AF792595

Modelling of Corroding Concrete Structures

RILEM BOOKSERIES

Volume 5

RILEM, The International Union of Laboratories and Experts in Construction Materials, Systems and Structures, founded in 1947, is a non-governmental scientific association whose goal is to contribute to progress in the construction sciences, techniques and industries, essentially by means of the communication it fosters between research and practice. RILEM's focus is on construction materials and their use in building and civil engineering structures, covering all phases of the building process from manufacture to use and recycling of materials. More information on RILEM and its previous publications can be found on www.RILEM.net.

For other titles published in this series, go to
www.springer.com/series/8781

Carmen Andrade • Giuseppe Mancini
Editors

Modelling of Corroding Concrete Structures

Proceedings of the Joint fib-RILEM Workshop held in Madrid, Spain, 22–23 November 2010

Springer

Editors
Carmen Andrade
IETcc_CSIC
Madrid, Spain

Giuseppe Mancini
Politecnico di Torino
Italy

ISSN 2211-0844
ISBN 978-94-007-0676-7 e-ISBN 978-94-007-0677-4
DOI 10.1007/978-94-007-0677-4
Springer Dordrecht Heidelberg London New York

© RILEM 2011
No part of this work may be reproduced, stored in a retrieval system, or transmitted in any form or by any means, electronic, mechanical, photocopying, microfilming, recording or otherwise, without written permission from the Publisher, with the exception of any material supplied specifically for the purpose of being entered and executed on a computer system, for exclusive use by the purchaser of the work.

Cover design: eStudio Calamar S.L.

Printed on acid-free paper

Springer is part of Sprincer Science+Business Media (www.springer.com)

Table of Contents

Foreword

Reinforced concrete is the most used construction material. It presents adequate performance and durability in many circumstances. However, premature deterioration, mainly from reinforcement corrosion, has been noticed. These failures have promoted the need to design the concrete for a pre-determined service life which is able to hold up in several exposure classes. The concrete assumes the absence of deterioration during its service life and therefore no rules exist on how to calculate the evolution of damage at structural level.

This lack of standards is especially dramatic when dealing with the assessment of existing structures, as the residual safety and the lifetime for intervention are urgent questions in deteriorating structures. Some Guidelines and Manuals exist but there is no accepted methodology, nor any models that could be used by engineers.

The subject demands the concurrence of material, corrosion and structural specialists as only from a multidisciplinary approach, new methods of calculation will be developed. Both RILEM and *fib* have members expert in these subjects. Rilem Technical Committee TC-213-MAI "Model assisted integral service life prediction of concrete structures with respect to reinforcement corrosion" has been working in a critical analysis of present models of corroding elements and the new Special Group fib-SAG "Assessment of Existing Structures" is starting the work to gather information in order to propose a Model Code on Assessing Existing Structures.

The present Workshop responds to the need of comprehensive but very specialized discussions on the main topics of the area in order to identify the aspects which demand more research and consensus.

Carmen Andrade
Giuseppe Manicini

Study of the Interaction between Loading and Rebar Corrosion in R.C. Ties

Luca Giordano, Giuseppe Mancini and Francesco Tondolo

Department of Structural and Geotechnical Engineering, Politecnico di Torino, Italy

Abstract. Rebar corrosion and reinforcement stress variation typically affect structures exposed to environmental actions and to significant loads, like bridges. In order to analyze the coupled effects of the two phenomena, an experimental campaign on reinforced concrete ties subjected to simultaneous loading and corrosion have been established. The cycling action is chosen in order to obtain a stress peak and a stress variation reproducing the live load effect in fatigue design situations. In the meanwhile, the environmental action, reproduced by means of an accelerated electrochemical procedure, simulates a high level of carbonation induced corrosion on the structure. Besides, other tests under static load were performed in order to compare the results obtained with the cyclic ones and to obtain a comprehensive analysis of the deterioration processes.

Transversal crack widths due to loading and longitudinal cracks openings due to corrosion are monitored. Results highlight the differences in terms of corroded and uncorroded specimens, static and cycling tests and also different loading amplitudes. Finally it can be observed as the growing of the damage is higher when a cycling action is combined with the chemical attack, so producing results unpredictable by means of static tests only.

Introduction

The durability of reinforced concrete (r.c.) structures is a subject of study since many years. It relates to the ability of a structure to maintain its characteristics such as safety level, efficiency and aesthetics during the service life time. One of the most important analyzed and studied effects that leads to the deterioration of concrete structures is the reinforcement corrosion [1, 2]. Service life of r.c. structures can be reduced by reinforcement corrosion. Corrosion reduces the rebar area primarily affecting the outer surface and gradually involving the inside. In addition, the oxide product expansion, caused by a volumetric increase, determines

C. Andrade and G. Mancini (eds.), *Modelling of Corroding Concrete Structures*, RILEM Bookseries 5, DOI 10.1007/978-94-007-0677-4_1, © RILEM 2011

transversal tensile stresses that can lead to a cracking in the direction of the bars [3]. Finally, with the growth of the corrosion penetration and the oxide products expansion, a decrease of bond between steel and concrete [4] can take place.
In roadway infrastructures, in addition to corrosion caused by environmental actions and aggressive agents, the permanent and variable loads due to traffic are present. Therefore a cyclic variation of the internal actions occurs, resulting in a stress variation and, if the structure is cracked, in a progressive evolution of the crack openings. These effects can determine unexpected degradation consequences, significantly different from those observed in static conditions. In the present work, the design and results of an experimental campaign on reinforced concrete ties subjected to: a) cyclic loading, b) cyclic loading and simultaneous corrosion, c) static loading and corrosion, are presented and discussed. An electrochemical corrosion process is used for the accelerated corrosion; the geometry of specimen and the loading procedure reproduces the usual conditions of bridges under fatigue loading.

Test Specimens Description

In order to study the effect of cracking and bond modification in r.c. structures, reinforced concrete ties loaded in tension with a length of 500 mm and a square section with side of 90 mm were used for this research.

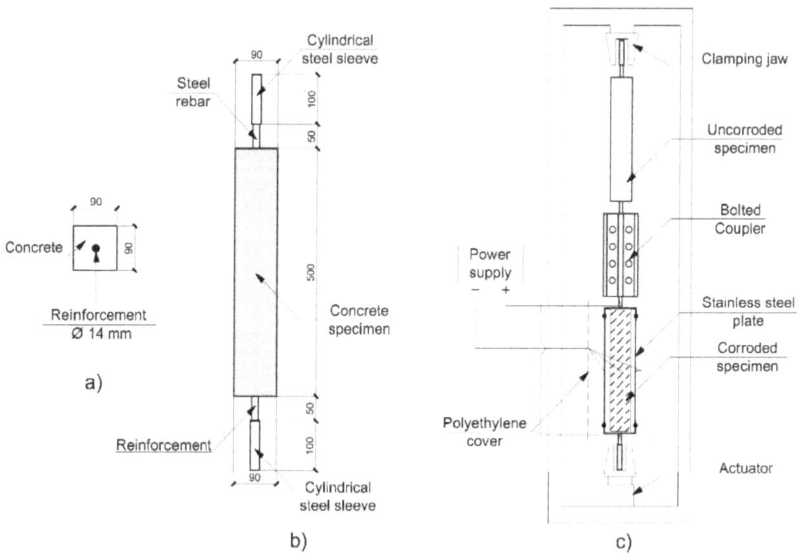

Figure 1. a) Transverse section, b) specimen, c) test setup

They were reinforced by a single ribbed bar of 14 mm diameter (see Fig. 1a) with a bond index of 0.079. They represent the tensed portion of a cantilever element of a bridge deck. A common Portland cement type was used for the casting of concrete; the mix proportions, by weight of cement, sand and gravel were 1:2.7:2.3 with a maximum aggregate size of 12 mm and a water/cement ratio of 0.4. The mean concrete compressive strength at 28 days, f_c, was 25.2 MPa, whereas the tensile strength f_{ct} was 2.7 MPa. In the concrete mix 3% of NaCl by weight of cement was added at casting so to obtain the depassivation of the reinforcement and enhance the artificial corrosion mechanism. The mechanical characteristics of steel were: elastic modulus E_s = 202 GPa, yield strength f_y = 507 MPa, ultimate strength f_t = 590 MPa, ultimate strain ε_u = 8%. The curing period was approximately 3 months where all the ties were unloaded and stored at 20 °C with RH 60%. The rebars were about 80 cm long with 10 cm coaxial steel sleeve at the extremities (see Fig. 2b) electrically insulated by means of epoxy resin. Two different types of test arrangement for the specimens were used: a coupled system of two specimens (Fig. 1c) and single specimen test. In the coupled system the upper specimen was subjected only to fatigue load, whereas the bottom one was also subjected to corrosion, instead with the second arrangement the specimen was subjected to both static load and corrosion.
The experimental research is still going on; to date twelve specimens were tested: five couples (cyclic loading with and without corrosion) and two single specimens (static loading and corrosion). The testing machine was an MTS device and the tests were conducted in the Laboratory "Franco Levi" of Department of Structural and Geotechnical Engineering at Politecnico di Torino.

Cyclic Loading and Corrosion Mechanism

According to the transversal analysis of different bridge decks subjected to the fatigue load defined by the European Standard for Load on Bridges [5], the amplitude of the cyclic loading equal to 50-70-90 MPa in the specimens were used. Furthermore, the maximum load level was chosen so to get, at the beginning of the test, ordinary crack width according to limitations of [6]. Using a frequency load of 3 Hz within 25 days of test about $6.5 \cdot 10^6$ cycles were reached.
In order to obtain corrosion, an electrochemical procedure was used. The details of the procedure and the assumption on the current density can be found in a previous work of the same authors [7] and are based on a consolidated literature [8]. According to that work, an electrical current density ($I_{c,t}$) of 200 $\mu A/cm^2$ was used. In this way it was simulated about 27 years of natural corrosion in exposure class XC4 during the 25 days of accelerated corrosion tests (the wetness parameter w_t = 0.75 for cycling wet and dry was ensured during the test).

Test Procedure

Five steps were followed for the coupled specimens:

1. a tensile load was progressively applied to the two specimens up to reaching transverse cracking. The maximum load level was chosen in order to obtain, for at least one crack (in the specimen subjected to corrosion) a value of the crack opening of 0.15, 0.20 and 0.25 mm respectively. The details of each specimen are summarized in Table I; the specimens subjected only to cyclic loading were named as FU, whereas specimens also subjected to corrosion were named FC. The number that follows the prefix indicates the maximum load level (expressed in kN) used in the test, while the last two digits are referred to the stress amplitude variation (expressed in MPa);

2. the specimens were then completely unloaded;

3. gauges for the measure of transversal cracks by means of mechanical strain gauges were positioned. Figure 2 shows number and location of the transversal cracks caused by the first loading. Only for corroded specimens, in a median position, four electrical strain gauges with base length of 50 mm were glued on the four surfaces, in order to measure the horizontal strain caused by corrosion:

4. the tensile load was increased to reach the value N_{max} registered during the first cycle (see Table 1);

5. a stress variation of 50/70/90 MPa, with sinusoidal shape and frequency of 3 Hz was applied to the bar. Simultaneously to the load application, the corrosion mechanism was initiated with the constant current density of 200 $\mu A/cm^2$ previously chosen (only for corroded specimens).

For the two specimens named SC40-00 and SC60-00 (see Table I) the load was maintained constant through the test while the rebars were corroded by using the electrochemical process.
In order to produce corrosion of reinforcement, the two extremities of the corroding element were connected to the positive terminal of an external power supply, while stainless steel plates were connected to the negative one (Fig. 1c). The contact between the stainless steel plates and the specimen was ensured by means of stainless steel wool in contact both with stainless steel plates and the concrete; in this way, no mechanical actions were produced between the specimen and the stainless steel plates and electrical continuity was obtained. For humidity level control, the specimen subjected to corrosion was encased into a polyethylene cover (Fig. 1c). As previously mentioned, each test took 25 days reaching about $6.5 \cdot 10^6$ cycles.

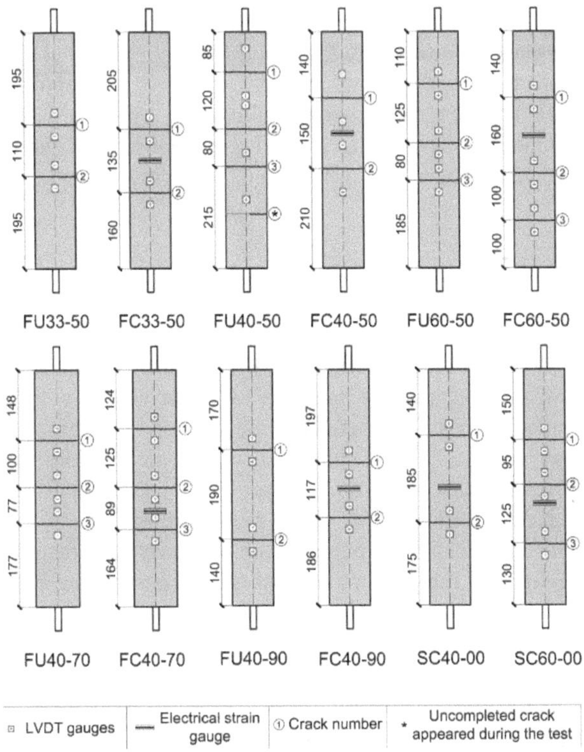

Figure 2. Test specimens, position of gauges, electrical strain gauges and transversal cracks

Specimen	Type of load	Corrosion?	$N_{max,}$ [kN]	W_{ini} [mm]	$\Delta\sigma$ [MPa]
FU33-50	Fatigue	No	33	-	50
FU40-50	Fatigue	No	40	-	50
FU60-50	Fatigue	No	60	-	50
FU40-70	Fatigue	No	40	-	70
FU40-90	Fatigue	No	40	-	90
FC33-50	Fatigue	Yes	33	0.15	50
FC40-50	Fatigue	Yes	40	0.20	50
FC60-50	Fatigue	Yes	60	0.25	50
FC40-70	Fatigue	Yes	40	0.20	70
FC40-90	Fatigue	Yes	40	0.20	90
SC40-00	Static	Yes	40	-	0
SC60-00	Static	Yes	60	-	0

Table I. Specimen of the experimental campaign

During each test, transversal crack openings were monitored by means of a mechanical strain gauge. Because of corrosion FC specimens, after an initial period of few days from the beginning of the test, showed a longitudinal cracking. Longitudinal crack opening were registered in some points along the specimen since their appearance by means of an optical strain gauge. All the measurements were daily performed at the maximum load level N_{max} (see Table I).

Transversal and Longitudinal Crack Opening

The tensile load introduced in the first step creates the transversal cracking. The crack openings associated to these cracks depend on the reinforcement percentage and dimensions of the specimen besides the load level (see Table I). Moreover their formation in terms of initial opening value and their evolution can be directly related respectively to the bond condition between the reinforcing bar and the surrounding concrete, and its evolution due to the mechanical action of fatigue and the effect of the rust formation. In fact, the deterioration of the mechanical connection determines higher slips between the two materials, resulting in a wider cracking [9]. Because of this, the transversal crack opening can be assumed as a reference for the internal structural condition of the contact between the two materials.

For each specimen, crack number and their location was different due to the scattered nature of the crack formation mechanism. As the number of the cracks influences the value of their width, the sum of the crack openings was chosen as a distinctive parameter for the cracking behavior of each specimen. This choice enables to compare the results obtained for the different specimens regardless the number of their cracks. In the same Figures, longitudinal cracks initiation and completion due to corrosion is reported.

Figure 3 shows the results obtained for three couples loaded at different maximum load level but subjected to the same stress amplitude. The aim is to highlight the effect of the N_{max} on crack width value. For the couple FC33-50 and FU33-50 no substantial differences between the increases of crack openings of the two specimens (Δ_{crack}) can be observed from the beginning to the end of the test. Consequently the damage due to the combined action of cyclic loading and corrosion is not significantly different from that detected for the specimen subject to only cyclic loading. For the couple FC40-50 and FU40-50, for the first 1.5×10^6 cycles, the crack openings remained practically unchanged for both the specimens; afterwards, while for FU40-50 no substantial differences were registered, in the corroded specimen a constant growing was detected, up to reaching, at the end of the test, a crack width increment of 56%, compared to the initial value.

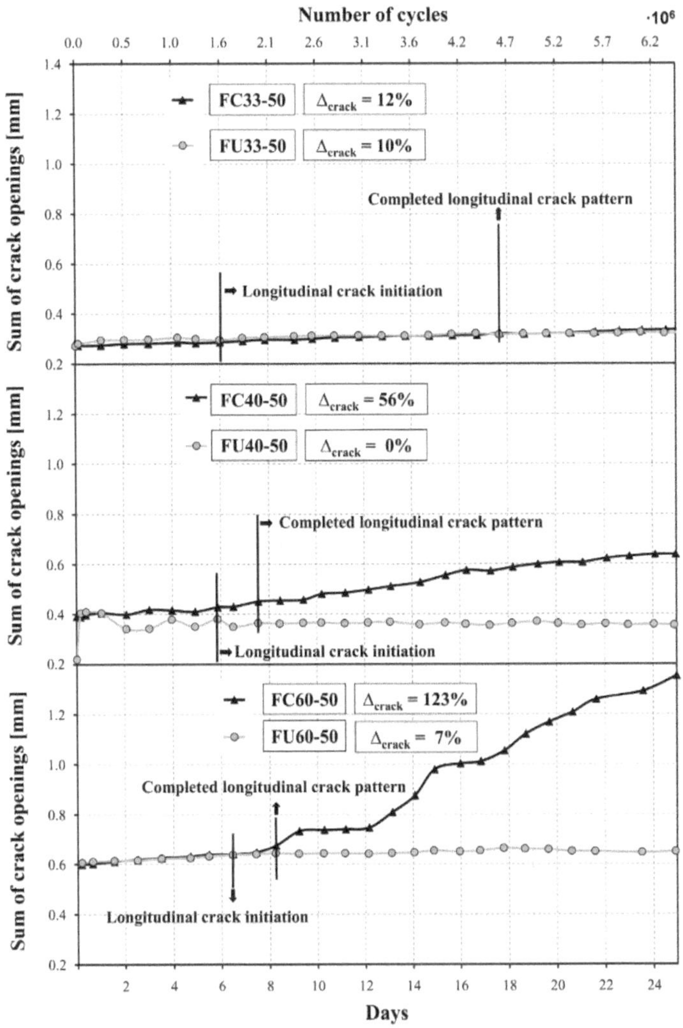

Figure 3. Sum of crack openings for the specimens subjected to a stress amplitude of 50 MPa

For the couples tested at 60 kN the initial value of the sum of crack widths for the two specimens was practically the same. During the test, the specimen FU60-50 showed a little but constant growing of the cracks that reached the value of 7%; in the corroded specimen, especially after the complete longitudinal crack pattern formation, a sudden increase is evidenced and the width at the end resulted higher than 100 % of the initial crack opening.

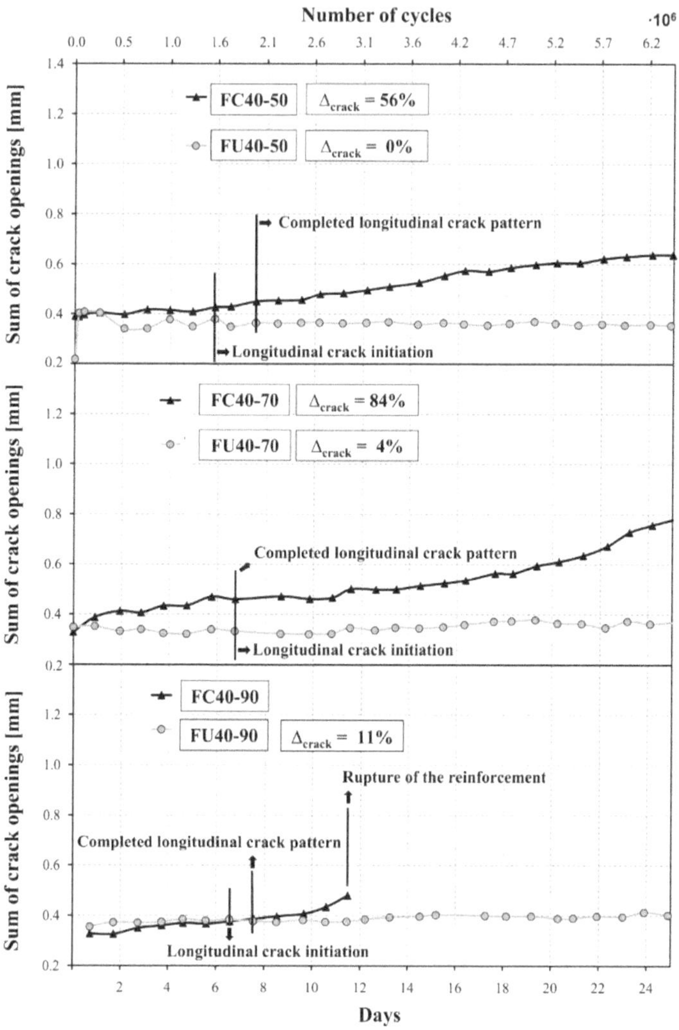

Figure 4. Sum of crack openings for the specimens subjected cyclic loading and $N_{max} = 40$ kN

In Figure 4 are collected the results of the three couples with the same N_{max} (40) but different stress amplitude. For comparison purpose, the results of couple FU40-50/FC40-50 are reported again. Moreover for the couple FU40-70/FC40-70, up to about $5 \cdot 10^6$ cycles (18 days of corrosion) the structural behaviour results to be quite similar to the couple with the same maximum load and a stress variation of 50 MPa. Afterwards, at 18 days of corrosion a significant slope increment can be seen, resulting in a final crack width increment for FC40-70 higher than that of specimen FC40-50. The couple tested at 40 kN with a stress amplitude of 90 kN showed a crack opening sum increment equal to 11% for the FU40-90 whereas, because of a localization of the corrosion, a rupture at the first crack location was registered for the specimen FC40-90. Before the collapse, as evidenced by Figure 4, an increase of the slope of the graphs revealed sudden bond deterioration. The lower value of the crack opening for FC40-90 at the beginning of the test, in comparison to the others specimens tested at 40 kN, can be probably due to the closeness of the two cracks (see Fig. 2).

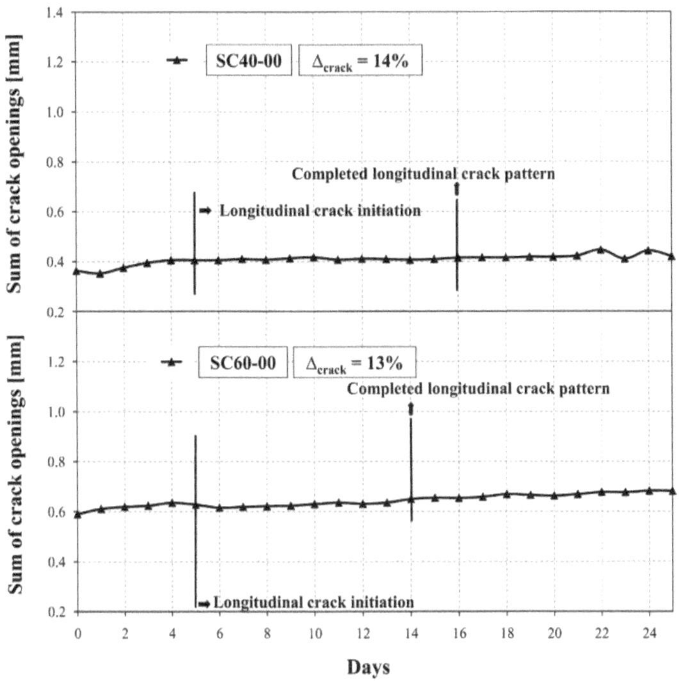

Figure 5. Sum of crack openings for the specimens subjected static loading

For the specimen SC40-00 it can be seen that the sum of cracks opening (Fig. 5) is substantially constant during the test, especially if compared with the crack width increments of the specimens with the same maximum load, but subject to fatigue. The comparison among the graphs in Figures 4 and 5 (graph on the top) demonstrates that the crack opening evolution for specimen tested at 40 kN, was essentially due to the interaction between cyclic loading and corrosion. The results obtained for the specimen SC60-00 confirmed that the coupled mechanical and chemical degradation is a key point for crack width evolution. In this case even if a constant increase of cracks opening was observed, they were much smaller than that obtained for the specimen FC60-50 and slightly wider than FU60-50.

For all the corroded specimens, an important role is played by the longitudinal cracking that is caused by corrosion. It is also influenced in its formation by the load level (see different elapsed time between the longitudinal crack patter initiation and completion among the corroded specimens), and also influences the transversal crack width evolution. In fact, the presence of this longitudinal discontinuity near the rebar, impair the confining action produced by the surrounding concrete, especially for high levels of stress in which bond is exerted by mechanical interlock among the reinforcement, by means of concrete struts radiating from the bar and the undamaged outer ring as described by some mechanical models [10].

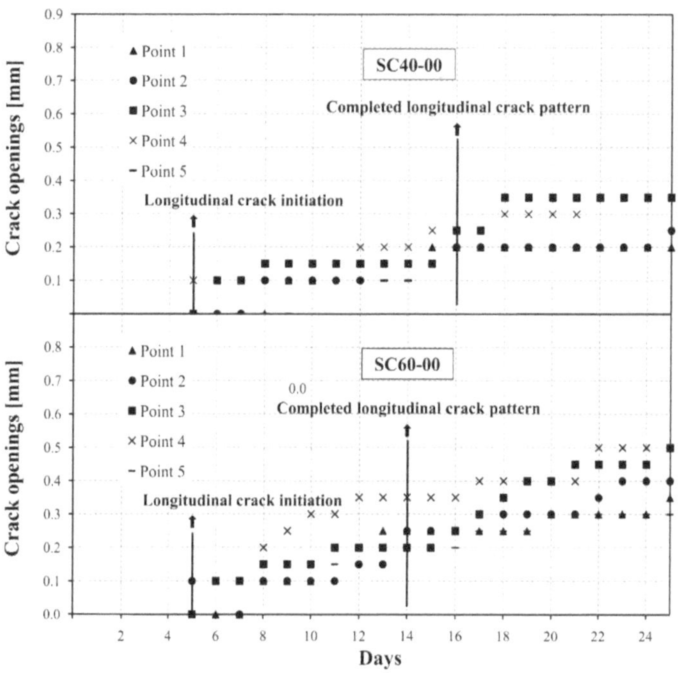

Figure 6. Sum Test results in specimens with corrosion and static load

As evidenced from the Figures, because of oxide formation and expansion, after about six days of corrosion, a longitudinal crack was formed. This crack was monitored from its first appearing to its complete formation all along the specimen until the end of the test. In Figures 6-8, longitudinal crack evolutions are reported for all the corroded specimens. Longitudinal crack can be considered as a reference for the level of corrosion of the reinforcement and their evolution is also dependent on the load level and the stress amplitude. In all the specimens subjected to corrosion (series FC and SC), the longitudinal crack appeared after about five or six days of test, independently from both the level and the type of load (cycling in series FC, static in series SC). On the opposite, the longitudinal crack was visible on the whole specimen after a number of cycles (or days) that depends on both the level and the type of load. In particular, in Figure 6 it can be seen how, if the maximum load levels increases from 40 to 60 kN, the time between the first longitudinal crack appearance and its full development (Δ_{tfd}), is slightly reduced from 11 to 9 days. This reduction can be associated with the higher tensile stress of concrete in longitudinal direction in the specimen subjected to 60 kN resulting in a reduction of the transversal bearing capacity.

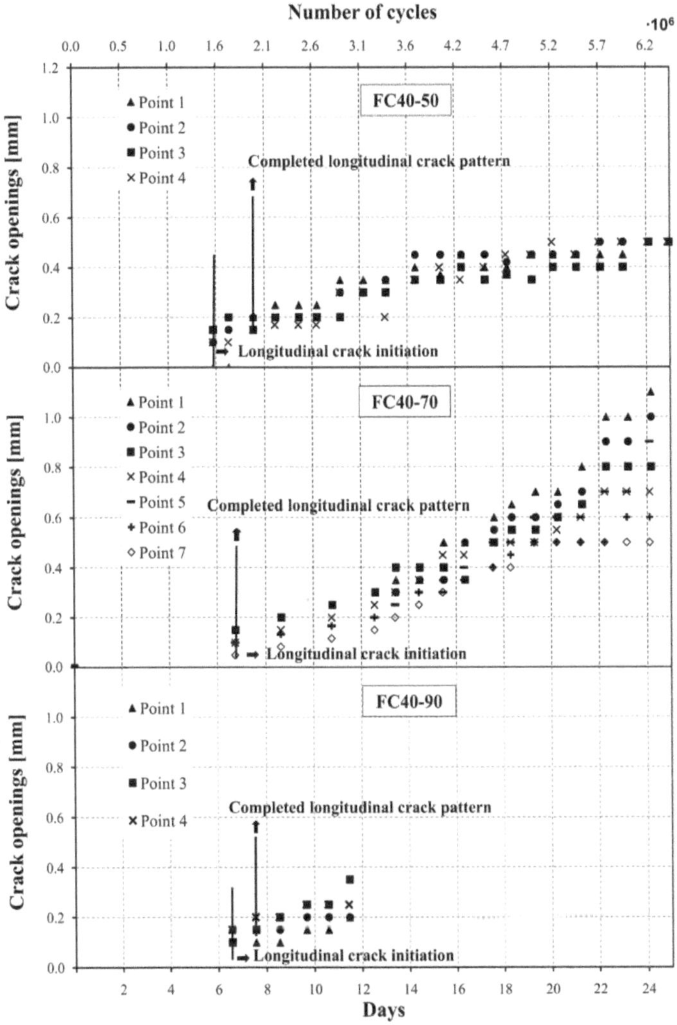

Figure 7. Test results in specimen with corrosion and fatigue load N_{max}

On the other hand, when a tensile stress variation is added to the static load, so determining a cyclic load, a deep change in structural behavior is registered. The comparison between the first graph of Figure 6 (static load with 40 kN) and the first one of Figure 7 (maximum cyclic load level of 40 kN and stress variation of the rebar $\Delta\sigma = 50$ MPa) allows to evaluate a significant reduction of Δ_{tfd}, that shifts from 11 to 2 days. Furthermore, if the stress variation also increases to 70 MPa, the Δ_{tfd} value becomes nil, as evidenced by the specimen FC40-70 (see Fig. 4). For specimen FC40-90 it is equal to one day.

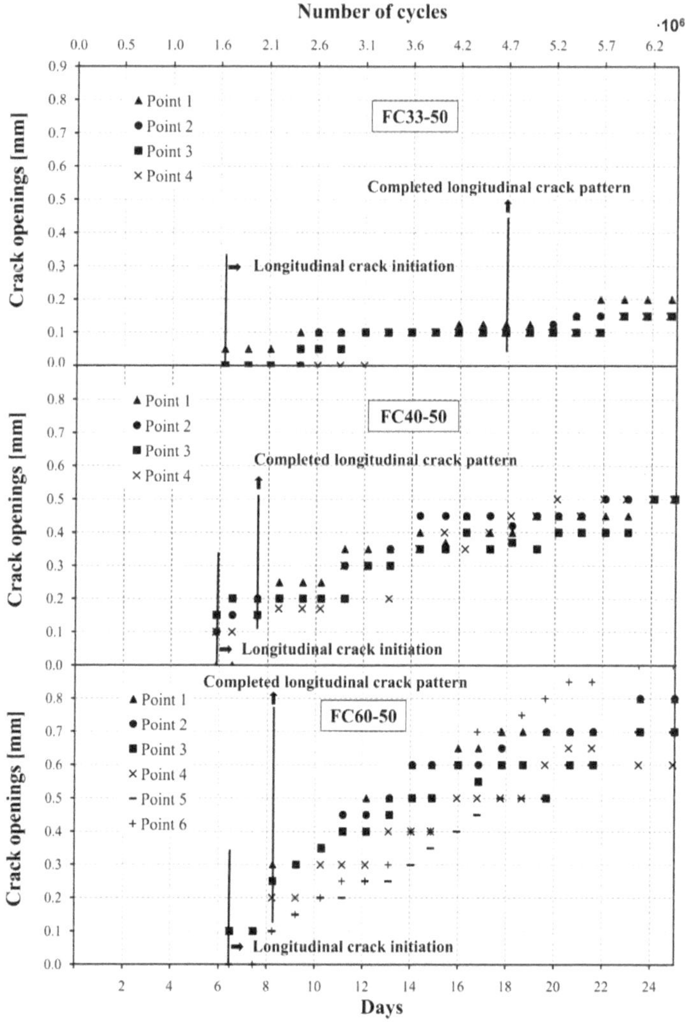

Figure 8. Test results in specimens with corrosion and fatigue load ($\Delta\sigma = 50$ MPa)

Anyway this suggests that the reduction of Δ_{tfd} is influenced by a combined effect of the two causes and then it could not be evaluated as the sum of each single contribution of deterioration. Finally, the three graphs of Figure 8 allow for the evaluation of the effect of the maximum load level variation with the same stress amplitude: a huge reduction of Δ_{tfd} is evidenced passing from 33 to 40 kN, whereas the increase from 40 to 60 kN is not followed by a reduction of the same parameter. This difference can be explained considering the variation of bond type that establishes for small load levels (adhesion type), and higher ones (interlock type).

Regarding the crack opening values, an important observation can be referred to the scattering of the experimental results: in Figure 8 it is evidenced as, with the growing of the maximum applied load, the scattering of the data measured along the specimen is higher. Similar considerations can be done about the influence of the stress variation comparing, for example, the experimental results of the specimens FC40-50, FC40-70 and FC40-90.

In terms of value of longitudinal crack openings, Figure 6,7 and 8 show how the type of load (cycling or static), the stress amplitude (50-70 90) load the load level (33-40-60) influence the final values. In particular starting from the specimens with static load (Fig. 6), it can be observed that, at the end of the test, an average value of crack opening of about 0.28 and 0.41 mm is present in the specimens loaded at 40 and 60 kN respectively.For the specimens loaded with a stress variation $\Delta\sigma$ = 50 MPa (see Figure 8), final mean values of crack width are equal to 0.16, 0.50 and 0.75 mm for the three levels of load considered (33, 40 and 60 kN). Moreover the three specimens tested at maximum level of 40 kN (Fig. 7) show mean openings equal to 0.20, 0.50 and 0.91 mm with the grow of the stress range (0, 50, 70 MPa). Unfortunately, the final value of the average crack widths is not reported because of the failure of FC40-90 but it can be observed as a value of 0.38 mm was read on a point since day 11. From those results it is highlighted as the expected value of crack opening on the surface is deeply influenced both by the mechanical action and by the stress range level, other than from the amount of corrosion that was kept constant during the various tests.

Conclusions

In the above described experimental work, the structural behavior of reinforced concrete ties subjected to both cyclic actions and accelerated corrosion is presented. During the tests both transversal cracks evolution and longitudinal crack formation for corroded specimens were monitored. For the transversal crack openings, the specimens subjected only to mechanical action show a crack width increase that could be considered weakly dependent on the maximum load level and is almost linear up to the considered number of cycles. In the specimens subjected to static action and corrosion, an increase can be observed even if it remains within the range of 14% of the initial value. For specimens subjected to cyclic action and corrosion, the growth of crack openings is dependent on the load level and the stress amplitude. In particular, for specimen tested with the same stress amplitude ($\Delta\sigma$ = 50 MPa) and maximum load level equal to 33-40-60 kN the cracks widths rise of 12-56 and 123% respectively. For specimen tested at the same maximum load level (40 kN) the influence of variation of the stress amplitude from 50 to 70 MPa is evidenced from the growth of the crack width to values from 56 to 84% of the initial value. The interaction between fatigue loading, the stress amplitude of 90 MPa and corrosion probably determined the early collapse of the specimen FC40-90 and then determining missing data. All these results confirm that the main effect of crack opening evolution is due to the interaction between corrosion and cyclic loading. Crack opening value is directly related to the bond between steel and concrete and so the previous results can be considered as findings about bond performance. The formation of longitudinal cracking caused by corrosion and its evolution, determined also from the loading condition were monitored. It can be assumed as an indicator of the corrosion level affecting rebars. Finally those observations underline the importance of a coupled mechanical and chemical test for a reliable evaluation of performance for structure under loading and corrosion.

References

[1] Rodriguez, J., Ortega, L. M., Casal, J., Diez, J. M. 1996. Assessing structural conditions of concrete structures with corroded reinforcement. *International congress Concrete in the Service of Mankind, Conference n. 5 Concrete Repair Rehabilitation and Protection*, Dundee, UK, pp. 14.

[2] Mangat, P. S., Elgarf, M. S. 1999. Flexural strength of concrete beams with corroding reinforcement. *ACI Struct J* 96(1): 149-158.

[3] Molina, F. J., Alonso, C., Andrade, C. 1993. Cover cracking as a function of rebar corrosion. *Mat Struct* 26: 532-548.

[4] Al-Sulaimani, G. J., Kaleemullah, M., Basunbul, I. A., Rasheeduzzafar 1990. Influence of corrosion and cracking on bond behavior and strength of reinforced concrete members. *ACI Struct J* 87(2): 220-231.

[5] EN 1991-2, 2005. Actions on structures - Traffic loads on bridges.

[6] EN 1992-2, 2006. Design of concrete structures Part 2: Concrete bridges.

[7] Giordano, L., Mancini, G., Tondolo, F. 2009. Experimental evaluation of corrosion effect on bond between steel and concrete in presence of cyclic action. *Key Eng Materials* 417-418: 350-353.

[8] Rodriguez, J., Ortega, L., Izquierdo, D., Andrade, C. 2006. Calculation of structural degradation due to corrosion of reinforcements. ECF 16, Alexandroupolis, Greece, 3-7 July, 527-536.

[9] Giordano, L., Mancini, G., Tondolo, F. 2009. Numerical interpretation of bond between steel and concrete in presence of corrosion and cyclic action. *Key Eng Materials* 417-418: 349-352.

[10] Tepfers, R. 1979. Cracking of concrete cover along anchored deformed reinforcing bars. *Mag of Concr Res* 31(106): 3-12.

Prediction of Corrosion Rate in RC Structures - A Critical Review

Mike Otieno, Hans Beushausen and Mark Alexander

Department of Civil Engineering, University of Cape Town, South Africa

Abstract. Corrosion rate is one of the most important input parameters in corrosion-induced damage prediction models for reinforced concrete (RC) structures. Its accurate assessment and/or prediction is therefore required if the damage prediction models are to be reliably used to predict both the rate and severity of damage and to plan for maintenance of these structures. However, it has not been assigned the level of importance it deserves especially with respect to its prediction. In most cases, instantaneous measurements or constant predicted corrosion rate values are used in damage prediction models hence neglecting its time-variant nature while in some cases, salient factors that affect corrosion rate such as cover cracking and concrete quality and not taken into consideration during the model development. The direct consequence of this may be under- or over-estimation of the severity and the time to corrosion-induced damage such as for example cover cracking, and hence service life of the structure. This paper presents a critical review of some of the available corrosion rate prediction models focusing mainly on chloride-induced corrosion. In addition, proposals for the improvement of these models are made.

Introduction

In the light of the paradigm shift to incorporate the corrosion propagation phase (t_p) in the service life of corrosion-affected reinforced concrete (RC) structures [1-3], several models have been developed to predict times to different corrosion-induced damages (limit states) such as cover cracking, loss of steel cross-section area, loss of stiffness, etc [4-7]. Even though these prediction models usually have several input parameters, it is clear that the rate of corrosion governs their outcome either in terms of the time to attainment of a pre-defined limit state or its severity at a given time. A sensitivity analysis by Li et al. [8] showed that corrosion rate (i_{corr}) is one of the most important input parameters in the corrosion-induced damage models.

C. Andrade and G. Mancini (eds.), *Modelling of Corroding Concrete Structures*, RILEM Bookseries 5, DOI 10.1007/978-94-007-0677-4_2, © RILEM 2011

The success in incorporating t_p in the service life of RC structures will therefore depend on how accurately and realistically the i_{corr} can be predicted. As discussed in later sections, its treatment with respect to prediction does not match the level of importance it deserves. Instead, more focus has been placed on the prediction of corrosion-induced damages, with little attention to the assessment and prediction of i_{corr}. The result of this has been the development of several corrosion-induced damage prediction models for each limit state (e.g. time to corrosion-induced cover cracking). This makes the task of selecting appropriate models difficult for the practicing engineer.

In contrast, the number of available/published i_{corr} prediction models is relatively small compared to that of the damage prediction models; this shows where the focus has been placed, as mentioned above. This paper presents a review of the available/published i_{corr} prediction models. It is divided into two main parts: the first part presents a review of individual models while the second part contains a summary of some of the salient aspects that should be incorporated in the models and a brief discussion of an ongoing study on the prediction of i_{corr}.

Before proceeding, it is important to note the following:

(i) i_{corr} is affected by many factors and can be expressed (using a factorial approach [9]) as follows:

$$i_{corr} = f(k_1, k_2, \ldots\ldots, k_n) \quad (1)$$

where k_1, k_2, ……, k_n represent the factors affecting corrosion propagation e.g. supplementary cementitious materials, moisture content, cyclic wetting and drying, sustained loading, loading history, concrete resistivity, concrete quality, cover depth, temperature, cracking, dissolved oxygen concentration and exposure conditions [10]. However, it is important to note that it may be impractical to explicitly incorporate each of these factors in a model. Some of these factors can be indirectly incorporated in a predection model e.g. concrete resistivity can be indirectly used to account for temperature, concrete quality and moisture content of the concrete.

(ii) i_{corr} prediction models can be developed based on one or a combination of the following approaches:
   - Electrochemical principles of corrosion of steel in concrete [e.g. 11],
   - Statistical analysis of experimental the test results [e.g. 12],
   - Electrochemical principles of corrosion and experimental testing [e.g. 13].

(iii) The following factors should be considered in the development of i_{corr} prediction models:
   - The model should be (as far as possible) representative of the actual i_{corr} characteristics in the RC structure in its service environment.

- It should account for variability in both i_{corr} and the model input parameters such as cover depth and concrete quality. Furthermore, this variability should be taken into account using adequate and representative data.
- If developed based solely on either electrochemical principles of corrosion or accelerated corrosion test results, its validation using natural corrosion data is important.
- The model should be capable of being adjusted to suit the prevailing concrete and exposure conditions of the RC structure.

With these in mind, the next sections will present a review of the available i_{corr} prediction models.

Existing Corrosion Rate Prediction Models

Alonso et al.'s model (1988)

The prediction model by Alonso et al. [14] is based on a statistical analysis of resistivity and accelerated carbonation-induced i_{corr} results. The experiments were carried out using 20 x 55 x 80 mm mortar specimens made with various binders (Portland cement (PC), Sulphate resistant PC, slag cement, Pozzolanic cement, fly ash (FA) cement and 70/30 PC/FA) with a w/b ratio of 0.50. Accelerated carbonation was achieved by exposing the specimens to CO_2 (in a CO_2-filled chamber (concentration not reported) assumed to be 100% CO_2) at 50-70% relative humidity (RH). i_{corr} was assessed by linear polarisation resistance (LPR) technique and later ascertained using the gravimetric method. The experimental results are presented in Figure 1.

The mathematical expression for the trend-line in Figure 1 was adopted as the prediction model, with concrete resistivity as the main input parameter, Eqn. (2):

$$i_{corr} = \frac{k_{corr}}{\rho_{ef}} \qquad (2)$$

where k_{corr} is a constant with a value of 3 x 10^4 μA/cm^2.kΩ-cm (slope of graph in Figure 1) and ρ_{ef} is the resistivity of the concrete at its actual degree of saturation. Alonso et al.'s [14] experimental results show a clear relationship between i_{corr} and concrete resistivity. The following can be noted with respect to this model:

- Accelerated tests were used without validation using natural corrosion test results. The use of small (mortar) specimens (20 x 55 x 80 mm) may have a size effect on the experimental results [15].
- In addition to concrete resistivity, i_{corr} in concrete can be affected by other factors such as presence of cracks, concrete cover depth, among others [10].

- Variability of i_{corr}-influencing parameters such as cover depth is not taken into accounted.
- It is not clear whether the i_{corr} predicted by the model is a constant i_{corr} or not.

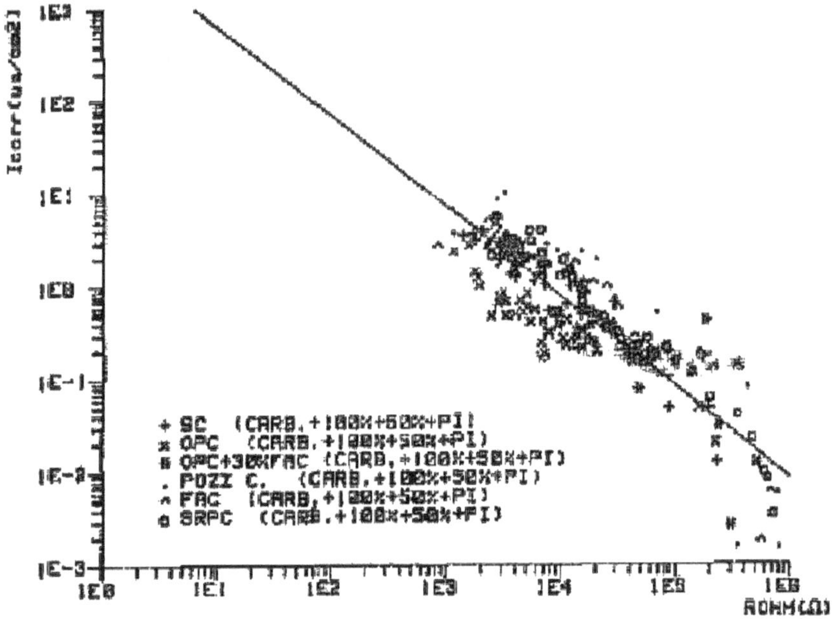

Figure 1: Corrosion rate vs. resistivity (for steel in carbonated mortars) [14]

However, even though shortcomings can be cited with respect to this model, it provides a clear indication of the possibility of using concrete resistivity as a potential durability indicator. The electrical resistivity of concrete provides indications on the pore connectivity and therefore, on the concrete resistance to penetration of liquid or gaseous substances; it can therefore be related to reinforcement durability [16]. The relationship between concrete resistivity and i_{corr} will be revisited later in this paper.

Yalcyn and Ergun's model (1996)

The model by Yalcyn and Ergun [17] was developed by studying the effect of chloride and acetate ions on i_{corr}. Corrosion was evaluated by measuring half cell potentials (HCP) and LPR. The model was developed based on results obtained from accelerated corrosion testing (admixed chlorides); with i_{corr} measurements taken up to a period of 90 days i.e. at 1, 7, 28, 60 and 90 days on cylindrical specimens of 150 mm diameter x 150 mm height. Only 90/10 PC/Pozzolanic cement was used. The model is presented as shown in Eqn. (3):

$$i_{corr} = i_o e^{-Ct} \quad (3)$$

where i_{corr} is the corrosion rate at time t, i_o is the initial corrosion rate and C is a corrosion constant that depends on the degree of concrete pore saturation, pH, permeability and the cover thickness; Yalcyn and Ergun proposed a value of C (evaluated from i_{corr} vs. time curves) as 1.1 x 10^{-3} day^{-1} for the different concrete samples they considered. Important notes about the model are as follows:

- The model predicts i_{corr} in uncracked concrete and may therefore not be applicable to cracked concrete; where both corrosion initiation and propagation may be affected (shortened or eliminated) [18].
- Factors which affect i_{corr} such as cover depth, temperature, concrete resistivity and cracking, among others [10], are not incorporated in the model.
- It is not clear whether the i_{corr} predicted by the model is a constant i_{corr} or not.
- The model as presented in Eqn. (3) assumes that the concrete and environmental conditions remain constant and that future i_{corr} is only dependent on time. However, these factors may vary from time to time and consequently affect i_{corr}.
- Taking into account that the model was developed using accelerated corrosion results, validation of the model should be carried out using data from natural corrosion tests.

Katwan et al.'s model (1996)

Katwan et al. [19] proposed an empirical model, based on the electrochemical noise (ECN) technique for the determination of i_{corr}. The measurements were made on full-scale RC beams (150 x 250 x 3000 mm) under dynamic (0.17 Hz) and static loading, and exposed to a corrosive environment (3.5% NaCl solution). Results from short and long-term tests suggested that for a given test condition, the i_{corr} can be predicted from the standard deviation (SD) of the HCP readings obtained using the ECN method.

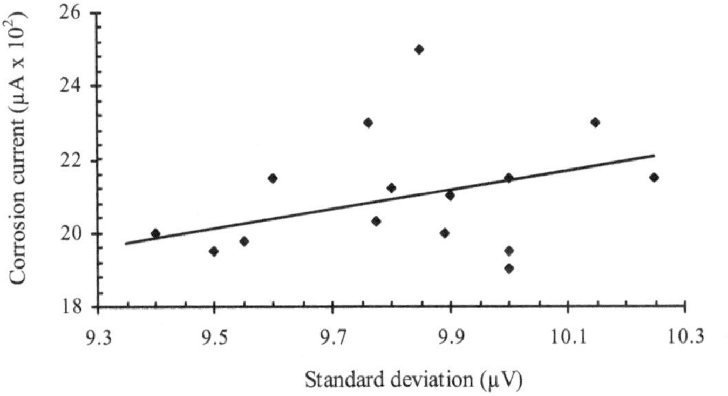

Figure 2: Corrosion current (μA) vs. standard deviation of HCP (μV) [19]

Although Katwan et al. [19] did not provide a mathematical expression for the relationship between corrosion current (I, μA) and SD of the HCP from the ECN technique, it is similar to that obtained by Page and Lambert [20] as shown in Eqn. (4):

$$\log I = 0.171 + 0.823 \log SD \tag{4}$$

The following can be noted with respect to this model:

- The use of HCP to assess corrosion has been criticised in the past for its instability depending on the prevailing measurement conditions; there is no general correlation between i_{corr} and HCP [21]. In most cases, for practical applications, HCP measurements need to be complemented with other corrosion assessment techniques e.g. LPR. Consequently, this limits the application of Katwan et al.'s model [19].
- The use of ECN technique in field of RC is not common (or well established) and therefore limits its practical applicability.
- In their paper [19], Katwan et al. reported that the SD was sensitive to change in the rate of sampling by the ECN technique but no recommendation was provided to account for this.
- It is not clear whether the i_{corr} predicted by the model is a constant i_{corr} or not.

Liu and Weyers' model (1998)

Liu and Weyers [12] developed an empirical model based on statistical analyses of experimental results obtained from a 5 year accelerated corrosion testing programme on 44 uncracked bridge deck slabs (1180 x 1180 x 216 mm) with covers of 25, 51 and 76 mm, w/b ratios of 0.41, 0.42, 0.43 and 0.45 and cement contents ranging from 337 to 382 kg/m^3; the objective being to obtain different accelerated corrosion rates. This was achieved by varying the amount of admixed NaCl (from 0 to 7.2 kg/m^3). The specimens were exposed to outdoor conditions during the test period. i_{corr} was measured using the LPR technique (3LP and Gecor devices were used). Liu and Weyers found out that the dynamic corrosion process of steel in in-service concrete is a function of the chloride content, temperature, resistivity of the concrete and active corrosion time. The outcome of statistical analyses of the results was a non-linear regression i_{corr} prediction model presented as shown in Eqn. (5):

$$i_{corr} = 102.47 + 10.09\ln(1.69Cl) - 39038.96(T^{-1}) - 0.0015R_c + 290.91t^{-0.215} \tag{5}$$

where i_{corr} is the 3LP corrosion rate (μA/cm^2), Cl is the total chloride content at the steel level (kg/m^3), T is the temperature at the steel surface (K), R_c is the resistivity of the cover concrete (Ω) and t is the corrosion time (years). The following can be noted with respect to Katwan et al.'s model [19]:

- Although concrete cover was a variable in the experimental set up, it is not an input parameter in the model. This was based on a statistical analysis of the results, but may also be partly due to the use of admixed chlorides. The absence of cover as an input parameter in the proposed model suggests that it has no influence on i_{corr}; this stands in contrast to results reported by various researchers indicating that concrete cover affects both corrosion initiation and its propagation [e.g. 22, 23].
- Similar to previous models discussed, it is not clear whether the i_{corr} predicted by the model is a constant i_{corr} or not.

Duracrete model (1998)

The Duracrete model [24] is an attempt to improve on Alonso et al.'s model [14] and proposes the incorporation of other i_{corr}-influencing factors by the introduction of coefficients/correction factors in the model as shown in Eqn. (6):

$$i_{corr} = \frac{k_{corr}}{\rho(t)} F_{cl}\, F_{Galv}\, F_{oxide}\, F_{Oxy} \tag{6}$$

where k_{cprr} is a constant regression parameter (10^4), F_{cl}, F_{Galv}, F_{oxide} and F_{Oxy} are factors to take into account the influence of chloride content, galvanic effects, continuous formation and ageing of oxides and availability of oxygen on i_{corr} respectively and $\rho(t)$ is the resistivity of concrete (Ω-m) at time t. $\rho(t)$ = $\rho_o\, f_e\, f_t\, (t/t_o)^n$; where ρ_o is the resistivity of concrete (Ω-m) at time t_o, n is a factor which takes into account the influence of ageing on ρ_o, f_e is a factor which modifies ρ_o to take into account the influence of the exposure environment and f_t is a factor which takes into account the influence of the resistivity test method. The following can be noted with respect to the Duracrete model:

- Even though the Duracrete model [24] attempted to improve on Alonso et al.'s model [14], it is still inherently susceptible to criticisms similar to those for Alonso et al.'s model as they are based on the same background.
- Although the above mentioned i_{corr}-influencing factors are suggested, no mention is made with respect to their quantification i.e. no guidelines are given on how to obtain or predict the values.
- It is not clear whether the i_{corr} predicted by the model is a constant i_{corr} or not.

Vu and Stewart's model (2000)

Vu and Stewart [25] developed an i_{corr} prediction model based on the assumption that O_2 availability at the steel surface (which depends on, among other factors, concrete quality, cover depth and environmental conditions such as temperature and relative humidity) are the governing factors. This was due to the consideration that for many locations in Australia, US, Europe and Asia, the average relative humidity (RH) is > 70%. For an ambient RH of 75% and a temperature of 20 °C, the influence of w/b ratio and cover depth (C, in cm) on corrosion rate up to 1 year ($i_{corr(1)}$, μA/cm^2) after the end of the corrosion initiation phase was expressed empirically as shown in Eqn. (7):

$$i_{corr(1)} = \frac{37.8(1-w/b)^{-1.64}}{C} \tag{7}$$

During the propagation phase i_{corr} is expressed as follows (Eqn. (8)):

$$i_{corr} = i_{corr(1)}\,0.85\,t_p^{-0.29} = \left(\frac{32.13(1-w/b)^{-1.64}}{C}\right) t_p^{-0.29} \tag{8}$$

where $t_p = t - t_i$; t is the time to which i_{corr} is to be predicted and t_i is the time to corrosion initiation. Plotting Eqn. (8) as shown in Figure 3 indicates that the model results in a constantly decreasing i_{corr} with time; this is contrary to the expected peak variations in i_{corr} depending on the prevailing concrete and exposure conditions, which vary from time to time. In addition, the initial i_{corr} is at the start of corrosion propagation is unexpectedly high.

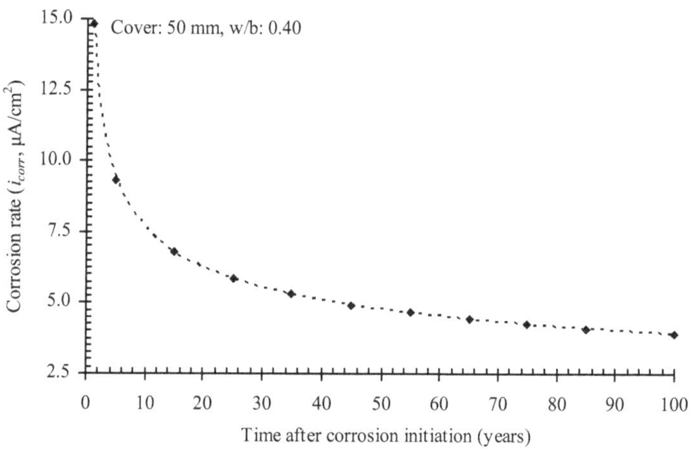

Figure 3: Variation of predicted i_{corr} with time as per Vu and Stewart's model [25]

Furthermore, it is necessary to note the following:
- The model does not take into account the variation in concrete quality with change in binder type, at a constant w/b ratio.

- The model may be limited to corrosion processes where cathodic reaction (oxygen availability) is the governing reaction e.g. PC concretes; it may not be applicable to those where concrete resistivity (anodic reaction) is the governing reaction e.g. blended cement concretes [22].

Later in 2005, Vu et al. [26] attempted to improve Vu and Stewart's model [25] to account for the time-variant nature (i.e. peak variation of i_{corr} with time) of i_{corr} by replacing the constants '0.85' and '0.29' with constants 'α' and 'β' respectively (Eqn. (9)) whose values depend on whether the i_{corr} is time-variant or not:

$$i_{corr} = \left(\frac{37.8(1 - w/b)^{-1.64}}{C} \right) \alpha \, t_p^{\beta} \tag{9}$$

The respective values for α and β and are given below:

Nature of i_{corr}	α	β	Reference
Time-invariant	0.85	-0.29	[25]
Time-variant	1.0	0	[26]

Scott's model (2004)

Scott's model [27] was developed using results from an experimental set-up comprising of cracked beam specimens (120 x 120 x 375 mm) with 0.2 and 0.7 mm crack widths, 20 and 40 mm concrete covers, a constant w/b ratio of 0.58 and a variety of binder types: 25/75 PC/GGBS, 50/43/7 PC/GGBS/Silica fume (SF), 50/50 PC/GGBS, 70/30 PC/FA, 75/25 PC/GGBS, and 93/7 PC/SF. The proposed i_{corr} prediction model was expressed as shown in Eqn. (10):

$$i_{corr} = \left(1.43 \frac{C_c}{f} + 0.02 \right) e^{\left[\left(\frac{40-x}{20} \right) 1.2 \left(\frac{C_c}{f} \right)^3 \right]} \tag{10}$$

where f is a slag correction factor, $f = 10^{(|0.5\text{-}S|\text{-}0.5+S)}$ (where S is the slag concentration expressed as a decimal e.g. 0.25 for 25%), C_c is the 90 day chloride conductivity index value (mS/cm) [28] and x is the concrete cover depth (mm). A plot of Eqn. (10) is presented in Figure 4.

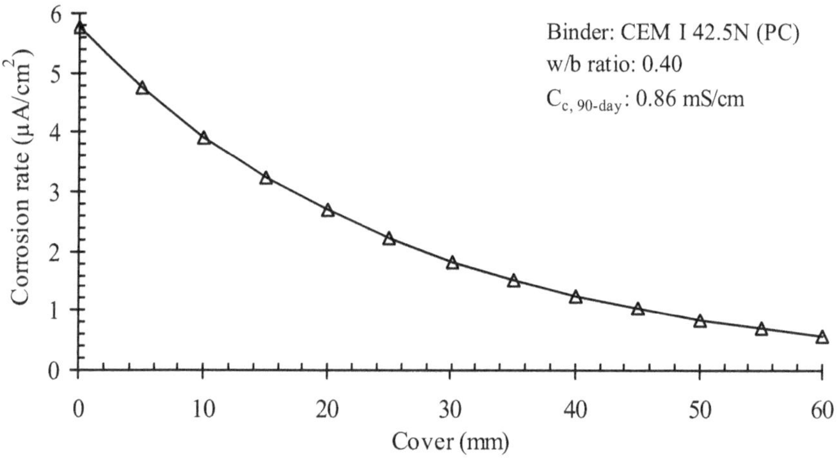

Figure 4: Plot of Scott's prediction model (values from [18])

The following are notable with respect to Scott's model:

- Although the model was developed using data from cracked specimens, it does not explicitly incorporate crack width as an input parameter. The applicability of this model to different scenarios other than those used during its development has not been conclusively ascertained in subsequent similar studies e.g. in Otieno et al.'s [18] study.
- Similar to other models already discussed, it is not clear whether the i_{corr} predicted by the model is a constant i_{corr} or not.
- Scott's model, similar to Alonso et al.'s [14], also underscores the possibility of using concrete resistivity (or its inverse conductivity) as a potential durability indicator and hence a service life prediction model input parameter.

Martinez and Andrade's model (2009)

Martinez and Andrade's model [29] can be used to predict an annual-averaged representative corrosion rate ($i_{corr,rep}$). The model, similar to Alonso et al.'s model [14], is based on concrete resistivity. The determination of $i_{corr,rep}$ (defined by Eqn. (11)) involves a series of steps described in [29]:

$$i_{corr,rep} = \frac{i_{corr,\sin gle} + i_{corr,\max}}{2} \qquad (11)$$

where $i_{corr,max}$ is the maximum predicted i_{corr} corresponding to a minimum resistivity (ρ_{min}) determined using in-situ cores, conditioned to 85% RH (for structures sheltered from rain) or vacuum water-saturated (for non-sheltered or submerged ones).

This model, similar to Vu and Stewart's model [25], assumes that corrosion is anodic- or resistivity-controlled, a phenomenon that may only be the case in high resistivity concretes e.g. in blended cement concretes [22]. Furthermore, even in situations where the corrosion process is anodic-controlled, differences may still arise for varied concrete qualities or in situations where the concrete cover is cracked [18, 22].

Models based on electrochemical principles of corrosion

These models rely mostly on solving the governing (*Laplace*) equation to determine the distribution of electric potential on the steel surface based on the assumptions of electrical charge conservation and isotropic conductivity [11, 30]. The governing equation can be presented as shown in Eqn. (12):

$$\frac{\partial^2 \phi}{\partial x^2} + \frac{\partial^2 \phi}{\partial y^2} = \nabla^2 \phi = 0 \qquad (12)$$

where x and y are planar Cartesian co-ordinates, ϕ is the electric potential and ∇ is the Laplacian/harmonic operator. The governing equation is solved using relevant boundary conditions (e.g. concrete resistivity, oxygen diffusion and pH of the pore solution), and suitable numerical methods such as finite element method, finite difference method and boundary element method. A model developed using this method cannot be presented as a single mathematical equation. The i_{corr} is determined from the potential distribution on the steel surface. The theory of electrochemical corrosion of steel is well established and can be found in the literature [11, 31]. Examples of models developed using this approach include those by Isgor and Razaqpur [11], Kim and Kim [30] and Dao et al. [32]. Specific details with respect to the principles of electrochemical corrosion and the determination of the input parameters and boundary conditions used to solve the governing (Laplace's) equation can be found in the literature [e.g. 11, 30, 31, 32].

From the on-going discussion, it is evident that models based on the application of electrochemical principles of corrosion, similar to others such as Katwan et al.'s model [19], rely on potential measurements. This has inherent disadvantages as discussed in previous sections. Therefore, validation of these models is important before they can be reliably used to predict i_{corr}. This will be covered in detail in the next section.

Summary of Salient Aspects to Be Considered When Developing a Corrosion Rate Prediction Model

A number of prediction models by different researchers, based on different approaches, have been presented, discussed and critiqued in the previous sections. The following section will present a summary of the salient issues related to these models.

Time-variant nature of corrosion rate

Corrosion rate in RC structures is affected by several factors (e.g. concentration of chlorides, concrete penetrability) that vary from time to time and with continued aging of the structure. Consequently, i_{corr} is also expected to show a similar variability and should be treated as such. This phenomenon has been appreciated by some researchers [e.g. 12, 33] but is yet to be successfully incorporated in prediction models. Researchers such as Yuan et al. [34] have proposed three distinct phases for a time-variant i_{corr} as (i) descent phase, (ii) steady phase and, (iii) ascent phase (Figure 5).

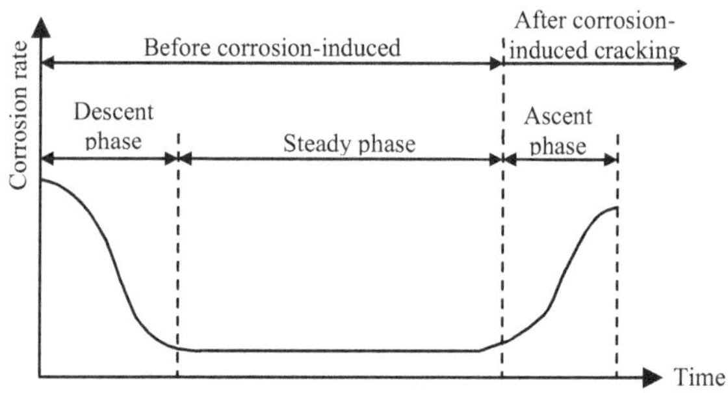

Figure 5: Phases of time-variant corrosion rate [34]

However, it is clear that such a trend cannot be used to generalize i_{corr} trend in RC structures. i_{corr} varies depending on the prevailing concrete and exposure conditions and hence will vary from one structure and exposure condition to another. Furthermore, the following can be noted:

- The proposed generalised trend may only be valid under the experimental set-up used by Yuan et al. Several other possible trends are possible depending on both the concrete and exposure conditions.
- In order to develop a more realistic generalised trend, the experimental set-up should be representative of the real RC structure and exposure conditions.
- It would be useful to the practicing engineer if approximate time periods and the corresponding corrosion rates for each phase are provided; this can also be used to plan for inspections and maintenance actions.

The variation of i_{corr} with time can be taken into account by incorporating the factors that affect it (as summarised in [10]) directly or indirectly as mentioned previously.

Influence of cover cracking on corrosion rate

Previous studies have shown that pre-corrosion cracking and crack characteristics can significantly influence both corrosion initiation and its propagation [18, 22]. The effect of cracking on corrosion may vary depending on concrete quality, concrete resistivity, crack width, crack density, crack self-healing and crack orientation. The effect of load-induced crack width, concrete quality (binder type and w/b ratio) and concrete resistivity on i_{corr}, obtained from an experimental study by Otieno et al. [18] can be summarised as presented in Figure 6.

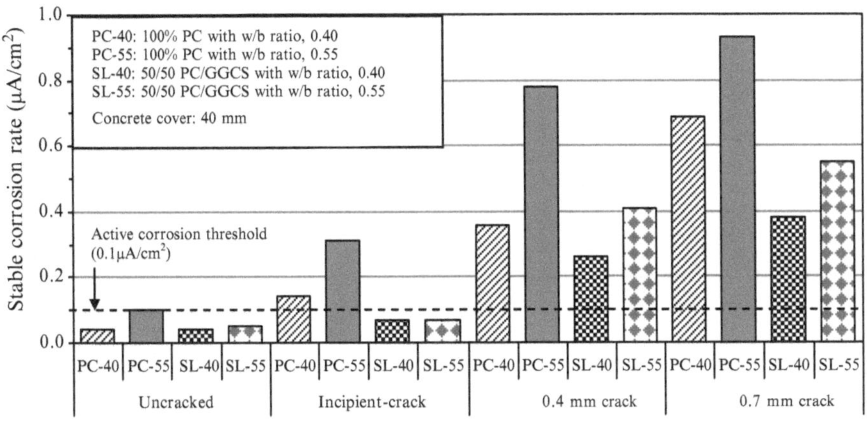

Figure 6: Effect of crack width and concrete quality on chloride-induced i_{corr} [18]

The study comprised of 100 x 100 x 500 mm beam specimens from four concrete mixes made using two w/b ratios (0.40 and 0.55) and two binder types (100% CEM I 42.5N (PC) and 50/50 PC/Corex slag (GGCS) blend. Details relating to GGCS can be found in [35]). A constant concrete cover of 40 mm was used. The beam specimens were pre-cracked (including the incipient cracks, being cracks that effectively had zero crack width) before being subjected to a cycle of 3-day wetting (with 5% NaCl) and 4-day air-drying under laboratory conditions (16-24 °C and 60-75% relative humidity) for a period of 32 weeks. From the study, the following conclusions were drawn:

- The initiation phase may either be substantially shortened or completely eliminated depending on crack width, concrete quality and resistivity.
- Cracking affects the i_{corr} during the propagation phase but the extent depends on the interaction between crack width, concrete quality and concrete resistivity.

To demonstrate the significance of the effect of crack width, concrete quality and resistivity on i_{corr}, Alonso et al.'s model [14], Eqn. (2), is compared with results obtained in a study by Otieno et al. [36] (
Figure 7).

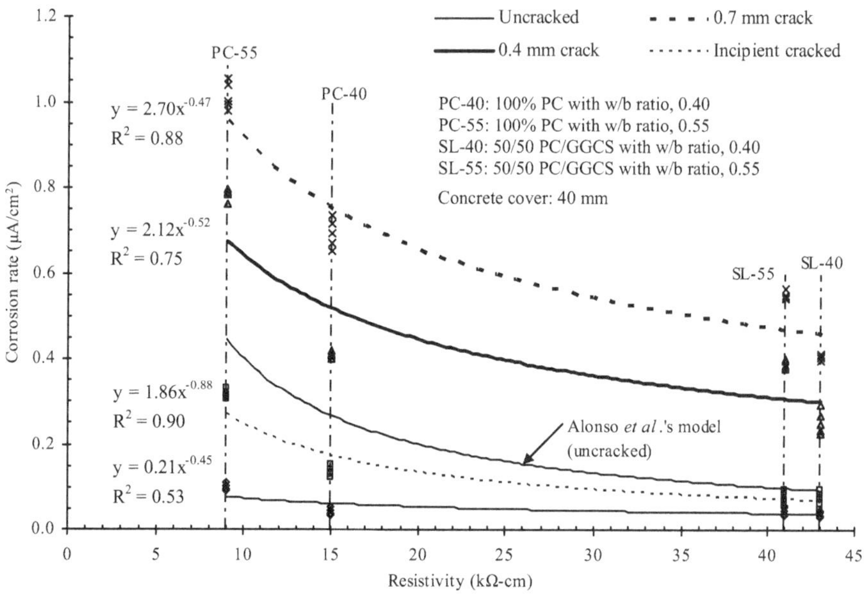

Figure 7: Effect of cracking, concrete quality and resistivity on i_{corr} (comparison of Alonso et al.'s model [14] and Otieno's results [37])

Even though the experimental set-ups in the two studies are different, this comparison shows that crack width, concrete quality and resistivity are significant factors when predicting i_{corr}.. Concluding from the data presented in
Figure 7, the relationship between i_{corr} and concrete resistivity (ρ) may be expressed as (Eqn. (13)):

$$i_{corr} = k\rho^{-m} \qquad (13)$$

where k and m are coefficients depending on crack width, concrete quality and/or resistivity. It is therefore important that cracking, among other factors that affect corrosion rate [10], are incorporated in i_{corr} prediction models. However, it is important to note that both Alonso et al. and Otieno et al.'s data were obtained from accelerated tests and should be verified with natural corrosion results.

Corrosion rate measurement techniques

Different assessment techniques and instruments can be used to quantify i_{corr} both in the laboratory and in the field. However, significant differences have been reported between different instruments when used to measure i_{corr} under the same conditions [12, 38]. For example, Liu and Weyers' [12] reported that the difference among three i_{corr} assessment techniques (gravimetric, 3LP and Gecor) was more than an order of magnitude; up to a factor of 17 with the 3LP being highest (Figure 8).

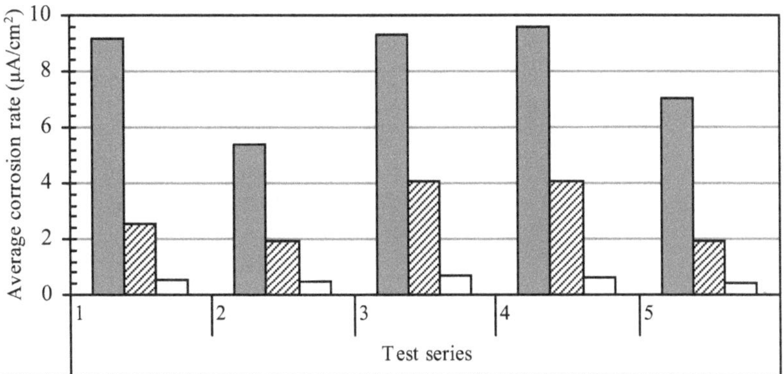

	1	2	3	4	5
3LP	9.17	5.39	9.32	9.56	7.01
Gravimetric	2.54	1.94	4.07	4.07	1.95
Gecor	0.58	0.45	0.72	0.64	0.42

Figure 8: Average i_{corr} from different assessment techniques [Data from 12]

It is therefore important that the assessment technique used to obtain the i_{corr} data used during the model development stage is clearly stated. In this way, adjustments can be made to obtain relevant i_{corr} values for use in the model. General guidelines for interpreting the results of the 3LP and Gecor are summarized in *Table I* [39, 40]. It must be noted that the interpretation guidelines for 3LP readings given in *Table I* may not be applicable to certain concrete and/or exposure conditions e.g. cracked concrete.

Table I: Gecor and 3LP interpretation guidelines [39, 40]

i_{corr} (μA/cm^2)	Gecor (corrosion state)	i_{corr} (μA/cm^2)	3LP (expected damage)
< 0.1	Passive	< 0.2	No damage expected
0.1 – 0.5	Low corrosion	0.2 – 1.1	Damage expected in 10-15 years
0.5 – 1.0	Moderate	1.1 – 11	Damage expected in 2-10 years
> 1.0	High corrosion	> 11	Damage expected in < years

Furthermore, in addition to the i_{corr} values differing from one measurement technique to another, they also differ from the actual (gravimetric) values. In the example presented here (Figure 8), the Gecor and 3LP i_{corr} values are conspicuously higher (up to 261%) and lower (up to 84%) respectively than the gravimetric results (which are usually used in the laboratory to depict actual i_{corr}).

In addition to different assessment techniques giving different results under the same measurement conditions, other techniques e.g. HCP measurement have been reported to give varied results depending on the measurement conditions. With respect to HCP, studies have shown that there is no general correlation between i_{corr} and HCP [21]. Furthermore, the HCP values can only be interpreted to give the likelihood or expected severity of corrosion in the structure. The use of such a measurement technique to obtain data for model development (for example as done by Katwan et al. [19]), may therefore not be ideal.

In general, i_{corr} results from different corrosion assessment techniques should be carefully interpreted and/or compared bearing in mind the technique employed.

Validation of models

Performing accelerated corrosion experiments has become a common technique to simulate corrosion-induced damage in RC structures, mainly because results can be obtained within a short period of time. However, this technique has been criticised for not being representative of the natural corrosion process and hence the results obtained from such tests may not be reliably extended to real structures [41, 42]. Taking into consideration that most of the i_{corr} prediction models are developed using results from accelerated experiments [e.g. 12, 14, 17, 18, 25, 27], it is important that to validate them using real/natural i_{corr} data. This is an important stage in ensuring that the model is representative of the actual performance of in-service RC structures.

Accounting for variability

Variability of model input parameters is important and should be modelled in such a way that realistic results, and hence decisions, can be derived from the model. In order to adequately account for the variability in i_{corr}, it is important that the variability of the i_{corr}-influencing factors such as cover depth, cracking and concrete quality are first taken into consideration. The outcome of such an approach should be expected i_{corr} trend-ranges/envelopes and not single ones.

However, accounting for variability should not only be taken to mean a statistical analysis of the i_{corr} data, regardless of how rigorous the process may be. It requires availability of not only a representative sample but also accurate and reliable data. . None of the available models covered in the previous sections is based on probabilistic analysis. The development of a probabilistic model has been hindered mainly by lack of adequate data to determine relevant statistical distributions for the various i_{corr}-influencing parameters [24]. Further, the available data must be compatible with each other especially with respect to the measurement techniques.

Taking into consideration the critique on the use of accelerated corrosion experiments discussed above, it may be debatable as to whether a probabilistic model developed using data from an accelerated corrosion testing regime is valid or not to be applied to real structures. The use of natural i_{corr} data would be ideal but in its absence, using data from accelerated tests appear acceptable subject to the model being validated in future as natural i_{corr} data becomes available.

Future Outlook (Ongoing Study)

The review presented in this paper was motivated by an ongoing study to develop a chloride-induced i_{corr} prediction model. The impetus for the study emanates from the need to quantify the corrosion propagation phase (t_p) of the service life of corrosion-affected RC structures; this can assist in the development of pro-active maintenance and repair strategies. Furthermore, the need to improve on some of the shortcomings of the existing models discussed in this paper must be pointed out. A brief summary of the ongoing study is presented in the next sections.

Objectives and expected outcomes of the study

The main objectives of the ongoing study are to:

1. Develop a finite element numerical i_{corr} prediction model for chloride-induced t_p in RC structures. The model will be developed based on the findamental electrochemical corrosion principles, and implemented using a commercially available finite element modelling software.
2. Validate the model using both laboratory and in-situ natural corrosion assessment data. This will involve (*i*) detailed comparison of the numerical model results with the in-situ and laboratory results, and (*ii*) modifying the model, if necessary, to fit into the in-situ corrosion rate results.
3. Integrate the model into an existing corrosion-induced damage model to obtain a comprehensive/complete service life prediction incorporating both the initiation and propagation phases of corrosion.

Important aspects of the study, in a bid to improve on some of the shortcomings of the available model discussed in this paper, are as follows:

(i) Combined effect of crack width, concrete cover and concrete quality on corrosion rate. Concrete quality will be quantified using the durability index tests [43] (chloride conductivity, oxygen permeability and water sorptivity).

(ii) Use of natural corrosion results to validate the numerical model (see next section for details).

(iii) The inherent variability of corrosion rate will be accounted for by carrying out a stochastic analysis on the corrosion-influencing parameters viz concrete cover, crack width and concrete quality.

It is envisaged that at the end of the study, with the above objectives met, the following outcomes will be realised:

(i) Provision of recommendations with respect to pro-active conservation (maintenance and repair) strategies that can be undertaken to either attain the initial design service life (incorporating t_p) or extend it.
(ii) Determination of corrosion acceleration factors to relate accelerated (laboratory) and in-situ/field corrosion rates; this will enable the determination of equivalent natural corrosion rates from accelerated corrosion rates.
(iii) Inclusion of principles for obtaining performance diagnosis and prognosis of RC structures from corrosion assessment techniques and prediction models.

Framework and scope of the study

The study focuses on the quantification of the effect of crack width, concrete quality (binder type and w/b ratio) and concrete cover on $i_{corr,Cl}$. The experimental variables include: binder type (100% CEM I 42.5N (PC), 50/50 PC/GGBS and 70/30 PC/FA), w/b ratio (0.40 and 0.55), concrete cover (20 and 40 mm) and crack width (uncracked, incipient-cracked, 0.4 mm and 0.7 mm). For each binder type, concrete cover, crack width and w/b ratio, three beam specimens (120 x 130 x 375 mm long) will be made.

The cracked specimens will be pre-cracked under 3-point flexural machine loading. Thereafter, the respective crack widths (except the incipient crack) will be maintained using the 3-point loading rig shown in Figure 9. All the 0.4 and 0.7 mm cracked specimens will remain in individual loading rigs for the duration of the experimental programme. Parallel experiments will be run in the laboratory (accelerated corrosion) and in the field (natural corrosion in a spray/splash marine zone) in an attempt to establish the correlation between accelerated laboratory and natural field corrosion rates.

A series of corrosion assessments will be carried out on the specimens: i_{corr}, HCP, concrete resistivity and visual inspections. In addition, relevant tests will be performed to characterise both the fresh and hardened concretes.

In summary, the ongoing study will directly or indirectly simultaneously incorporate the following i_{corr}-influencing factors in the resulting prediction model: concrete quality, crack width, sustained loading, temperature, moisture content and cover depth. Ultimately, an improved and validated i_{corr} model is expected to be developed at the end of the study.

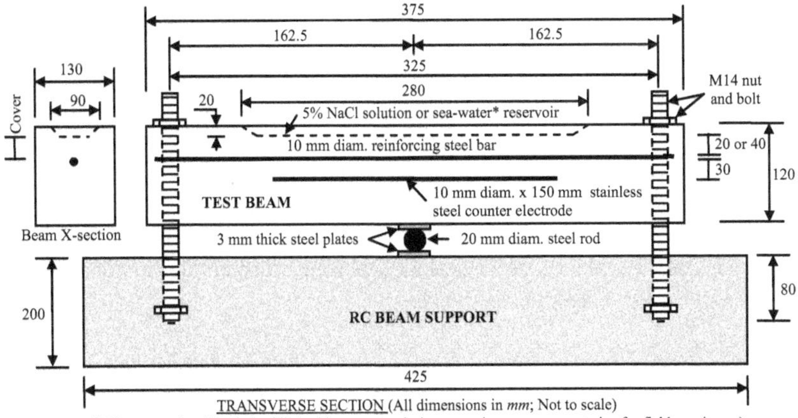

Figure 9: Beam loading set-up for 0.4 and 0.7 mm cracked specimens

Closure

Different aspects of various corrosion rate prediction models were reviewed in this paper. In summary, the following points can be made:

(i) Corrosion rate is one of the most important input parameters in corrosion-induced damage prediction and should therefore receive sufficient attention with respect to its assessment and prediction.

(ii) Corrosion-influencing factors such as cover cracking, resistivity, concrete quality and cover depth should be (explicitly) incorporated in corrosion rate prediction models.

(iii) Realistic prediction models and/or clear definition of corrosion rate and its assessment are needed to predict the actual performances of corrosion-affected RC structures. In addition, a clear definition of acceptable corrosion-induced damage levels should be made a priori.

(iv) Model validation using data from natural corrosion is a vital stage in model development especially for numerically developed models and those developed using data from accelerated corrosion tests. Therefore, both field/laboratory data should be combined with modelling to develop reliable corrosion rate prediction models.

(v) The inherent variability of corrosion rate and the factors influencing it should be accounted for in prediction models using probabilistic methods.

(vi) The ongoing (laboratory and field) study presented in this paper will attempt to address some of the salient corrosion-influencing factors presented by taking into account cover cracking, cover depth and concrete quality (binder type, w/b ratio). The variability of these factors will also be accounted for using a probabilistic approach.

References

[1] Raupach, M. (2006) Models for the propagation phase of reinforcement corrosion - an overview. *Materials and Corrosion,* 57(8), pp. 605-612.

[2] Redaelli, E., Bertolini, L., W., Peelen & Polder, R. (2006) FEM-models for the corrosion propagation period of chloride-induced reinforcement corrosion. *Materials and Corrosion,* 57(8), pp. 628-634.

[3] Warkus, J., Brem, M. & Raupach, M. (2006) BEM-models for the corrosion propagation period of chloride-induced reinforcement corrosion. *Materials and Corrosion,* 57(8), pp. 636-641.

[4] Zhang, R., Castel, A. & Francois, R. (2009) Serviceability limit state criteria based on steel–concrete bond loss for corroded reinforced concrete in chloride environment. *Materials and Structures,* Vol. 42, pp. 1407-1421.

[5] Torres-Acosta, A. A., Navarro-Guitierrez, S. & Teran-Guillen, J. (2007) Residual flexure capacity of corroded reinforced concrete beams. *Engineering Structures,* Vol. 29, pp. 1145-1152.

[6] Liu, Y. & Weyers, R. E. (1998) Modelling the time-to-corrosion cracking in chloride contaminated reinforced concrete structures. *ACI Materials Journal,* 95(6), pp. 675-681.

[7] El Maaddawy, T. & Soudki, K. (2007) A model for prediction of time from corrosion initiation to corrosion cracking. *Cement & Concrete Composites,* 29(3), pp. 168-175.

[8] Li, C. Q., Melchers, R. E. & Zheng, J. J. (2006) An analytical model for corrosion-induced crack width in reinforced concrete structures. *ACI Structural Journal,* 103(4), pp. 479-487.

[9] Cox, D. R. (1958) *Planning of experiments*, John Wiley and Sons, Inc., New York.

[10] Otieno, M. B., Beushausen, H.-D. & Alexander, M G. (2010) Corrosion propagation in reinforced concrete structures - state of the art review and way forward. *Proceedings of the 6th international conference on Concrete under severe conditions - Environment and loading, CONSEC'10.* June 7th -9th 2010, Merida, Yucatan, Mexico. Accepted for publication.

[11] Isgor, O., B. & Razaqpur, A. G. (2006) Modelling steel corrosion in concrete structures. *Materials and Structures,* Vol. 39, pp. 291-302.

[12] Liu, T. & Weyers, R. W. (1998) Modelling the dynamic corrosion process in chloride contaminated concrete structures. *Cement and Concrete Research,* 28(3), pp. 365-379.

[13] Song, X. & Liu, X. (2000) Experiment research on corrosion of reinforcement in concrete through cathode-to-anode area ratio. *ACI Materials Journal,* 97(2), pp. 148-155.

[14] Alonso, C., Andrade, C. & Gonzalez, J. A. (1988) Relation between resistivity and corrosion rate of reinforcements in carbonated mortar made with several cement types. *Cement and Concrete Research,* 18(5), pp. 687-698.

[15] Azad, A. K., Ahmad, S. & Al-Gohi, B. H. A. (2010) Flexural strength of corroded reinforced concrete beams. *Magazine of Concrete Research,* 62(6), pp. 405-414.
[16] Andrade, C. (2009) Types of models of service life of reinforcement: the case of the resistivity. *Proceedings of the 7th Asia Pacific Structural Engineering and Construction Conference & 2nd European Asian Civil Engineering Forum (APSEC / EACEF 2009).* 4-6 August 2009, Langkawi, Malaysia, Kedah, Malaysia pp. 30-35.
[17] Yalcyn, H. & Ergun, M. (1996) The prediction of corrosion rates of reinforcing steels in concrete. *Cement and Concrete Research,* 26(10), pp. 1593-1599.
[18] Otieno, M. B., Alexander, M. G. & Beushausen, H. D. (2010) Corrosion in cracked and uncracked concrete - influence of crack width, concrete quality and crack re-opening. *Magazine of Concrete Research,* doi: 10.1680/macr.2008.62.00.1.
[19] Katwan, M. J., Hodgkiess, T. & Arthur, P. D. (1996) Electrochemical noise technique for the prediction of corrosion rate of steel in concrete. *Materials and Structures,* 29(5), pp. 286-294.
[20] Page, C. L. & Lambert, P. (1986) Analytical and electrochemical investigations of reinforcement corrosion. *Contractor Report 30,* Transport and Road Research Laboratory (TRRL), Crowthorne.
[21] Broomfield, J. P. (2007) *Corrosion of steel in concrete - understanding, investigation and repair (2nd Edition)*, Taylor & Francis, Oxford, United Kingdom.
[22] Scott, A. N. & Alexander, M. G. (2007) The influence of binder type, cracking and cover on corrosion rates of steel in chloride-contaminated concrete. *Magazine of Concrete Research,* 59(7), pp. 495-505.
[23] Bentur, A., Diamond, S. & Berke, N. (1997) *Steel corrosion in concrete, Fundamentals and Civil Engineering Practice*, E & FN Spon, London. pp. 41-43.
[24] Duracrete (1998) Probabilistic performance based durability design: modelling of degradation. Document, D. P. BE95-1347/R4-5, The Netherlands,
[25] Vu, K. & Stewart, M. G. (2000) Structural reliability of concrete bridges including improved chloride-induced corrosion models. *Structural Safety,* 22(4), pp. 313-333.
[26] Vu, K., Stewart, M. G. & Mullard, J. (2005) Corrosion-Induced Cracking: Experimental Data and Predictive Models. *ACI Structural Journal,* 102(5), pp. 719-726.
[27] Scott, A. N. (2004) The influence of binder type and cracking on reinforcing steel corrosion in concrete. *PhD Thesis*, Department of Civil Engineering, University of Cape Town.
[28] Streicher, P.E. & Alexander, M. G. (1995) A chloride conduction test for concrete. *Cement Concrete Research,* 25(6), pp. 1284–1294.

[29] Martínez, I. & Andrade, C. (2009) Examples of reinforcement corrosion monitoring by embedded sensors in concrete structures. *Cement & Concrete Composites,* Vol. 31, pp. 545-554.
[30] Kim, C. Y. & Kim, J. K. (2008) Numerical analysis of localized steel corrosion in concrete. *Construction and Building Materials,* 22(6), pp. 1129-1136.
[31] Revie, R. W. & Uhlig, H. H. (2008) *Corrosion and Corrosion Control: An introduction to corrosion science and engineering, 4th Edition*, John Wiley & Sons Inc., New Jersey, USA.
[32] Dao, L. T. N., Dao, V. T. N. & Ann, K. Y. (2010) Modeling Steel Corrosion in Concrete Structures - Part 2: A Unified Adaptive Finite Element Model for Simulation of Steel Corrosion. *International Journal of Electrochemical Science,* Vol. 5, pp. 314-326.
[33] Melchers, R. E. & Li, C. Q. (2006) Phenomenological Modeling of Reinforcement Corrosion in Marine Environments. *ACI Materials Journal,* 103(1), pp. 25-32.
[34] Yuan, Y., Ji, Y. & Jiang, J. (2009) Effect of corrosion layer of steel bar in concrete on time-variant corrosion rate. *Materials and Structures,* Vol. 42, pp. 1443-1450.
[35] Mackechnie, J. R., Alexander, M. G. & Jaufeerally, H. (1993) Structural and durability properties of concrete made with Corex slag, *Research monograph No. 6*, Department of Civil Engineering, University of Cape Town.
[36] Otieno, M. B., Alexander, M. G. & Beushausen, H. D. (2010) Suitability of various measurement techniques for assessing corrosion in cracked concrete. *Accepted for publication in the ACI Materials Journal.*
[37] Otieno, M. B. (2008) Corrosion Propagation in Cracked and Uncracked Concrete. *Masters Dissertation*, Department of Civil Engineering, University of Cape Town.
[38] Breysse, D., Klysz, G., Derobert, X., Sirieix, C. & Lataste, J. F. (2008) How to combine several non-destructive techniques for a better assessment of concrete structures. *Cement and Concrete Research,* 39, pp. 783-793.
[39] Clear, K. C. (1992) Measuring Rate of Corrosion of Steel in Field Concrete Structures. *Transportation Research Board,* Record No. 1211, Washington DC, pp. 28-37.
[40] Broomfield, J. P., Rodriguez, J., Ortega, L. M & Garcia, A. M. (1993) Corrosion rate measurements in reinforced concrete structures by a linear polarization device. *Proceedings of the International symposium on condition assessment, protection, repair, and rehabilitation of concrete bridges exposed to aggressive environments, ACI Fall Convention.* 9th –10th November, Minneapolis, MN. pp. 644-651.
[41] Alonso, C., Andrade, C., Rodriguez, J. & Diez, J. M. (1998) Factors controlling cracking of concrete affected by reinforcement corrosion. *Materials and Structures,* Vol. 31, pp. 435-441.

[42] Andrade, C., Alonso, C. & Molina, F. J. (1993) Cover cracking as a function of bar corrosion: Part I-Experimental test. *Materials and Structures,* Vol. 26, pp. 453-464.

[43] Alexander, M G., Ballim, Y. & Mackechnie, J. R. (2001) Use of durability indexes to achieve durable cover concrete in reinforced concrete structures. *Materials Science of Concrete,* VI, pp. 483-511.

Proposal of Corrosion Rate Analytical Model of Reinforced Concrete with Crack

Shinichi Miyazato[1] and Yusuke Hasegawa[2]

[1]Kanazawa Institute of Technology, Japan
[2]Kajima-Renovate Co. Ltd., Japan

Abstract. The initiation stage can be estimated by analyzing the penetration of chloride ions in concrete. In addition, it is necessary to predict the corrosion rate after the propagation stage. On the basis of this background, in this paper, a basic model that estimates a steel corrosion rate in reinforced concrete with a crack is proposed. Therefore, first, a model that can calculate the microcell and macrocell corrosion rates is constructed. The input data of this model are the anodic polarization curve, cathodic polarization curve, polarization resistance, and electrical resistivity of concrete. This model is further verified using a mortar specimen and a concrete specimen. The experimental parameters are the water-cement ratio, crack width, and humidity. Consequently, it can be confirmed that the analyzed and measured values are equal. Moreover, the engineering value of this model is confirmed. Based on the above results, the following conclusions can be drawn: 1) a model that can be used for analyzing the steel corrosion rate of the reinforced concrete with the crack is proposed. 2) By carrying out the experiment using the mortar specimen and the concrete specimen with a crack, we can confirm the validity of the model. 3) The influences of the water-cement ratio, crack, and humidity on the steel corrosion rate can be confirmed by using this model.

Introduction

A considerable amount of steel reinforced concrete is used for social infrastructure. Therefore, it shall be placed in service by safety comfortably. The steel reinforced concrete structure that is exposed to severe environmental conditions contributes to the society in a number of ways. In such cases, the reinforcing steel bar corrodes, particularly at the cracked area, because of chloride ions. In the case of a performance-based design, it is necessary to check the long-term performance of a concrete structure by considering such deterioration. Therefore, the long-term deterioration prediction of a concrete structure is attempted. Appropriate materials

C. Andrade and G. Mancini (eds.), *Modelling of Corroding Concrete Structures*,
RILEM Bookseries 5, DOI 10.1007/978-94-007-0677-4_3, © RILEM 2011

can be selected for such a structure and a section of the structure can be designed efficiently, while considering the performance of the structure during a service period, if the deterioration speed of the steel reinforced concrete is predicted quantitatively in the future.

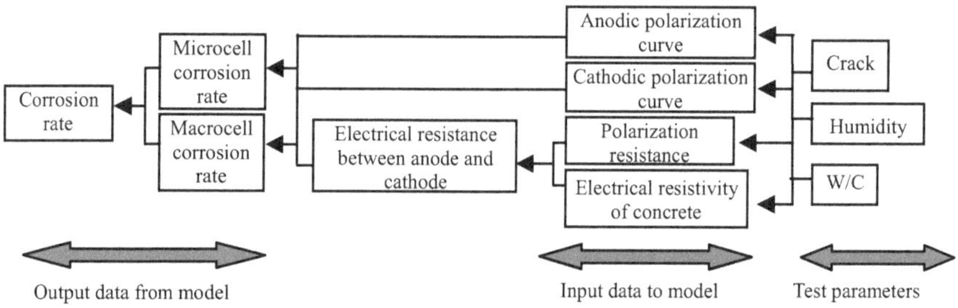

Figure 1. Relationship between factors that influence corrosion rate of steel in concrete

Thus far, the deterioration in the initiation stage can be predicted. That is, chloride ion ingress in a concrete can be analyzed, and the corrosion starting time can be predicted. However, it is necessary to further simulate the corrosion rate to predict the deterioration in the propagation stage.

On the basis of the above background, in this study, a basic model that estimates the corrosion rate in the reinforced concrete with a crack is proposed. Therefore, the construction of this model is discussed in Section 2, and it is verified by using a mortar specimen in Section 3 and by using a concrete specimen in Section 5. Further, the engineering value of this model is confirmed in Section 4.

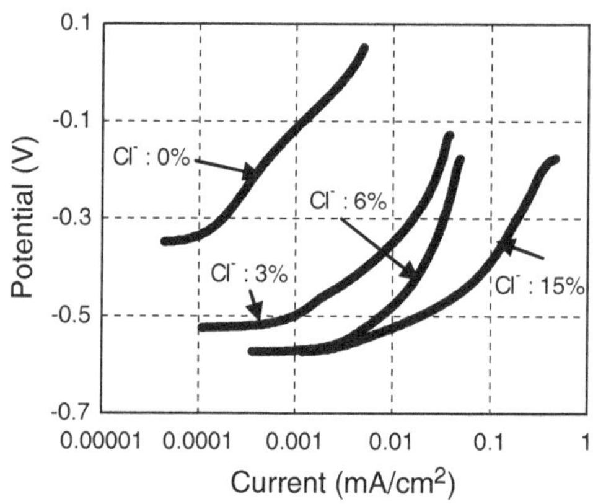

Figure 2. Example of anodic polarization curves of steel in saturated hydraulic calcium oxide aqueous solution whose chloride ion concentration differed

Construction of Model

Flow of model

The relationship of the factors, which influence the corrosion rate of the steel in concrete, is shown in Figure 1. The corrosion rate, which is the most important information for determining durability, is shown on the left. Further, this corrosion rate can be calculated by totaling the microcell and macrocell corrosion rates. Herein, the microcell is the narrow electric circuit between the anode and cathode, whereas the macrocell is the electric circuit in which the anode and the cathode are widely separated. Next, these corrosion rates can be electrochemically calculated using the polarization curves. This polarization curve shows the polarization characteristics of the electrode and is changed by the quantities of the reactant, etc. For example, the anodic polarization curves of the steel in a saturated hydraulic calcium oxide aqueous solution, whose chloride ion concentration differs, is shown in Figure 2. Moreover, the electrical resistance between the anode and cathode such as the polarization resistance of the steel and the electrical resistivity of concrete must be considered. Therefore, the output of the model is the corrosion rate, and the anodic polarization curve, cathodic polarization curve, polarization resistance, and electrical resistivity of concrete are the input data.

Setting of steel element and definition of corrosion cell as used in this paper

The reinforcing steel bar is divided into elements to analyze the macrocell and microcell corrosion rates. That is, one steel bar is considered to be a continuum of multiple steel elements, as shown in Figure 3. Further, the corrosion cell formation pattern is a "microcell," when the anode and the cathode exist in the single steel element. However, when the anode and the cathode are formed between different steel elements, the corrosion cell formation pattern is a "macrocell."

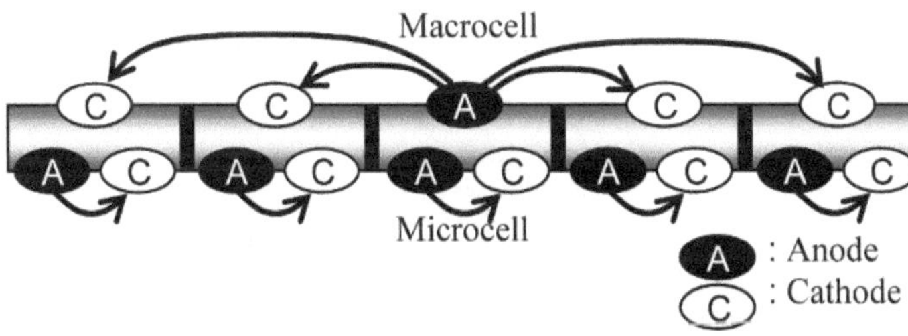

Figure 3. Divided rebar and corrosion cell formation pattern (Macrocell, Microcell)

Analysis of corrosion current

In Figure 4, the relationship between the polarization curves and the corrosion currents is explained. It is assumed that steel element-1 exists in concrete without the crack, while steel element-2 has a crack, as shown in Figure 4(a). Herein, the chloride ions and oxygen permeate easily from the external environment to steel element-2, in comparison with steel element-1. Therefore, both the anodic polarization curves that the chloride ions influence and the cathodic polarization curves that the oxygen influences are different in the case of steel elements-1 and 2.

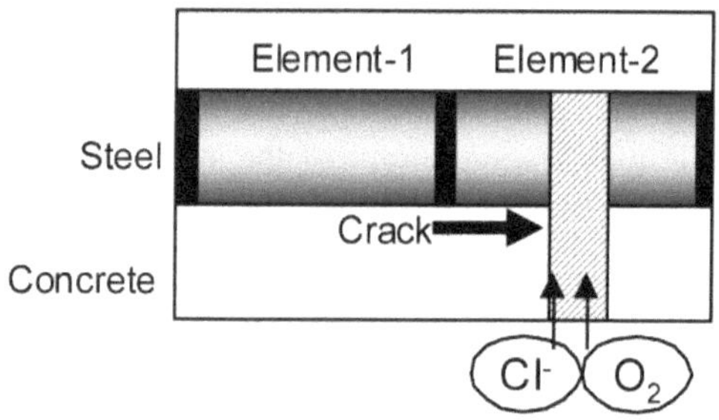

(a) Location of crack and steel element

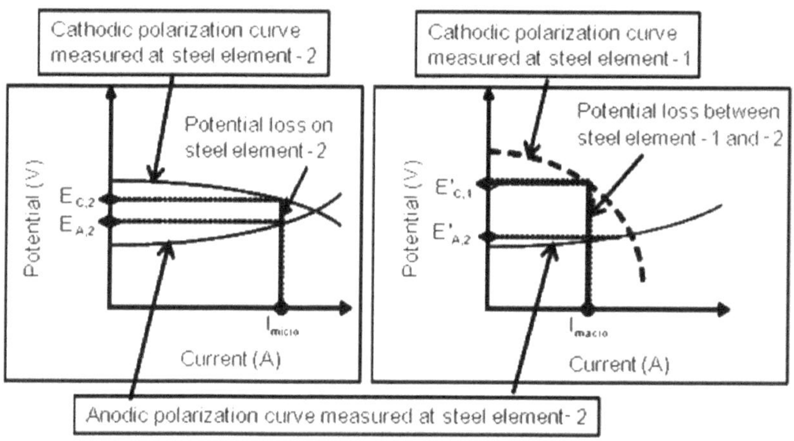

(b) Microcell formed only in steel element-2

(c) Macrocell formed between steel elements-1 (Cathode) and -2(Anode)

Figure 4. Relationship between polarization curves and corrosion currents

The intersection point of the anodic and cathodic polarization curves is electrochemically balanced. Therefore, the corrosion rate is controlled at this point. Herein, it is necessary to consider the potential loss by the resistance. Therefore, the polarization resistance is considered in the case of the microcell, as shown in Figure 5(a). That is, the microcell corrosion current, which satisfies Eqn. (1), flows on the left side from the intersection point of the anodic and cathodic polarization curves (the corrosion current becomes small), as shown in Figure 4(b).

$$E_{C,2} - E_{A,2} = (Rp_2 - Rp_2) \times I_{micro} \tag{1}$$

where Rp_2 represents the polarization resistance of steel element-2 (Ω).

On the other hand, it is also necessary to consider the potential loss with an increase in the polarization resistances and the electrical resistivity of concrete in the case of the macrocell, as shown in Figure 5(b). That is, the macrocell corrosion current, which satisfies Eqn. (2), flows as shown in Figure 4(c).

$$E'_{C,1} - E'_{A,2} = (Rp_1 - Rc_{1,2} - Rp_2) \times I_{macro} \tag{2}$$

where Rp_1 represents the polarization resistance of steel element-1 (Ω), and $Rc_{1,2}$ represents the electrical resistivity of concrete between steel element-1 and steel element-2 (Ω).

The corrosion current calculated by the above method is $I_{(i,j)}$, when steel element-i is the anode and steel element-j is the cathode.

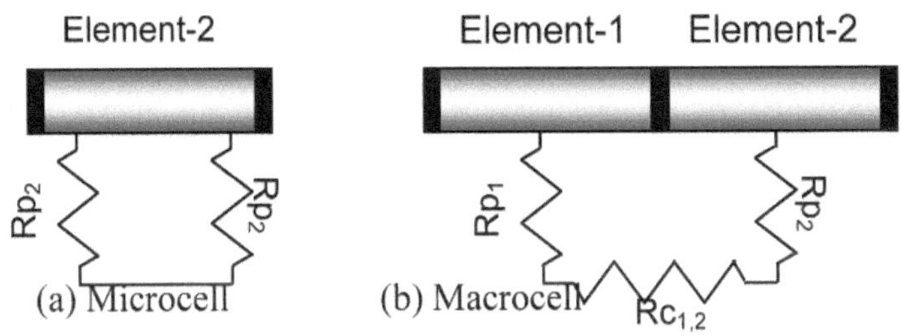

Figure 5. Resistance in corrosion cell as shown in Figure 4

Experiment for Verification of Model Using Mortar Specimen

Experimental procedure

Material used and mixture proportion. Ordinary Portland cement, land sand, and a round steel bar were used as listed in Table I. The water-cement (W/C) ratios of mortar were 0.30 and 0.50. S/C was 2.5. Further, 5.0% of high-range water reducing admixtures was added when the water-cement ratio was 0.30.

Table I. Materials used

	Type	Spec
Cement	Ordinary Portland Cement	Density : 3.16g/cm^3 Specific surface : 3270cm/g^2
Fine aggregate	Land sand	Density:2.59g/cm^3 Water absorption:2.83% Fineness modulus:2.58
Water	Tap water	
Steel bar	Round bar	Yield point:235N/ mm^2 over, φ:9mm

Experimental and analytical cases. The experimental and analytical cases are listed in Table II. The water-cement ratio of two levels, the crack width of three levels, and the humidity in the exposure air of two levels were studied.

Table II. Experimental and analytical cases of mortar specimen

No.	Crack width (mm)	Humidity (RH) (%)	W/C
1	No crack	90	0.50
2	0.2		
3	0.4		
4		60	
5		90	0.30

Specimen configuration. The used specimen is shown in Figure 6. The steel bar is composed of five elements for the purpose of analyzing the macrocell and microcell corrosion currents [1].

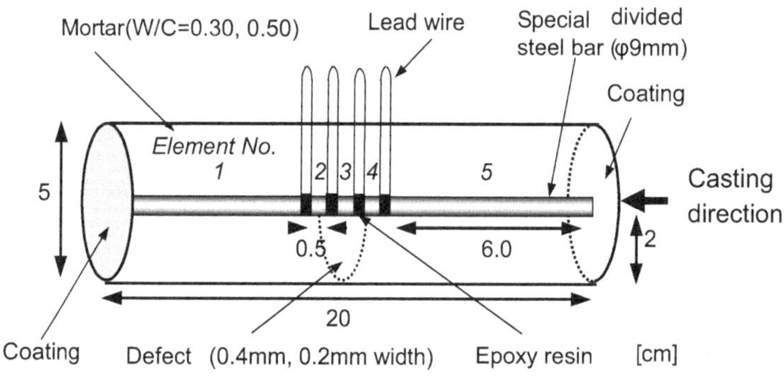

Figure 6. Mortar specimen configuration

The produced method of this specially divided steel bar is explained. Three 0.5-cm long steel elements and two 6.0-cm long steel elements are cut. Lead wires are soldered at both ends of each element before joining these elements with epoxy resin having a high insulating capacity. The short elements are used for examining in detail the corrosion near the defects.

The casting procedure is explained. Mortar is cast twice to form the artificial cracks. First, after the steel bar is placed at the center of the mold, the mortar is cast to half of the mould. Herein, the slit with a width of 0.2 mm or 0.4 mm is inserted to simulate a crack before the mortar is cast to another part of the mould. For specimen nos. 3, 4, and 5, as shown in Table II, the 0.4-mm-wide acrylic board is inserted and is pulled out after 6 h. In the case of specimen no. 2, a 0.2-mm-wide tissue is inserted. Further, specimen no. 1 does not have any defect. All specimens are subjected to the initial curing in the wet environment at 20°C and 90% RH for 28 days after removal from the mold. Then, all specimens are exposed to a chloride-accelerated environment at 20°C for 49 days. This environment is a two-process cycle for 3.5 days: one process is the exposure to a saltwater shower (NaCl 3.0 wt %) at 90% RH for 1 day, and the other is the exposure to an air at 90% or 60% RH for 2.5 days.

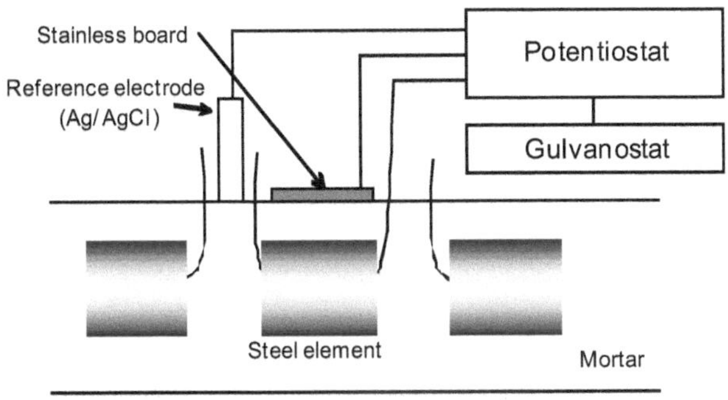

Figure 7. Measurement method of polarization curves

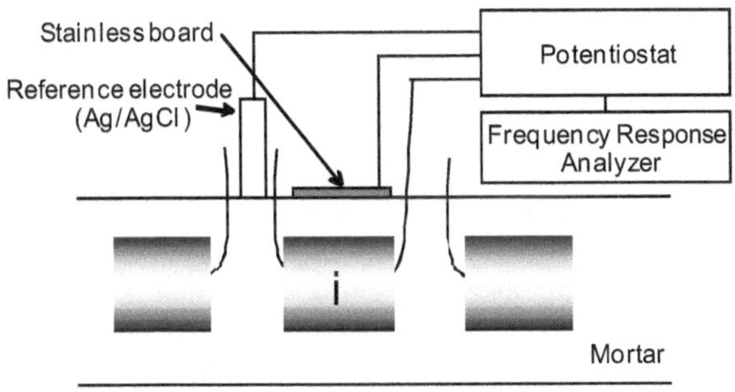

Figure 8. Measurement method of polarization resistance

Measurement of input data for the model. The anodic polarization curve, cathodic polarization curve, polarization resistance, and electrical resistivity of mortar were measured as input data to the model.

The measuring method of the anodic/cathodic polarization curves is shown in Figure 7. Ag/AgCl was used as the reference electrode. The potential of the steel element was changed to noble or ignoble at 1 mV/s [2], and the flowing current was recorded. Further, the polarization curve, which reacts only on the steel surface, was obtained considering the IR drop. Therefore, the potential losses by the polarization resistance on the steel surface and the electrical resistivity of mortar between the stainless board and the steel were considered.
The polarization resistance on the steel surface was measured by using the alternating current impedance method, as shown in Figure 8. First, the lead wire, which connected the steel elements, was cut off. In this condition as soon as possible, the polarization resistance of each steel element was measured within the range from 10 mHz to 10 kHz, thereby giving a voltage with a 50-mV amplitude. The examples of the measured Bode diagram and the Call-Call plotting are shown in Figure 9.

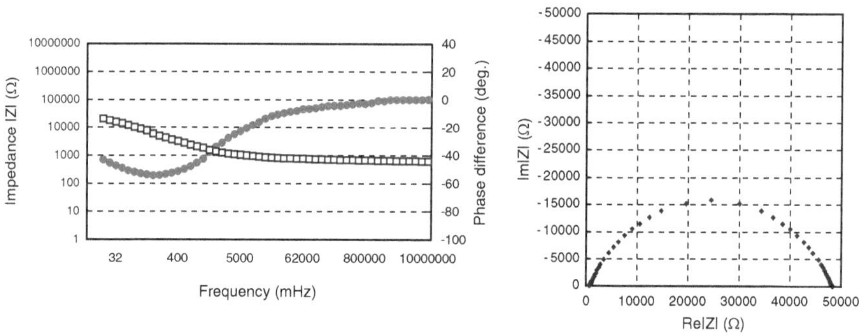

(a) Bode diagram (b) Call-Call plotting

Figure 9. Example of graphs measured for polarization resistance

The mortar resistance between two steel elements was measured by using the alternating current impedance method, as shown in Figure 10.

Measuring method of corrosion current as comparison data. Referring to the literatures [1, 3-5], we measured the macrocell and microcell corrosion currents by comparing the theoretical values with the output data of the model.

The measuring method of the macrocell corrosion current is shown in Figure 11. The non-resistance ammeter was connected between adjoining steel elements, and the macrocell corrosion current was measured. Next, the current that flowed from the steel elements on both sides was totaled. By dividing this current by the surface area of the steel element, we calculated the macrocell corrosion current density (I_{macro}) as shown in Eqn. (3).

$$I_{macro} = \frac{I_{i-1.i} + I_{i+1.i}}{S_i} \quad (3)$$

where I_{macro} represents the macrocell corrosion current at steel element i (A/cm^2), $I_{i.j}$ represents the corrosion current from steel element i to j (A), and S_i represents the surface area of steel element i (cm^2).

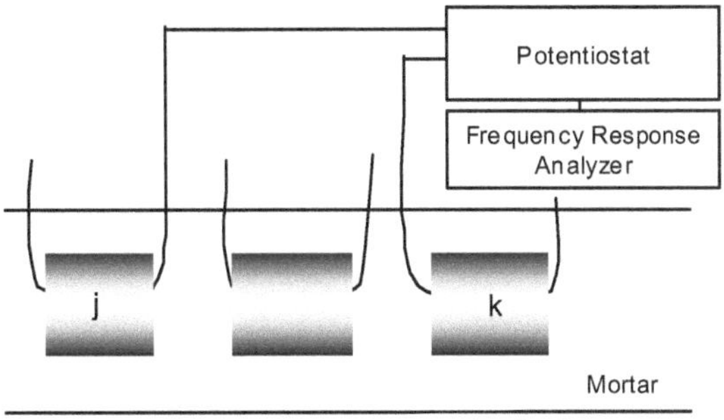

Figure 10. Measurement method of mortar resistance between steel elements j and k

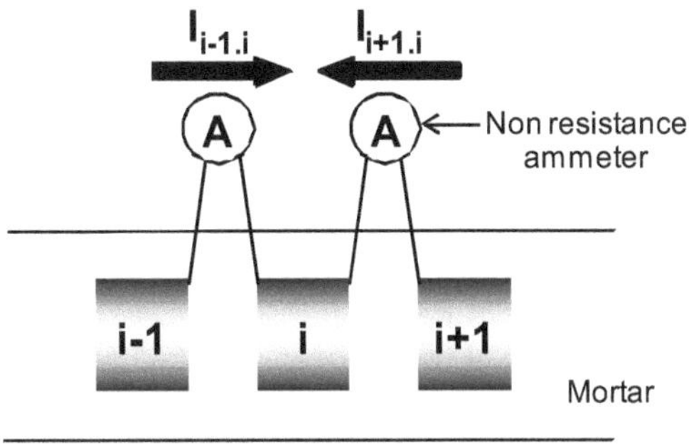

Figure 11. Measurement method of macrocell corrosion current

The microcell corrosion current density (I_{micro}) was obtained by using Eqn. (4) from the literatures [6-8].

$$I_{micro} = \frac{K}{Rp_i} \quad (4)$$

where I_{micro} represents the microcell corrosion current density at steel element i (A/cm^2), Rp_i represents the polarization resistance at steel element i measured as shown in Figure 8 (Ω cm^2), and K is a constant (=0.0209 V) [6].

Further, the sum of the measured macrocell and microcell corrosion currents was the total corrosion current.

Measured input data

Anodic polarization curve. The anodic polarization curves are shown in Figure 12. This figure shows that the anodic reaction is easy when the curve is located in the right area. According to Figure 12(a), the anodic corrosion current flows easily when there is a crack in the structure. According to Figure 12(b), the anodic corrosion current flows easily when the humidity is high. According to Figure 12(c), the anodic corrosion current flows easily when the water-cement ratio is high.

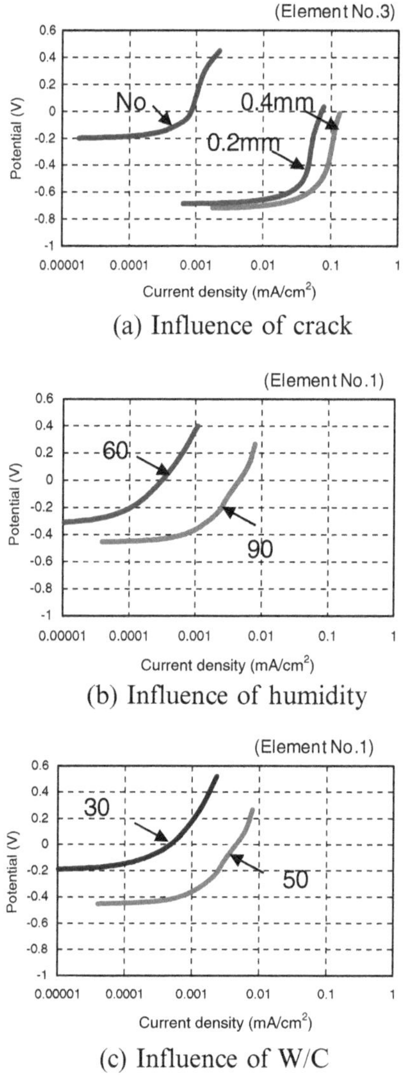

Figure 12. Example of measured anodic polarization curves

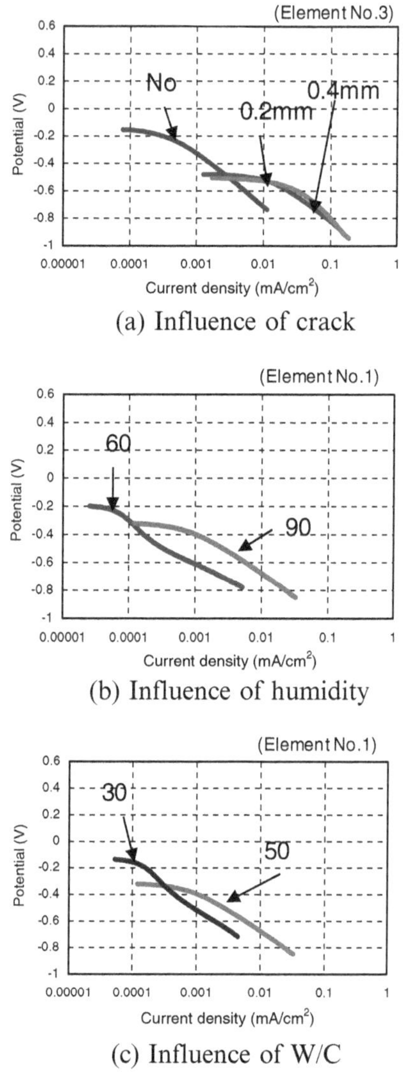

Figure 13. Example of measured cathodic polarization curves

Cathodic polarization curve. The cathodic polarization curves are shown in Figure 13. This figure shows that the cathode reaction is easy when the curve is located in the right area. According to Figure 13(a), the cathodic corrosion current flows easily when there is a crack. According to Figure 13(b), the cathodic corrosion current flows easily when the humidity is high. According to Figure 13(c), the cathodic corrosion current flows easily when the water-cement ratio is high.

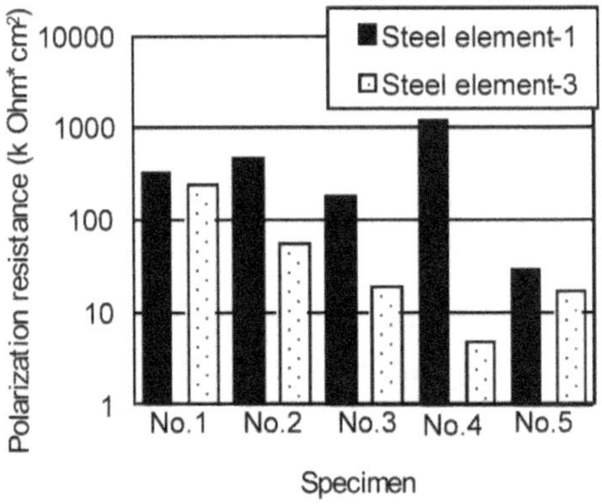

Figure 14. Example of polarization resistance (Results of steel element-3 of meeting with the crack and steel element-1 in the sound mortar are shown)

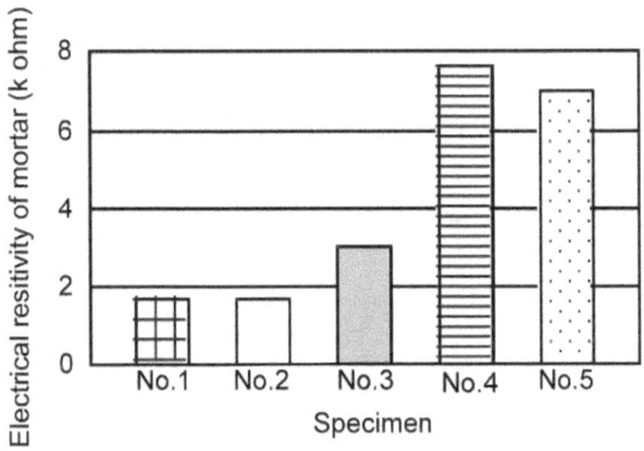

Figure 15. Example of electrical resistivity of mortar (Electrical resistance between steel elements-1 and -3 is shown as a representative value.)

Polarization resistance. The polarization resistances of steel elements-1 and -3 are shown in Figure 14. According to this figure, the polarization resistance is small at steel element-3 across the crack, in comparison with steel element-1, which is located in the sound mortar.

Electrical resistivity of Mortar. The electrical resistivity between steel elements-1 and -3 is shown in Figure 15. According to this figure, the electrical resistivity decreases when the crack width is narrow. In contrast, the electrical resistivity increases when the humidity is low. Further, the electrical resistivity increases when the water-cement ratio is low.

Analytical result

Example of analysis procedure. The analysis procedure for specimen no. 3 is discussed. All input data are shown in Figure 16 and Tables III and IV.

Table III. Example of input data of polarization resistance

Steel element No	Polarization resistance (kΩ cm^2)
1	184.6
2	196.6
3	18.3
4	69.8
5	82.7

Table IV. Example of input data of electrical resitivity of mortar (kΩ)

Steel element No	1	2	3	4	5
1	-	4.0	2.8	3.5	2.5
2	4.0	-	3.0	3.0	4.2
3	2.8	3.0	-	3.2	3.0
4	3.5	3.0	3.2	-	2.0
5	2.5	4.2	3.0	2.0	-

On the basis of the information given in Section 2, it is concluded that the anodic and cathodic polarization curves in the any steel element are piled up. As shown in Figure 17, the anodic and cathodic polarization curves at the identical steel element are combined when the microcell corrosion is analyzed. Further, two times the polarization resistance considers the potential loss as the electrical resistance between the anode and the cathode. On the other hand, as shown in Figure 18, the anodic and cathodic polarization curves in the different steel elements are combined when the macrocell corrosion is analyzed. Further, the sum of both the polarization resistances and the electrical resistivity of the mortar consider the potential loss. As a result of such analysis for all steel elements, Table V shows the corrosion current that flows among all steel elements.

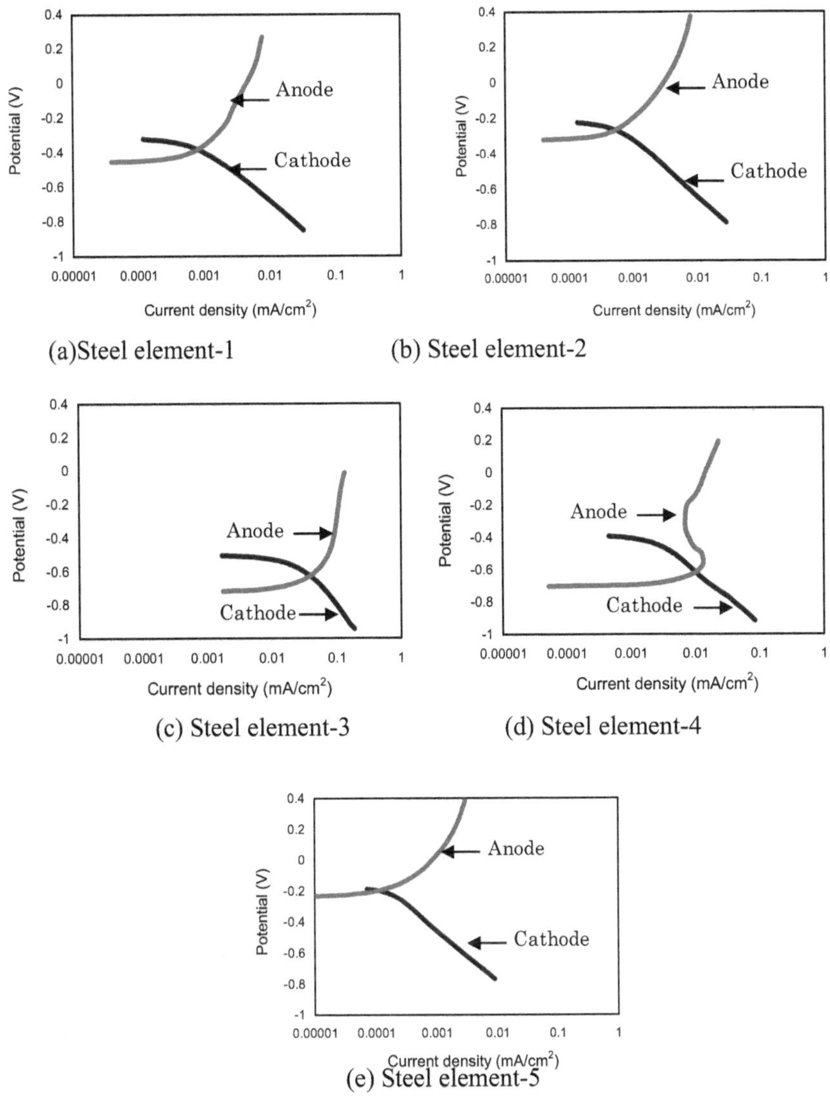

Figure 16. Example of input data of polarization curves

Table V. Example of analyzed corrosion current (μA)

		Cathodic steel element no.				
		1	2	3	4	5
Anodic steel element No	1	7	1	0	1	6
	2	0	0	0	0	1
	3	10	2	7	4	15
	4	5	1	2	2	5
	5	0	0	0	0	1

$$\text{Macrocell corrosion current density} = \sum_{j \neq i} I_{(i,j)} \div S_i \qquad (5)$$

$$\text{Microcell corrosion current density} = I_{(i,i)} \div S_i \qquad (6)$$

where S_i indicates the surface area of steel element no. i (cm^2).

The macrocell corrosion current density at steel element no. i is calculated as shown in Eqn. (5). On the other hand, the microcell corrosion current density at steel element no. i is calculated as shown in Eqn. (6). Further, the total corrosion current density is calculated as the sum of the macrocell and microcell corrosion current densities. The corrosion current density of each steel element is shown in Table VI. The corrosion current density of 100 $\mu A/cm^2$ can be converted into the corrosion rate of 1.16 mm/year.

Table VI. Example of analyzed corrosion current density ($\mu A/cm^2$)

Steel element No	microcell	macrocell	total
1	0.4	-0.4	0.4
2	0.1	-2.5	0.1
3	5.0	17.0	22.0
4	1.4	5.7	7.1
5	0.1	-1.3	0.1

Note: The plus values mean the corroding. Also, the minus macrocell corrosion current shows the cathodic current. Therefore, when the total corrosion current is calculated, the minus values of the macrocell current are considered '0'.

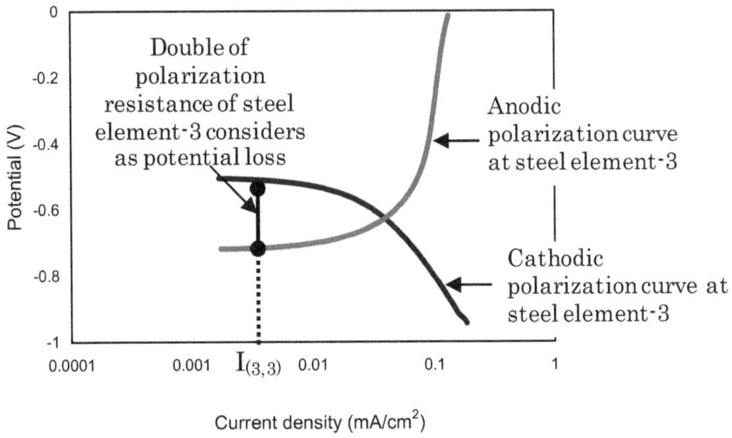

Figure 17. Example of analyzed microcell corrosion current ($I_{(3,3)}$, Steel element-3)

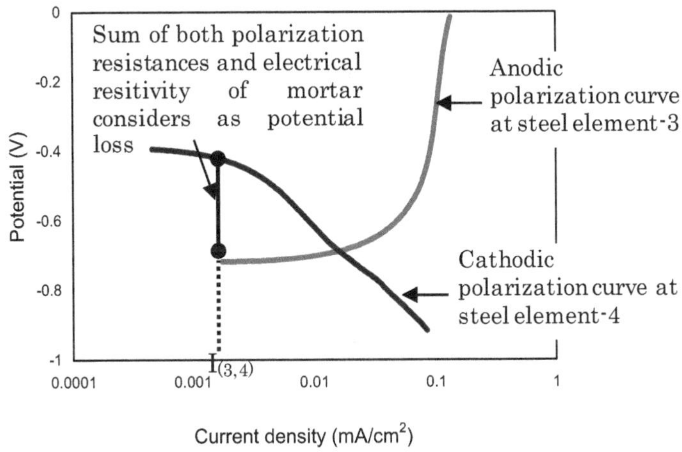

Figure 18. Example of analyzed macrocell corrosion current ($I_{(3,4)}$, Steel element-3 is anode and -4 is cathode)

The analytical result of the corrosion current density distribution is shown in Figure 19. According to this figure, it can be confirmed that both the macrocell and microcell corrosion current densities may become maximum at the crack.

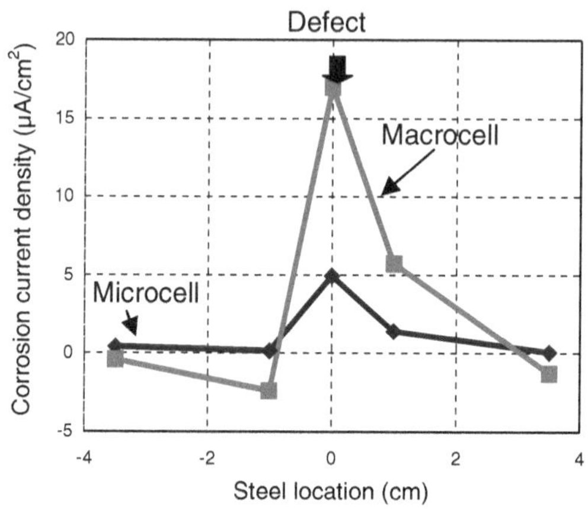

Figure 19. Example of analyzed corrosion current density

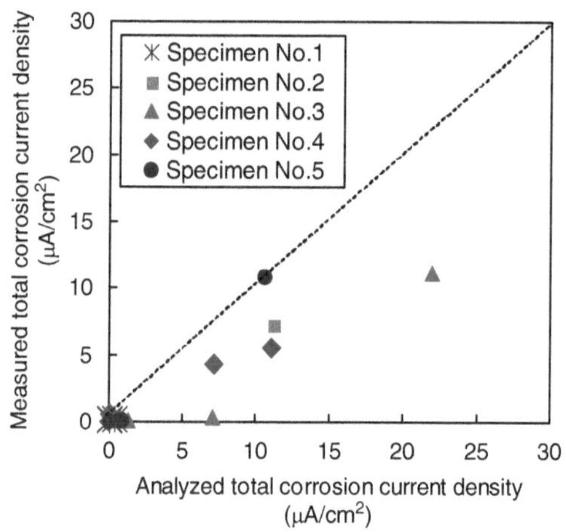

Figure 20. Comparison between analyzed and measured total

Comparison of experimental and analytical results. In Figure 20, the analyzed and measured values of the total corrosion current density at the all steel elements in all specimens are compared. According to this figure, it can be confirmed that the order of the analyzed and measured values may be equal. That is, 1) if the total corrosion current density is low, the analyzed and measured values become 0.5 $\mu A/cm^2$ or less and corrosion does not progress. 2) The differences between the analyzed and the measured values are less than three times if the total corrosion current density is high. Still, the influence of each parameter (existence of crack, humidity, and water-cement ratio) on the analyzed corrosion rate is compared in the next section to verify that this model is appropriate from an engineering point of view.

Influence of Various Conditions on Analyzed Corrosion Rate

Existence of crack

The influence of the existence of the crack on the total corrosion current density is shown in Figure 21. According to this figure, it can be confirmed that when the crack exists, the corrosion current density is high, and the local corrosion progresses. This phenomenon has been confirmed according to the previous experimental result [3, 9, 10] and the existing structure survey [11]. Therefore, it becomes clear that the influence of the existence of the crack on the corrosion rate can be reproduced analytically by this model.

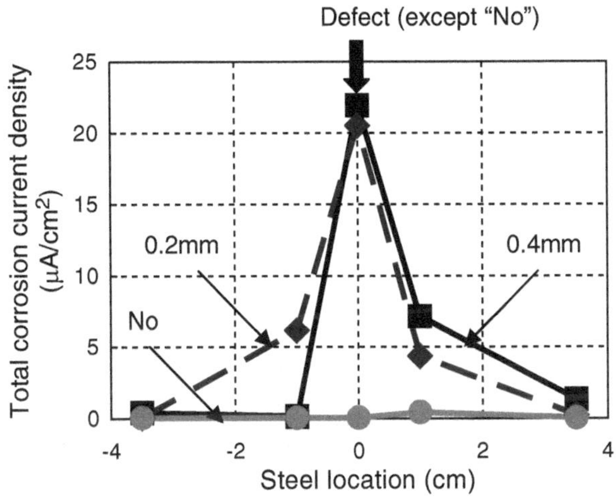

Figure 21. Influence of crack on corrosion rate

Humidity

The influence of the humidity on the total corrosion current density is shown in Figure 22. According to this figure, it can be confirmed that the corrosion current density of 90% R.H. is higher than that of 60%. This phenomenon has been confirmed according to previous experimental results [12, 13]. Therefore, it is clear that the influence of the humidity on the corrosion rate can be reproduced analytically by this model.

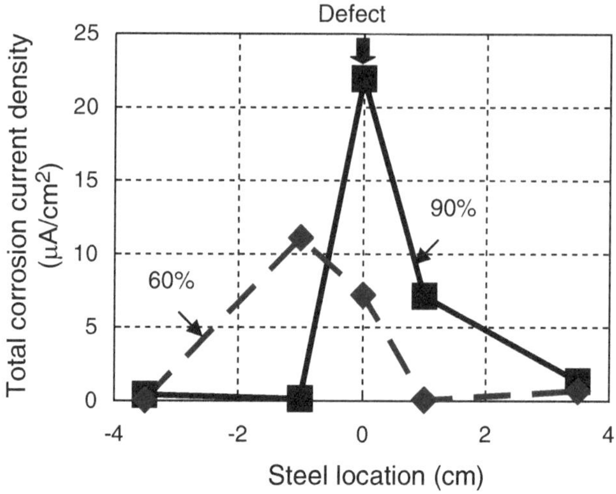

Figure 22. Influence of humidity on corrosion rate

Water-cement ratio

The influence of the water-cement ratio on the total corrosion current density is shown in Figure 23. According to this figure, the corrosion current density in the case of 0.50 W/C is higher than that in the case of 0.30 W/C. This phenomenon has been confirmed by previous experimental results as well [3]. Therefore, it is clear that the influence of the water-cement ratio on the corrosion rate can be reproduced analytically by this model.

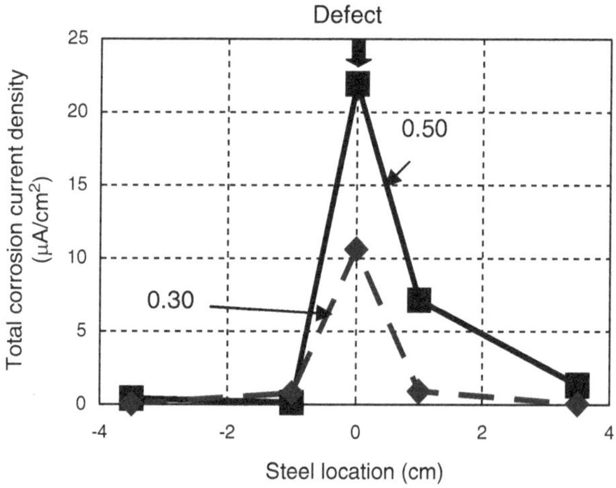

Figure 23. Influence of W/C on corrosion rate

Experiment for Verification Using Concrete Specimen

Experimental procedure

Material used and mixture proportion. Ordinary Portland cement, land sand, crushed stone, and deformed steel bar were used as shown in Table VII. The water-cement ratios were 0.30, 0.50, and 0.70. The mixture proportions are shown in Table VIII.

Table VII. Materials used

	Type	Spec
Cement (C)	Ordinary Portland Cement	Density:3.16g/cm^3, Specific surface:3270cm^2/g
Chemical admixture	Super plasticizer (SP)	Naphthalene-sulfonic acid
	Water reduce agent (AD)	Lignin sulfonic acid
Fine aggregate (S)	Land sand	Density:2.59g/cm^3, Water absorption:2.83%, Fineness modulus:2.58
Coarse aggregate (G)	Crushed stone	Density:2.64g/cm^3, Water absorption:1.40%, G.Max:20mm
Water (W)	Tap water	
Steel bar	Deformed bar	Yield point:235N/mm^2 over, D:10mm

Table VIII. Mixture proportions of concrete

W/C	s/a (%)	Unit weight (kg/m^3)				(g/m^3)	
						Ch. Ad.	
		W	C	S	G	SP	AD
0.30	45	179	597	704	864	4770	—
0.50	49	185	370	848	886	—	111
0.70	53	191	273	954	848	—	821

Experimental and analytical cases. The experimental and analytical cases are shown in Table IX. The water-cement ratio of the three levels and the crack in the three levels were studied.

Table IX. Experimental and analytical cases of concrete specimen

No.	Crack width (mm)	W/C
6	No crack	0.50
7	0.1	
8	0.4	0.30
9		0.50
10		0.70

Specimen configuration. The used specimen is shown in Figure 24. The steel bar was composed of ten elements. The crack was generated by the 3-point bending load after the initial curing in water at 20°C for 28 days. The crack width was 0.4- mm or 0.1-mm. Then, all specimens were exposed to the chloride-accelerated environment at 20°C for 91 days. That is, the salt solution was sprayed (NaCl 3.0 wt%, 600 ml for cracked plane) twice a day, while exposing the sample to wet air (RH 90%).

Measuring method of input data to model. The anodic polarization curve, cathodic polarization curve, polarization resistance, and electrical resistivity of concrete were measured for obtaining the input data with the same method as that discussed in Section 3. Further, the macrocell and microcell current densities were measured for the comparison data; this measurement was also carried out by using the same method that as discussed in Section 3.

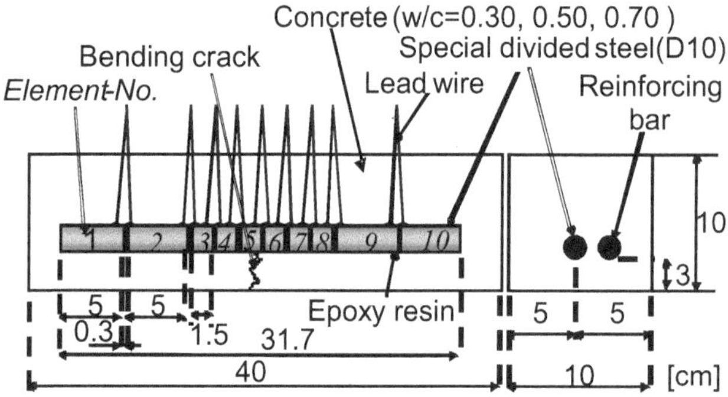

Figure 24. Concrete specimen configuration

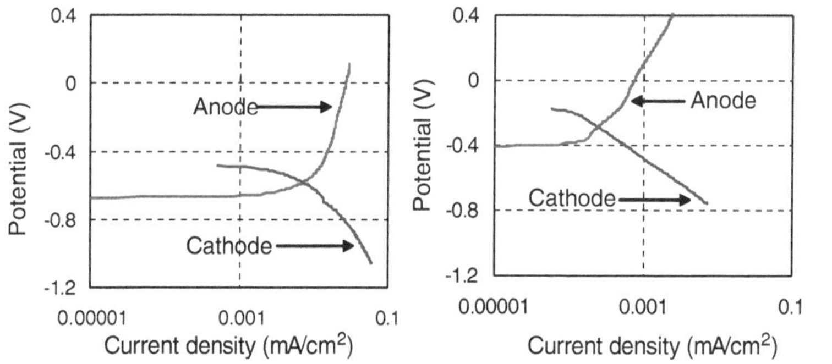

(a) Steel element-4 (Defect) (b) Steel element-9 (Sound)

Figure 25. Example of input data of polarization curves (Specimen no. 9)

Analytical result

Example of analysis procedure. The analysis procedure for specimen no. 9 is discussed. Examples of input data are shown in Figure 25, Table X, and Figure 26.

Table X. Example of input data of polarization resistance (Specimen no. 9)

Steel element No.	Polarization resistance (kΩ cm^2)
1	294.5
2	864.4
3	40.4
4	27.4
5	715.9
6	491.5
7	536.2
8	491.3
9	1791.7
10	1640.6

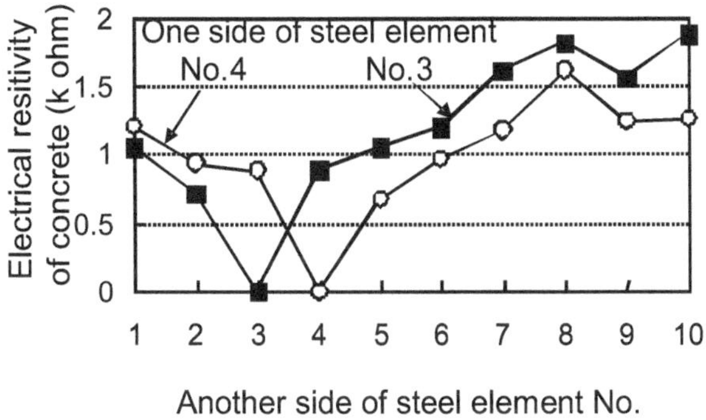

Figure 26. Example of input data of electrical resitivity of concrete (Specimen no. 9)

On the basis of the information given in Section 2, it can be concluded that the anodic and cathodic polarization curves in the any steel element are piled up. As shown in Figure 27, the anodic and cathodic polarization curves at the identical steel elements are combined when the microcell corrosion is analyzed. On the other hand, as shown in Figure 28, the anodic and cathodic polarization curves in different steel elements are combined when the macrocell corrosion is analyzed. As a result of such analysis of all steel elements, the corrosion current density of each steel element is listed in Table XI.

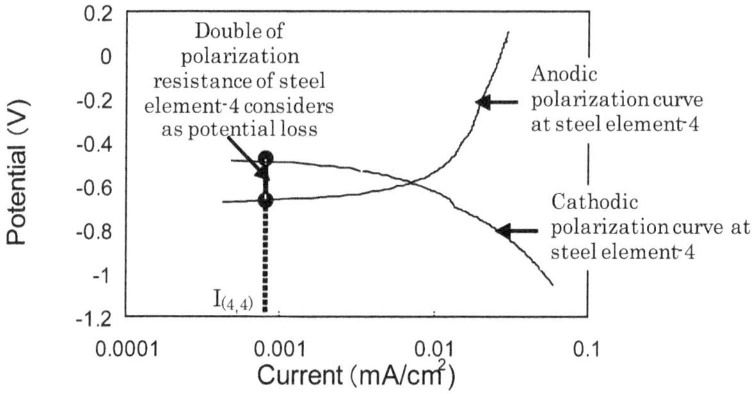

Figure 27. Example of analyzed microcell corrosion current ($I_{(4,4)}$, Steel element-4)

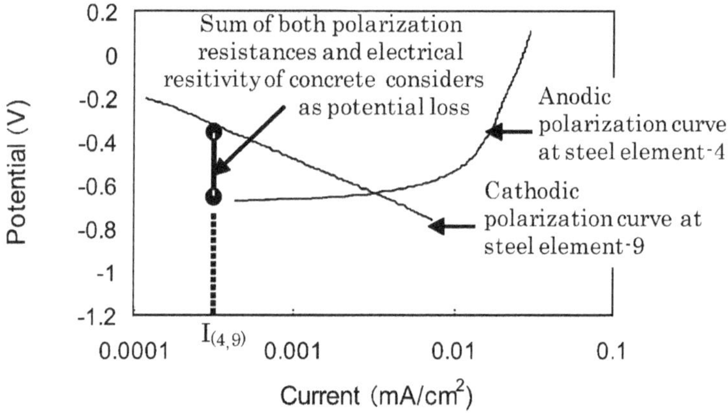

Figure 28. Example of analyzed macrocell corrosion current ($I_{(4,9)}$, Steel element-4 is anode and -9 is cathode)

Table XI. Example of analyzed corrosion current density (μA/cm^2)

Steel element no.	Microcell	Macrocell	Total
1	0.3	-1.2	0.3
2	0.1	-0.5	0.1
3	0.7	3.7	4.4
4	2.2	7.2	9.4
5	0.1	-0.8	0.1
6	0.1	2.0	2.1
7	0.1	-1.8	0.1
8	0.2	-1.2	0.2
9	0.5	-0.4	0.5
10	0.1	-0.7	0.1

The analytical result of the corrosion current density distribution is shown in Figure 29. According to this figure, both the macrocell and the microcell corrosion current densities may become maximum at the crack.

Comparison of experimental and analytical results. In Figure 30, the analyzed and measured values of the total corrosion current density at all the steel elements in all the specimens are compared. According to this figure, the order of the analyzed and measured values may be equal. In addition, the differences between the analyzed and the measured values are considered, and the accuracy will be improved.

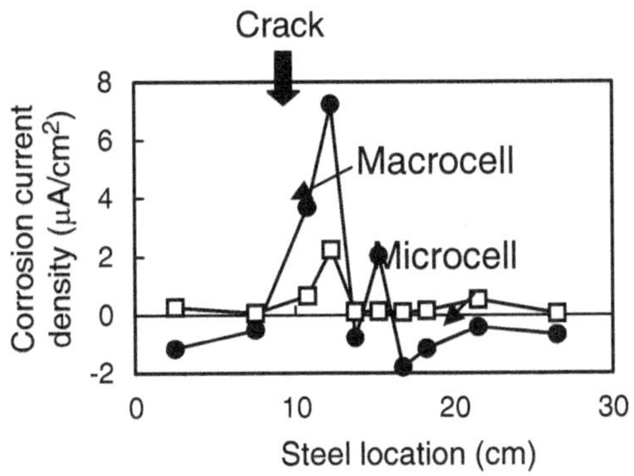

Figure 29. Example of analyzed corrosion current density distribution

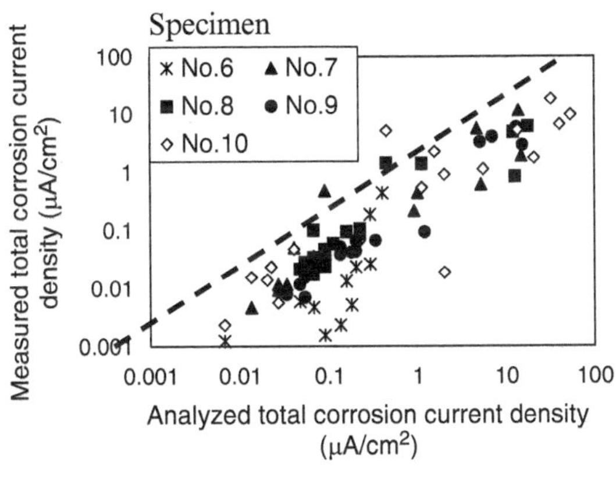

Figure 30. Comparison between analyzed and measured total corrosion current densities

Conclusions

The conclusions are as follows:

1) A model, which can analyze the corrosion rates of the reinforced concrete with the crack, was proposed. The input data of the model were the anodic polarization curve, cathodic polarization curve, polarization resistance, and electrical resistivity of concrete.
2) By carrying out an experiment using a mortar specimen and a concrete specimen with a crack, we confirmed the validity of the proposed model.
3) The influences of the crack, humidity, and water-cement ratio on the corrosion rate were confirmed by this model.

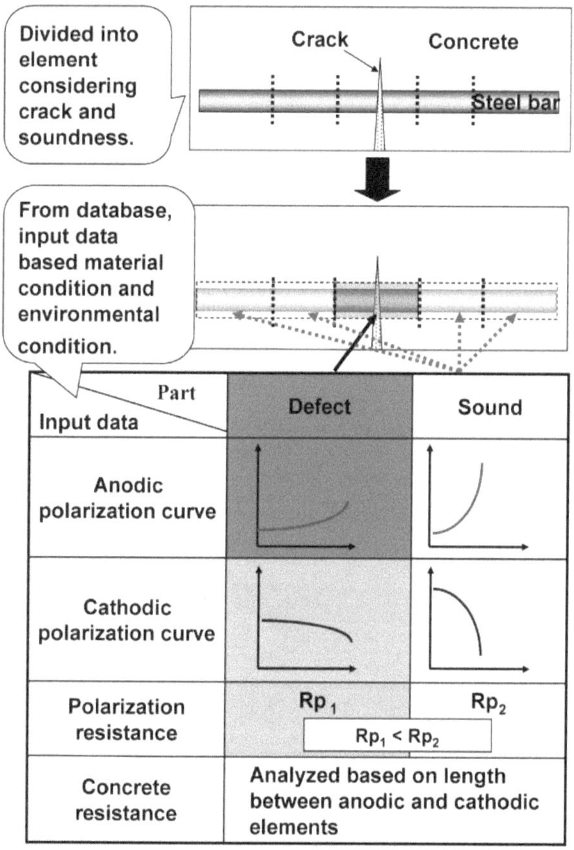

Figure 31. Application of model to existing structure

In this study, input data such as the polarization curves, polarization resistance, and electrical resistivity of concrete were obtained by carrying out an experiment using a specimen that was embedded in a specially divided steel bar. In the future, these values should be stored in a database for every material condition and every environmental condition. As a result, input data are arranged in such a manner that they can be obtained without carrying out an experiment. Further, as shown in Figure 31, for the general structure (that is, the undivided steel bar), the reinforcing bar is divided into appropriate lengths (1-10 cm), and an application of this model is attempted.

Acknowledgments

The author would like to appreciate the cooperation of Dr. K. Yokozeki and Mr. T. Oyamoto, Researchers at Kajima Corporation, and the Japan Concrete Institute - Technical Committee on "Analysis Model Supporting Verification of Long-Term Performance of Concrete Structures in Design" (Chairman: Dr. K. Takewaka, and Coordinator: Dr. T. Noguchi).

References

[1] S. Miyazato and N. Otsuki, (2010), Journal of Advanced Concrete Tech., vol. 8, n. 2, pp. 135-144.

[2] H. Hamada, Y. Sagawa, T. Ikeda, and R. Morikawa, (2010), In: Several factors affecting the anodic polarization curve of steel bars embedded in mortar, Proceedings of the 6th International Conference on Concrete under Severe Conditions, vol. 1, pp. 201-208, P. Castro, Borges et al. (Ed.), Merida.

[3] N. Otsuki, S. Miyazato, N. B. Diola, and T. Suzuki, (2000), Material Journal of American Concrete Institute, vol. 97, n. 53, pp. 454-464.

[4] S. Miyazato, N. Otsuki, and A. Konagai, (2001), In: The experimental and theoretical investigation of macrocell current measurement method using special divided steel bar, Proceedings of the Japan Concrete Institute, vol. 23, n. 2, pp. 547-552 (in Japanese).

[5] S. Miyazato, (2004), In: Relationship between corrosion rate and weight loss of special divided steel bar embedded in mortar and example test, Proceedings of the 31th JUCC Congress on Cement and Concrete, pp. 23-28 (in Japanese).

[6] T. Tsuru, R. Maeda, and S. Haruyama, (1979), Corrosion Eng., pp. 638-644 (in Japanese).

[7] C. Andrade, I. R. Maribona, S. Feliu, J. A. Gonzalez, and S. Feliu, Jr., (1992), Corrosion Science, vol. 33, n. 2, pp. 237-249.

[8] N. Sato, (1995), Corrosion Science, vol. 37, n. 12, pp. 1947-1967.

[9] A. W. Beeby, (1983), Concrete International, n. 2, pp. 35-40.

[10] P. Schiebl, and M. Raupach, (1997), Material Journal of American Concrete Institute, vol. 94, n. 1, pp. 56-62.
[11] S. Miyazato, (2010), Monitoring of macrocell corrosion rate in existing structures, Proceedings of the 6th International Conference on Concrete under Severe Conditions, vol. 1, pp. 169-182, P. Castro, Borges et al. (Ed.), Merida.
[12] S. Nagataki, N. Otsuki, A. Moriwake, and S. Miyazato, (1996), Journal of Materials, Concrete Structures and pavement, n. 544, pp. 109-119 (in Japanese).
[13] C. Andrade, I. R. Maribona, S. Feliu, J. A. Gonzalez, and S. Jr. Feliu, (1992), Corrosion Science, vol. 33, n. 2, pp. 237-249.

Reinforcement Corrosion Rate in Cracked Areas of RC-Members Subjected to Sustained Load

Davor Grandić[1] and Dubravka Bjegović[2]

[1]Faculty of Civil Engineering, University of Rijeka, Rijeka, Croatia
[2]Faculty of Civil Engineering, University of Zagreb, Zagreb, Croatia, and Institute IGH, Zagreb, Croatia

Abstract. The results of measuring the reinforcement corrosion rate of reinforced concrete beam and slab specimens subjected to sustained load are presented in this paper. Beam and slab specimens were subjected to corrosion of reinforcement under controlled conditions in an environmental chamber, where corrosion was initiated and accelerated in cycles of spraying with sodium chloride solution and drying the specimens. Cracks in concrete were caused by sustained load. During the experiment reinforced concrete samples were cracked, except near the supports, in area with small bending moments. A significant effect of increasing of corrosion rate due to presents of cracks was observed. The corrosion rates in uncracked areas of specimens were significantly smaller than in the cracked areas of the same specimens. The corrosion rate was measured by using galvanostatic impulse method. The measuring points were distributed along the beam and slab specimens. In this paper the influence of cracks on the time of initiation of corrosion and corrosion rate is analysed. Also, the relation between the measured values of the half-cell potential and corrosion rate is investigated. The results presented in the paper are part of the comprehensive experimental research.

Introduction

Many studies, when investigating durability of reinforced concrete structures exposed to corrosion of reinforcement in chloride environment do not consider the influence of cracks which occur due to loading of structures. Cracks in loaded reinforced concrete elements are a common occurrence, as consequence of small tensile density of concrete. Concrete cracking regularly occurs at service load, i.e. at the regular use of reinforced concrete structures. Cracks in reinforced concrete structures cannot be generally avoided, but with suitable reinforcing its width can be decreased. Hence, the influence of cracks on durability of reinforced structures

C. Andrade and G. Mancini (eds.), *Modelling of Corroding Concrete Structures*, RILEM Bookseries 5, DOI 10.1007/978-94-007-0677-4_4, © RILEM 2011

cannot be generally disregarded.

The influence of cracks occurred due to loading of reinforced concrete elements on the time of initiation and corrosion rate was investigated by a small number of authors. Chung Quing Li [1] conducted an experimental study of the influence of cracks occurred due to loading to chlorides penetration and time of initiation of corrosion of reinforcement. Corrosion was induced and accelerated exposing reinforced concrete elements to sodium-chloride solution and wetting/drying periods in the environmental chamber. Specimens were designed by varying the concrete composition, type and quantity of cements and width of protective cover of concrete to reinforcement steel. It was established that when the crack width is greater or equal to 0.1 mm, it significantly accelerates the penetration of chlorides in concrete, depassivation and occurrence of corrosion of reinforcement despite the thickness of the concrete cover to the depth of the reinforcement, type and quantity of cement and composition of concrete [1, 2].

Otsuki et al. [3] investigated the influence of bending cracks of elements and water-cement ratio on chloride-induced corrosion of reinforcing bars. In that study, reinforced concrete specimens were exposed to reinforced corrosion accelerated by wetting/drying cycles for 4 weeks. Part of the specimens has been exposed to chloride induced environment outside the laboratory for 13 weeks (without acceleration of corrosion of reinforcement). It was established that the corrosion rate at the points where bending cracks occurred are approximately 3 times greater than the corrosion rate measured in cases when there were no bending cracks [3].

De Shutter Geert [4] investigated durability of marine concrete structures damaged by early thermal cracks. He suggested using the method of factor of influence of cracks on acceleration of carbonization and penetration of chlorides based on results from experiments conducted on small prisms of mortar.

Yoon et al. [5] investigated the reinforcement corrosion on reinforced concrete specimens (beams) exposed to bending. Bending has caused occurrences of cracks in tensile area of the specimens. Corrosion was induced by outer source of direct current. Three levels of loading of reinforced concrete specimens were examined: without loading, 45% and 75% loading failure of reinforced concrete specimens. It was established that the corrosion rate of reinforcement increase with the increasing of the level of loading the specimens [5].

Specified studies [1-5] were conducted aiming to establish influence of cracks on acceleration of penetration of chlorides [1, 2, 4, 5], on the time of corrosion initiation [1, 2, 5], and on increasing corrosion rates [3, 5]. In these experimental studies the time of exposure to corrosion environment was relatively short so that corrosion could not manage to develop significantly [1, 2, 3], or its acceleration was induced by outer source of direct current [5], or it was conducted on specimens of mortar, and not on specimens with reinforced concrete elements [4].

Except the previously mentioned experiments in which accelerated penetration of chlorides have been used the long-term experiments on reinforced concrete members exposed to corrosive influence of chloride environment are also conducted [6, 7].

Zhang et al. are presented the results of an extensive research of behaviour of beams exposed to reinforcement corrosion in chloride environment [6]. In the research [6] the beams were loaded with sustained load during the time of chloride exposure. On the basis of presented distribution of local steel cross-section reduction along tensile bars of the corroded beams exposed to reinforcement corrosion in chloride environment for 14 and 23 years it can be seen that the greatest cross-section reduction is at the middle of the span, in area with maximal bending moments. The observed differences in cross-section reduction along the beam are less obvious for the beam exposed for 23 years. The distribution of cross-section reduction is determined from reinforcement mass loss by weighting the samples.

In [7] a detailed investigation of chloride induced corrosion damage was performed on a 40 years old reinforcement concrete beam exposed to marine environmental. The beam was cracked on the upper side due to prestressing with two tendons in the lower part of the beam. On the basis of corrosion rate deducted from polarisation resistance measurements and visual inspection of reinforcement after removing the concrete cover it was observed that reinforcement cross-section loss in tensile zone of the beam is significantly higher than in the compressive zone.

This paper presents and analyses results of a comprehensive experimental research [8, 9]. In [9] the partially processed experimental results and an analysis of observed structural deterioration of beam and slab specimens, at that time still ongoing experimental procedure, were given.

The experiment in [8] was conducted on a total of 24 specimens of reinforced concrete beams and slabs which were simultaneously exposed to reinforcement corrosion and sustained loading. The reinforcement corrosion of beams and slabs were induced and accelerated in chloride environment in the environmental chamber. The acceleration of corrosion was achieved by cycles of spraying the specimens with salt water and drying them at increased temperature. Corrosion of reinforcement embedded in beam and slab specimens is initiated and accelerated by alternate wetting with salt water and drying cycles in an environmental chamber under controlled conditions to simulate the corrosion process with corrosion characteristics on the surface of the steel bar similar to that which occurs in reinforced concrete members in natural chloride environment (pitting corrosion). The experimental results obtained from this research are therefore more representative than the results of researches in which the corrosion of reinforcement is accelerated by using an impressed direct current technique [10].

Exposing the specimens to corrosion environment in environmental chamber lasted 383 days. Diameters of the reinforcing bars and coefficients of reinforcing of beams and slabs were chosen proportional to dimensions of the specimens. Sustained loading was chosen in the way that it agrees to the real level of loading of structures in use, and the average width of cracks under loading were 0,1 mm. The corrosion was measured on measuring points which were distributed along the beam and slab specimens (in area with cracks and without them). This way it was possible to observe the influence of cracks to reinforcement corrosion.

Experimental Program

A comprehensive experimental research was conducted, where data about advancement of chloride corrosion of reinforcement, and damaging effects of corrosion on reinforced concrete elements simultaneously exposed to sustained load was collected [8]. In this paper a part of the already mentioned experimental program which relates to inducing the acceleration of reinforcement corrosion in specimens of beams and slabs, and measuring corrosion parameters is presented.

Test specimens

The beam specimens selected for the experiment had a cross-section of 8×12 cm and length of 200 cm. They were reinforced with two bars having a diameter of 8 mm in the tensile area, two bars of 6-mm diameter in the compressive area of cross-section, and with stirrups of 6 mm in diameter spaced at intervals of 8 cm. Concrete cover of concrete to the depth of reinforcement (stirrups) was 1.0 cm (Figure 1).

Slab specimens had a cross-section 50×8 cm, and length of 200 cm. They were reinforced in the tensile zone of the cross-section with four bars of 6 mm in diameter. Secondary reinforcement of slabs consisted of 9 bars of 6 mm in diameter. The concrete cover to the reinforcement was 1.0 cm (Figure 2). Bars of 6 mm nominal diameter were of cold worked ribbed reinforcing steel, while the ones of nominal diameter of 8 mm were of hot rolled ribbed reinforcing steel.

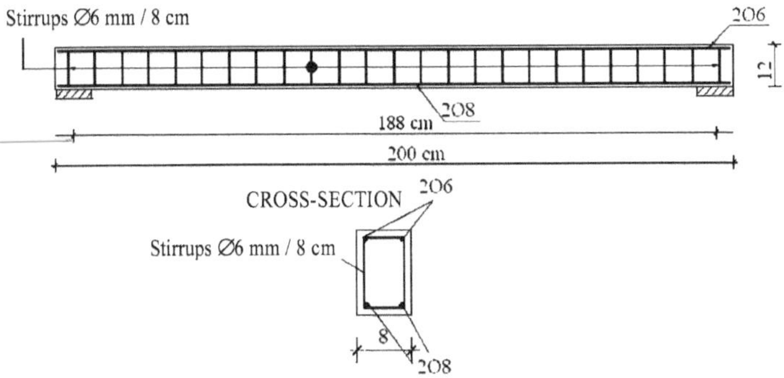

Figure 1. Beam specimen

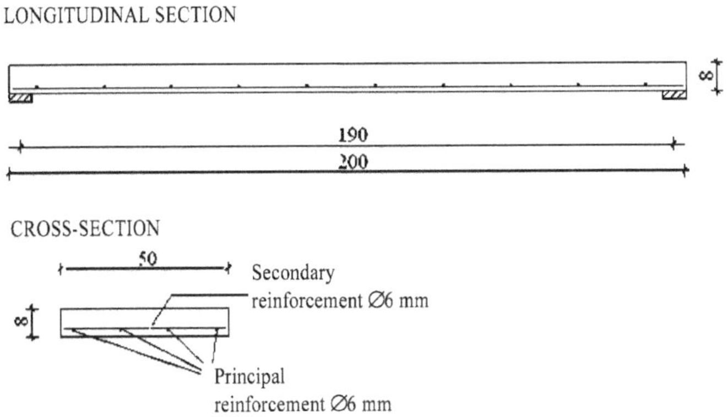

Figure 2. Slab specimen

Material properties were determined by testing on minimally three specimens. Mean values of tested specimens are hereby specified. By testing the warm rolled steel for reinforcing, the mean value of yielding strength 589 MPa and the mean value of tensile strength 684 MPa were determined. Mean value of strain at maximum force of hot rolled reinforcing steel (A_{gt}) is 6.38%. Cold worked reinforcing steel has the mean value of yielding strength 573 MPa and mean value of tensile strength 607 MPa. Mean value of strain at maximum force of cold worked reinforcing steel (A_{gt}) is 2.13%.

In Table I the composition of fresh concrete is specified. Compressive strength of 28 day-old concrete was tested on concrete cubes with edge length of 150 mm, and mean value of testing results was 35.2 MPa.

Table I. Composition of fresh concrete

Binder	Portland cement with 18% of slag
aggregate	Washed separated natural gravel and sand
Quantity of cement on 1 m^3 of concrete	380 kg
Water-cement ratio	0.52
Quantity of aggregate on 1 m^3 of concrete	1790 kg
Biggest grain of aggregate	16 mm

The following transport-related (penetrability) properties of concrete were tested: capillary water absorption according to EN ISO 15148 [11], gas permeability according to EN 993-4 [12], and resistance to penetration of chloride ions according to ASTM 1202 [13].

Testing determined the mean value of the coefficient of capillary absorption is 0.62 $kgm^{-2}h^{2}$ and mean value of specific coefficient of permeability is $1.353 \cdot 10^{-13} cm^2$. Resistance to penetration of chloride ions is evaluated [13] over the quantity of electricity which passed through the rolled specimens of nominal diameter of 102 mm and nominal high of 51 mm. Mean quantity of passing electricity through the specimens was 1249 coulombs.

Specified results of testing showed that the specimens of beams and slabs were composed of concrete of mean quality according to transport-related properties [13, 14].

Three levels of reinforcement corrosion were foreseen with the programme of experiments. After the approximate reaching of corrosion levels, testing of specimens of beams and slabs to failure was conducted (Table II), as well as establishing the real corrosion state of reinforcement on specimens of corroded reinforcement removed from beams and slabs. Value P_{corr} in Table II is the depth of corrosion in relation to the original surface of uncorroded reinforcement.

For each degree of corrosion, a series of four specimens of reinforced concrete beams and slabs were foreseen. Four control specimens of beams and slabs, which were not exposed to reinforcement corrosion, were also foreseen in the programme of experiments. After reaching individual degrees of reinforcement corrosion - one specimen of beams and slabs from the series were destroyed, the reinforcement was removed in order to conduct a detailed examination, measuring and testing of corroded reinforcement.

Table II. Number of beam and slab specimens* and levels of corrosion with proposed depth of reinforcement corrosion

<table>
<tr><td>Level of corrosion</td><td>0</td><td>I</td><td>II</td><td>III</td></tr>
<tr><td>Depth of corrosion</td><td>Control specimens (not exposed to corrosion): P_{corr} = 0 mm</td><td>P_{corr} 0.05mm</td><td>0.1mm P_{corr}< 0.2mm</td><td>P_{corr}> 0.2mm</td></tr>
<tr><td rowspan="2">Number of specimens</td><td>1 specimen at an age of 28 days tested to failure</td><td colspan="3">1 each of beam and slab specimens in the climate chamber for testing corrosion state of reinforcement and chloride penetration</td></tr>
<tr><td>3 specimens loaded in laboratory conditions (20 2 C, RH 48%) on which deflections were measured until experiment completion and after that tested to failure</td><td colspan="3">3 each of the specimens in climate chamber exposed to the following actions:
- sustained static load,
- cycles of spraying with chloride solution and drying (reinforcement corrosion); after the specimens had reached a particular corrosion stage (I, II, III) they were tested to failure</td></tr>
</table>

(The number of test specimens was the same for the beams and slabs, i.e. a total of 16 beams and 16 slabs)

Initiation and acceleration of the process of reinforcement corrosion

Initiation of reinforcement corrosion in reinforced concrete beams and slabs and its acceleration after it had started, was conducted by repeating the cycles which consisted of wetting specimens by spraying with salt water and drying them in environmental chamber. The concentration of sodium-chloride (NaCl) melted in salt water used to spray specimens was 3.8% in relative to the water mass. A cycle in duration of 3 days started with spraying the specimens with salt water, second day the specimens were exposed to air temperature which was approx. 20°C, relative humidity was 60 to 70%. The third day the environmental chamber was heated for 3 hours, so that the highest temperature was around 50°C or somewhat higher, ventilating at the same time with fans, and relative humidity came down to 20%.

Measuring of reinforcement corrosion parameters

During the exposure to reinforcement corrosion specimens of beams and slabs were exposed to constant sustained loading in steel frames (Figures 3, 4). Sustained loading of specimens caused cracks in concrete with average width of 0.1 mm.

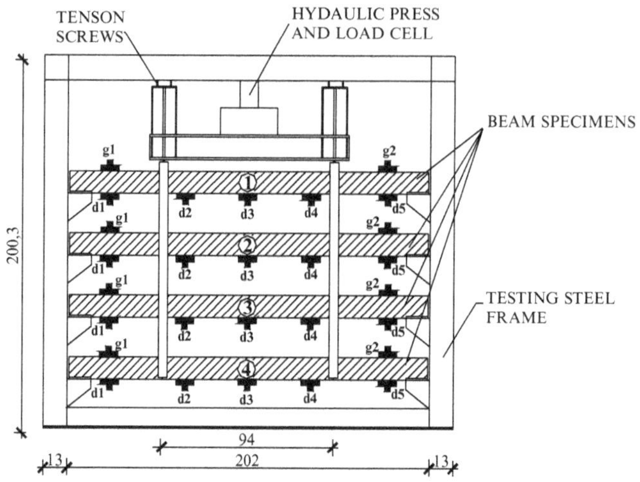

① to ④ ... marks of beam specimen

g1, g2 ... corrosion measuring points on the top side of the beams

d1 to d5 ... corrosion measuring points on the bottom side of the beams

Figure 3. Beam specimens in the testing frame and corrosion measurement points

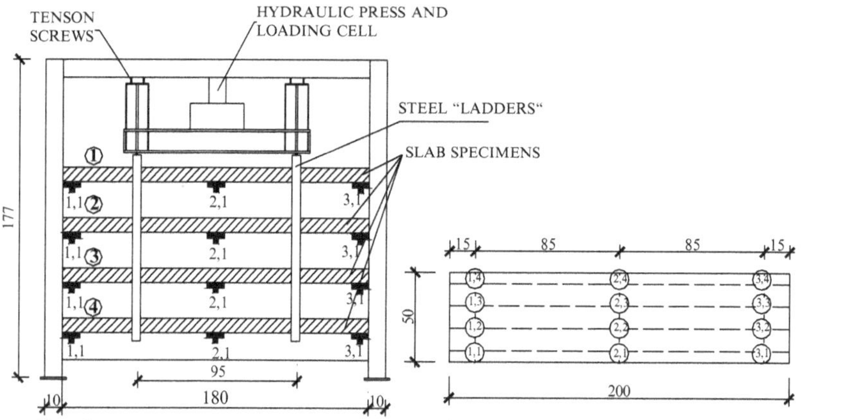

① to ④ ... marks of slab specimen

1,1 to **3,1** ... corrosion measuring points in front view

1,1 to **3,4** ... corrosion measuring points on the bottom side of the slab

Figure 4. Slab specimens in the testing frame and points of corrosion measurement

Periodically, the reinforcement corrosion parameters (corrosion rate, half-cell potential and electrical resistance), deflections, and cracks caused by loading and corrosion were measured and observed on specimens of reinforced concrete beams and slabs. During the exposure of beam and slab specimens to reinforcement corrosion which lasted 383 days, a total of 18 phases of measurements of corrosion parameters were conducted. At the beginning of exposure of specimen beams and slabs to corrosion, the parameters of corrosion were measured once a week, four weeks every two weeks, then every three weeks and at the end once a month. GalvaPulse^{TM} device was used for those measurements, based on galvanostatic impulse method [15, 16].

On every beam specimen the corrosion was measured at two points on the upper surface of the beam and at five points on the bottom surface. On the slab specimen, corrosion was measured at 12 points on the bottom surface of the slabs. The beam and slab specimens in the testing frames and the arrangement of points where corrosion was measured on the specimens are shown in Figures 3 and 4.

Measured Values of Corrosion Parameters

Corrosion rates of reinforcement in beam and slab specimens

Due to a great number of measurement results, mean measured value of corrosion rate for each series of specimens of beams and slabs $i_{corr,m}$ (A/cm^2) is shown (Figures 5 and 6). Mean measured values of corrosion rate $i_{corr,m}$ (A/cm^2) is determined as the mean value of corrosion rate measured on measuring points which have the same mark and position on individual specimens in the series (Figures 3 and 4).

Measured values from the bottom surface of the beam and slab specimens are shown. At the bottom, tensile side of the specimen, cracks occur in the area of the greatest bending moment. Marks GI, GII, GIII and PI, PII, PIII on Figures 5 and 6 mark series of beams and slabs in which the different level of reinforcement corrosion were reached (levels I, II and III).

According to the criteria for evaluating measurements with the galvanostatic impulse method specified in the FORCE Institute Report [17] (Figure 3), high measured values of corrosion rate of reinforcement were reached.

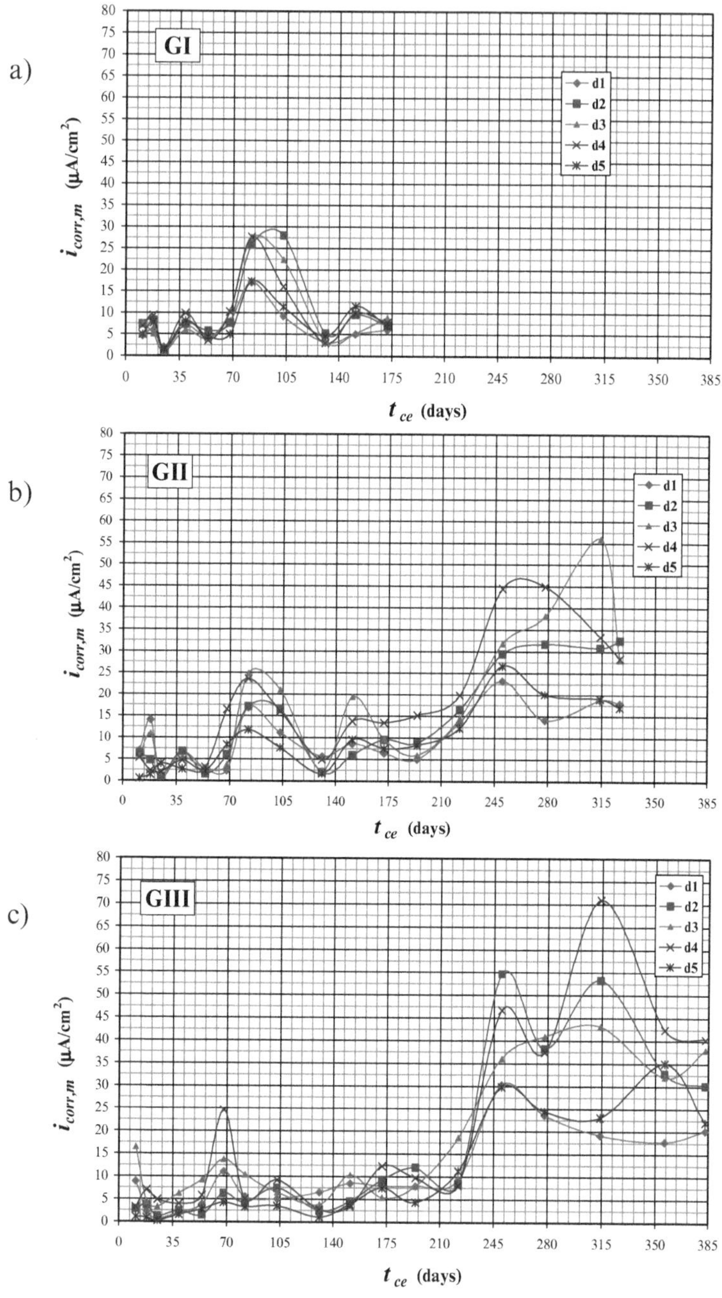

Figure 5. Corrosion rates in beams: a) series GI; b) series GII; c) series GIII

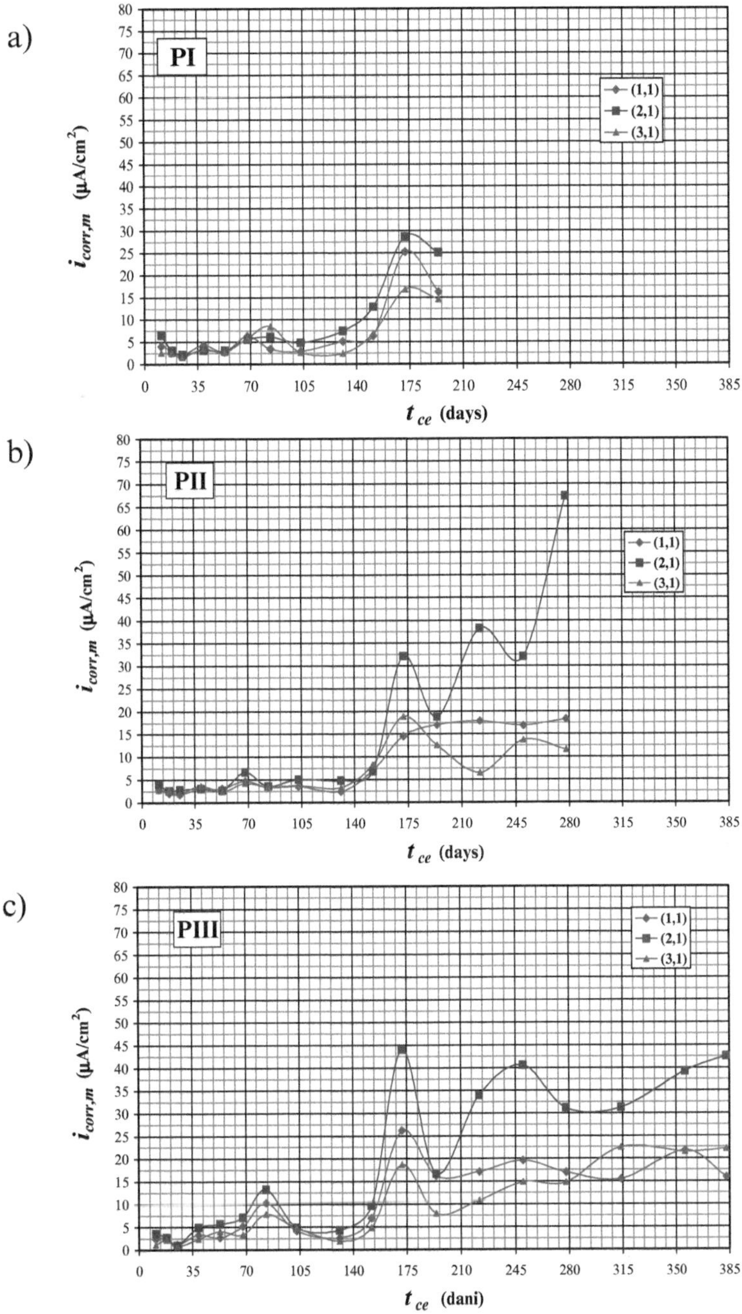

Figure 6. Corrosion rates in slabs: a) series PI; b) series PII; c) series PIII

Table III. Evaluation of measured values of corrosion rates [17]

Corrosion rates i_{corr} (A/cm^2)	Evaluation of corrosion rates
$i_{corr} > 15$	High
$5 < i_{corr}$ 15	Moderate
$1{,}0 < i_{corr}$ 5	Low
$i_{corr} < 1{,}0$	Negligible

Relation of half-cell potential and corrosion rate in beams

Half-cell potentials (HCP) in beam and slab specimens were measured with GalvaPulse™ device in relation to Ag/AgCl electrode. Mean value of half-cell potential and corrosion rate determined from results of measuring all specimens of beams which were comprised by measuring corrosion parameters in each measurement phase. Figure 7 shows the described mean measured values in areas of maximum bending moment (measuring points d2, d3 and d4), i.e. in cracked area of the beams.

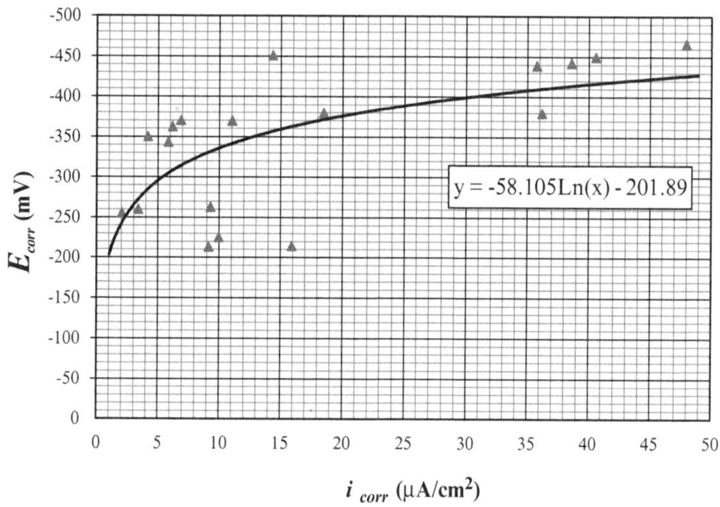

Figure 7. Relation between HCP and corrosion rates in cracked area of beams

Figure 8 shows mean measured values which relate to the measuring points near the supports (d1 and d5), i.e. in uncracked area of the beams.

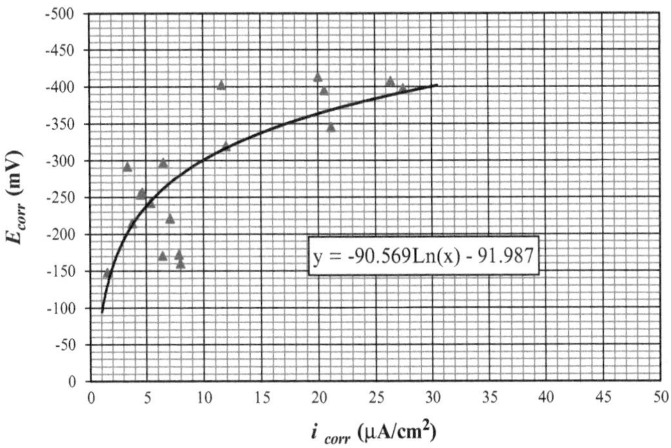

Figure 8. Relation between HCP and corrosion rates in uncracked area of beams

Relation of HCP and corrosion rate in slabs

Mean values of HCP and corrosion rate determined from results of measuring all specimens of slabs which were comprised by measuring corrosion parameters in each measurement phase are shown in Figures 9 and 10.

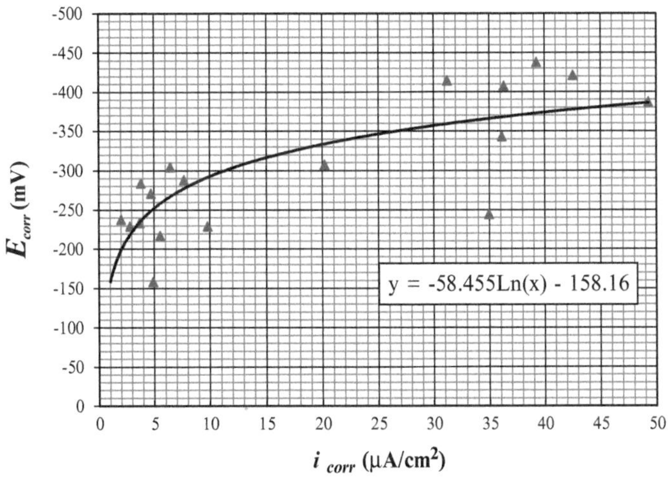

Figure 9. Relation between HCP and corrosion rates in cracked area of slabs

Figure 9 shows mean measured values in areas with maximum bending moment (measured in points in the middle of span of slabs) i.e. in cracked area. Figure 10 shows mean measured values which relate to the measured points near the support, i.e. uncracked area of slabs.

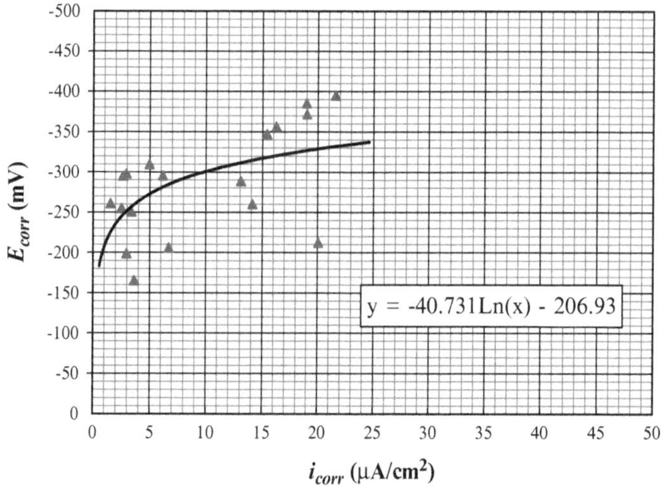

Figure 10. Relation between HCP and corrosion rates in uncracked area of slabs

Depth of Reinforcement Corrosion According to Results of Measurements of Corrosion Rate

Mean depth of corrosion in measuring area of the sensors P_{corr} (further in text: depth of corrosion) is calculated, according to Poulsen [18], based on measured values of corrosion rate according to following expression [8]:

$$P_{corr}(\tau) = 11.6 \cdot \int_0^{\tau} i_{corr}(t)dt, \qquad (1)$$

where $P_{corr}(\ )$ is the mean depth of corrosion of the reinforcement in microns, $i_{corr}(t)$ is the corrosion rate A/cm^2 obtained by periodical measurements on reinforced concrete specimens, t is time of duration of reinforcement corrosion in years, and is the observed moment of time at which the mean depth of corrosion $P_{corr}(\ )$ is calculated. The expression (1) is solved with numerical integration.

Average value of mean depth of corrosion in measured area of sensors ($P_{corr,m}$) in time were determined as arithmetical mean depth of corrosion P_{corr} calculated on the basis of measured corrosion rates in one series of beams or slabs according to the expression (1). For values P_{corr} which were calculated during reaching individual degrees of corrosion, except the mean value, also the standard deviation of depth of corrosion in one series of specimens was determined, which is shown by distribution of depth of corrosion along reinforced concrete specimens of beams and slabs (Figures 11 and 12). For each series of specimens of beams and slabs, mean value of coefficients variation V_m are shown. As has been shown on Fig 3 and 4, d1 to d5 and (1,1) to (3,1) are corrosion measuring points.

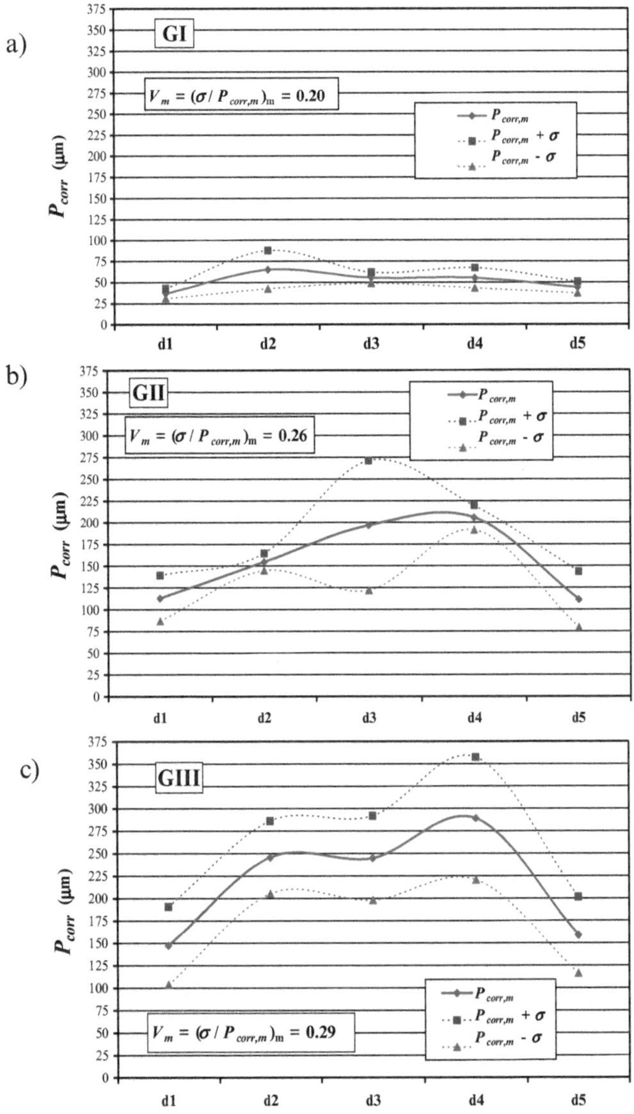

Figure 11. Depth of corrosion along beams: a) series GI; b) series GII; c) series GIII

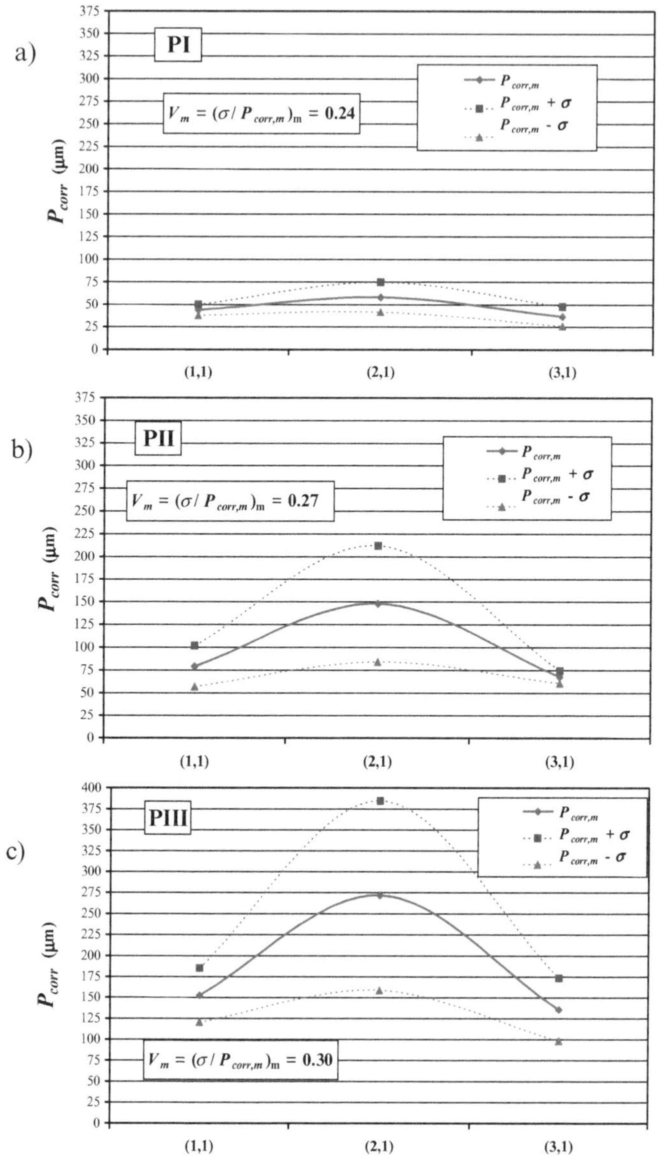

Figure 12. Depth of corrosion along slabs: a) series PI; b) series PII; c) series PIII

Observations and Discussion

Based on the results of measuring, the following is observed:

1. In cracked areas of beams corrosion advances 1.6 times faster than in uncracked areas near the supports. Reinforcement corrosion of slab specimen in cracked areas advance twice as fast than in uncracked areas near the support. In the paper [3] it is said that the corrosion advances 3 times faster due to the cracks. The difference can be explained with different methods of measuring the corrosion rates.

2. Measured values of corrosion parameters have big dispersion regarding the mean value. However, it is shown [8] that there is good correlation between decrease of diameter determined on the basis of mean measured value of corrosion rates and decrease of diameter determined by measurements conducted on corroded bars removed from the specimens.

3. From the relation of measured values of HCP and corrosion rates (Figures 7 to 10), it is seen that at the corrosion rates smaller than 10 A/cm^2 the potentials in uncracked area of beams are greater from the potentials in cracked area. At the corrosion rate greater than 10 A/cm^2, i.e. between 10 and 28 A/cm^2 for beams and 10 to 22 A/cm^2 for slabs, the differences of relations between corrosion rates and potentials in cracked and uncracked area of beams are insignificant.

4. Measured value of HCP rise with the corrosion rates. HCP E_{corr} = -256 mV (90% probability of active corrosion according to ASTM C876 [19]) correspond to following corrosion rate: 2.54 A/cm^2 (Fig 7) 6.12 A/cm^2 (Figure 8), 5.35 A/cm^2 (Figure 9) and 3.35 A/cm^2 (Fig 10). According to the criteria for evaluating measurements of corrosion rates with the help of galvanostatic impulse method (Table III), specified values of corrosion rates are found in the area of low to moderate corrosion rates, which points to great probability of active reinforcement corrosion [18, 20].

Conclusion

Sustained load to which the specimens of beams and slabs were exposed to, caused in tensile area of specimens cracks upright to tensile reinforcement of approximate width of 0.1 mm. These cracks have significant effect to increasing of measured corrosion rates. In the points of beams and slabs the corrosion rates were 1.6 to 2 times higher from the corrosion rates measured in areas of specimens without cracks. Previous studies of other researches have also pointed to the effect of cracks to faster initiation of corrosion and increasing of HCP and corrosion rates. There are not many such studies, and the experiment described in this paper is specific, because of the comprehensive procedure of corrosion rates that were conducted exchanging the wetting/drying cycles, where the natural corrosion process of reinforcement was simulated in chloride environment. Similar process happens in nature e.g. at reinforced concrete in marine environment in areas splashed by waves. The time the specimens were in chloride conditions in the environmental chamber, the specimens were subjected to sustained load. Specimens were designed so that they had real reinforcement ratio, and the diameters of reinforcement were adjusted to the dimensions of the specimens. From the analysis of relation of measured corrosion rates and HCP, it is to be concluded that the corrosion rate, which also points to active corrosion, corresponds to HCP, for which, according to ASTM, there is 90% probability of existence of active reinforcement corrosion (E_{corr} = -256 mV).

Adverse influence of cracks to penetration of chlorides and acceleration of corrosion of reinforcement steel should be further investigated in order to improve norms for calculation and design of concrete structures aiming at designing and constructing safe and durable reinforced concrete and prestressed structures. What should be taken in account is the fact that due to cracking of concrete the efficiency of protective cover of concrete to the reinforcement steel in protection of reinforcement from corrosion is significantly damaged.

Acknowledgments

This research was performed within scientific project "The Development of New Materials and Concrete Structure Protection Systems" (082-0822161-2159), funded by Croatian Ministry of education, science and sport.

References

[1] Li, C. Q.: *Initiation of Chloride-Induced Reinforcement Corrosion in Concrete Structural Members–Experimentation,* ACI Structural Journal, 98 (2001) 4, 502-510.
[2] Li, C. Q.: *Initiation of Chloride-Induced Reinforcement Corrosion in Concrete Structural Members–Prediction,* ACI Structural Journal, 99 (2002) 2, 133-141.
[3] Otsuki, N.; Miyazato, S.; Diola, B. D.; Suzuki, H.: *Influences of Bending Crack and Water-Cement Ratio on Chloride-Induced Corrosion of Main Reinforcing Bars and Stirrups,* ACI materials Journal, 97 (2000) 4, pp. 454-464.
[4] Shuter, De G.: *Durability of Marine Concrete Structures Damaged by Early Age Thermal Cracking*, International RILEM Workshop on Life Prediction and Aging Management of Concrete Structures, Cannes, France, 2000, 177-186.
[5] Yoon, S.; Wang, K.; Weiss, W. J.; Shah, S. P.: *Interaction between Loading, Corrosion, and Serviceability of Reinforced Concrete*, ACI Materials Journal, 97 (2000) 6, 637-644.
[6] Zhang, R.; Castel, A.; François, R.: *Serviceability Limit State criteria based*

on steel-concrete bond loss for corroded reinforced concrete in chloride environment, Materials and Structures, 42 (2001), 1407-1421.

[7] Poupard, O.; L'Hostis, V.; Catinaud, S.; Petre-Lazar, I. : *Corrosion damage diagnostic of reinforced concrete beam after 40years natural exposure in marine environment*, Cement and Concrete Research, 36 (2006), 504-520.

[8] Grandić, D.: *Calculation procedures for evaluating remaining load bearing capacity and serviceability of corroded reinforced concrete structures*, Doctoral Thesis, Faculty of Civil Engineering, University of Zagreb, Croatia, 2008. (in Croatian)

[9] Grandić, D.; Bjegović D.: *Structural Deterioration due to Chloride-Induced Reinforcement Corrosion,* Supplementary Papers – Seventh CANNMET/ACI International Conference on Durability of Concrete, Montreal, Canada 2006, 173-188.

[10] Yuan, Y.; Ji, Y.; Shah, S. P.: *Comparison of Two Accelerated Corrosion Techniques for Concrete Structures*, ACI Structural Journal, 104 (2007) 3, 344-347.

[11] EN ISO 15148:2002 *Hygrothermal performance of building materials and products - Determination of water absorption coefficient by partial immersion.*

[12] EN 993-4: 1995 Methods of test for dense shaped refractory products – Part 4: *Determination of permeability to gases.*

[13] ASTM Designation: C 1202 - 91: *Standard Test Method for Electrical Indication of Concrete's Ability to Resist Chloride Ion Penetration.*

[14] Ukrainczyk, V.; Bjegović, D.: *Materials testing in the insurance system of durability of concrete structures,* Civil engineering yearbook 1995, Croatian Society of Civil Engineers, Zagreb, Croatia, 1995., pp. 209-286. (in Croatian)

[15] Brite-Euram III "Smart Structures", Contract No. BRPR-CT98-0751: Integrated Monitoring Systems for Durability Assessment of Concrete Structures, Project Report, September 2002.

[16] Klinghoffer, O.: *In-Situ Monitoring of Reinforcement Corrosion by Means of Electrochemical Methods*; Nordic Concrete Research, No. 16, 1/95, Oslo, 1995.

[17] Klinghofer, O.; Berge, H.-E.: *Methods of Test of Reinforcement Material Properties Based on Mix of Stainless Steel and Carbon Steel*, Nordest Report, TR 528, Approved 2003-04.

[18] Klinghoffer, O.; Frřlund, T.: *Rebar Corrosion Rate Measurements for Service Life Estimates,* ACI Fall Convention 2000, Toronto, Canada, Committee 365 "Practical Application of Service Life Models".

[19] ASTM Designation: C876 - 09: *Standard Test Method for Half-Cell Potentials of Uncoated Reinforcing Steel in Concrete.*

[20] Poursaee, A.: *An Analysis of the Factors Influencing Electrochemical Measurements of the Condition of Reinforcing Steel in Concrete Structures*, Ph.D. thesis in Mechanical Engineering, University of Waterloo, Ontario, Canada, 2007.

SBRA Model for Corrosion Initiation of Concrete Structures

Pratanu Ghosh[1], Petr Konečný[2] and Paul J. Tikalsky[3]

[1]Department of Civil and Environmental Engineering, University of Utah, USA
[2]Faculty of Civil Engineering, VŠB - Technical University of Ostrava, Czech Republic
[3]Department of Civil and Environmental Engineering, University of Utah, USA

Abstract. This paper presents a finite element based probabilistic corrosion initiation model using Simulation Based Reliability Assessment (SBRA). The model is focused on the effect of diffusions and ingress of chloride ions in bridge decks with excessive cracks to investigate the performance of different corrosion resistant steel reinforcements. The objective of this research is based on the formulation of probabilistic corrosion initiation model with the inclusion of distribution of High Performance Concrete (HPC) diffusion coefficients computed from fundamental electrochemistry and the variation of surface chloride concentration from the field data of Virginia bridge decks as well as other random variable parameters. This study shows the variability and sensitivity on estimation of the time to onset of corrosion using Monte Carlo technique. In addition, the estimation of corrosion free service life for the preliminary design of concrete structures in harsh chloride environments will be indicated.

Introduction

One of the most significant types of distress in many bridge decks is the corrosion of reinforcing steel from the ingress of chloride salts applied to melt snow and ice. This can lead to loss of structural capacity and promote reduced service life, thus leading to increased life-cycle costs. Though models for chloride ingress and corrosion development have been studied (see e.g. [1, 2, 3, 4, 5, 6, 7, 8, 9, 10, 11]), there are still many issues that must be addressed to become useful engineering tools, especially with regards to randomness of pertinent input variables. Tikalsky et al. (2005) [12], Konečný et al. (2007) [13] used Monte Carlo technique in

C. Andrade and G. Mancini (eds.), *Modelling of Corroding Concrete Structures*, RILEM Bookseries 5, DOI 10.1007/978-94-007-0677-4_5, © RILEM 2011

corrosion model to show the variability on estimation of the time to onset of corrosion and the required depth of concrete cover to extend the service life up to 100 years. Shim, 2006 [14] demonstrated that the resulting probability distribution is most sensitive to the chloride diffusion coefficients. Lounis et al., 2004 [15] presented a research approach for the probabilistic modelling of the chloride induced corrosion of carbon steel reinforcement in concrete structures. They also considered the uncertainties of the governing parameters such as concrete diffusivity, surface chloride concentration, concrete cover depth, and threshold chloride level. The parameters were modelled as random variables and the distribution of the corrosion time and probability of corrosion rate are determined by using Monte Carlo simulation. This method was incorporated to predict the level of corrosion in the top layer of reinforcing carbon steel of a highway bridge deck that was exposed to chlorides from de-icing salts for forty years. The prediction of the proposed model correlated very well with the field data, which clarifies the usefulness of the probabilistic models to characterize the corrosion response and the actual condition of reinforced concrete structures.

All of the previously mentioned probabilistic corrosion model lack the distribution of performance based diffusion coefficients data of various high performance cementitious materials or performance of several corrosion resistant steel reinforcements. Some research needs to be performed to incorporate the steady state diffusion coefficients of HPC materials to investigate the performance of several steel reinforcements to delay chloride induced corrosion initiation.

Figure 1. Crack on the Grinded Bridge Deck (left) and Scratched Epoxy-coating (right)

This paper utilizes a finite element based probabilistic corrosion initiation model. The model is focused on a reinforced concrete bridge deck 2-D chloride ingress that accounts for the scatter of input random variables with regards to diffusion, surface cracks and reinforcing steel protection type (i.e. black bar, epoxy coating bar, and galvanized bar). The model combines a finite element model [13] and the

SBRA method that is introduced in [16] and proposed for application with chloride ingress by Tikalsky et al. [12] and [17]. The model is focused on the effect of cracks on chloride ingress in bridge decks to investigate the performance of different corrosion resistant steel reinforcements. The objective of this research is based on the formulation of probabilistic corrosion initiation model with the inclusion of distribution of HPC diffusion coefficients computed from fundamental electrochemistry, the variation of surface chloride concentration from the field data of Virginia bridge decks [18] and the variation of chloride threshold for reinforcing steel [19]. This study shows the variability and sensitivity on estimation of the time to onset of corrosion using Monte Carlo technique.

Chloride Ingress Induced Deterioration

Service life

If one considers the corrosion process driven by chloride ingress as the governing durability issue then service life $t_{service}$ can be written according to [20] as

$$t_{service} = t_{initiation} + t_{propagation} \tag{1}$$

where $t_{initiation}$ is the period before the onset of corrosion and $t_{propagation}$ is the time for corrosion to reach an unacceptable damage level once it has started. The initiation time period is primarily influenced by the concrete diffusion coefficients, concrete cover, surface chloride concentration, temperature, level of saturation and the required concentration at the level of reinforcing steel to initiate corrosion. If there is epoxy coating on the reinforcing steel, one must consider the concentration near holidays, if any are present. The model adopted in this paper focuses only on this initiation period. The propagation period begins once a sufficient concentration of chlorides has reached the reinforcing steel to dissolve the passivation layer and initiate active corrosion.

Mechanism of chloride ingress and Transportation Model

Diffusion is the primary variable by which chlorides penetrate to the level of reinforcing steel to initiate corrosion. The effects of hydraulic permeation and capillary absorption are small in comparison in most cases and are neglected herein. It is widely accepted that Fick's 2nd law of diffusion can represent the rate of chloride penetration into concrete as a function of depth and time, see [21]. The solution (referred to as the Crank's Solution) of the governing differential equation is given as Eqn. (2) [22].

$$C_{x,t} = C_0\left[1 - erf\left(\frac{x}{\sqrt{4D_c t}}\right)\right] \tag{2}$$

where $C_{x,t}$ is the concentration of chlorides (percent by mass of total cementitious materials) at time t (years) and depth x (meters), C_0 is the concentration of chlorides (% by mass of total cementitious materials) at the surface directly inside the concrete and D_c is the apparent diffusion coefficient (m^2/year) Eqn. (2) is widely used for chloride ingress models but does not account for cracks and must be modified to account for time dependent changes in material property or boundary conditions.

Performance assessment

Severity of the chloride ingress can be assessed by comparing the chloride threshold value C_{th} with the chloride concentration at the exposed areas of reinforcing steel. This value will depend on the type and preparation of the reinforcing steel and the constituents of the concrete as well as other factors. The performance function, RF_t, of a bridge deck is expressed as the time-dependent exceedance of the corrosion initiation threshold by the location dependent chloride concentration, $C_{xy,t}$. The performance function characterizing the above described limit state is expressed as:

$$RF_t = C_{th} - C_{xy,t} \tag{3}$$

Probabilistic time-dependent analysis can be thought of as a comparison of the joining extrema of the chloride concentration $C_{xy,t}$ and threshold C_{th} random realizations. The performance of the system was estimated by the probability of chloride threshold exceedance (corrosion initiation) at a specific age $P_{f,t}$.

$$P_{ft} = P(C_{th} - C_{xy,t} < 0) \tag{4}$$

2-D Finite Diffusion model with crack effect

The 2-D finite element model generated in ANSYS based on the Ficks 2nd law of diffusion focuses on the movement and accumulation of chloride ions to the level of reinforcing steel during the initiation period of corrosion. The analysis is 2-D in order to cover the interaction between the effect of crack on the chloride ion ingress and the holidays in the epoxy coating.

This proposed model focuses on the chloride transportation in reinforced concrete bridge decks with cracks and on the estimation of chloride ion concentration in particular locations on the embedded reinforcing steel bars or damaged areas of epoxy-coated bars. Although transverse reinforcing steel bars are susceptible to corrosion, the presented model is focused on longitudinal steel because longitudinal bars will cross the crack while only one transverse bar (closest to the crack) is affected by rapid chloride ingress through crack. A 2-D finite element method is chosen for the solution of Fick's second law (Eqn. (2)) to meet the analysis requirements. The commercial software ANSYS was used [23]. The assumptions for the applied model are explained in [13]. The 4-noded thermal solid element PLANE55 was used for the chloride diffusion analysis because of the analogy between the mechanisms of thermal flow and ion diffusion. The 0.25 m thick deck was vertically divided into 25 elements that are 10 mm by 10 mm in size. This resolution provided sufficient accuracy. Sample graphical output is shown in Fig. 2.

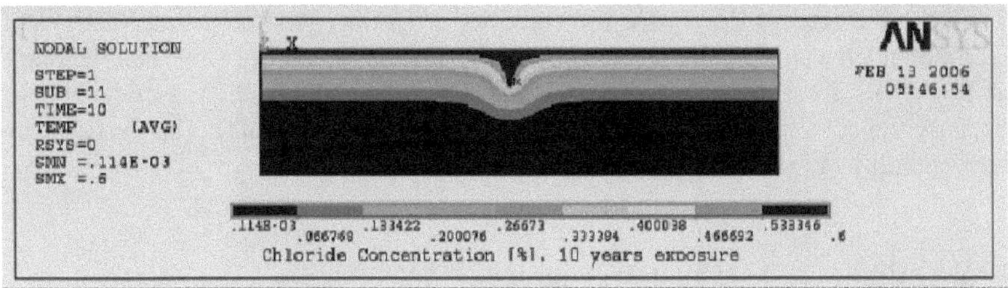

Figure 2. Chloride ion concentration in concrete slab with crack, t = 10 years

One can observe the effects of a crack that allow the chloride ions to move in both directions and reach the rebar level more rapidly. The highest concentration surrounding one of the epoxy-coating damaged areas was selected as the critical case where corrosion would occur first whereas the highest concentration on the reinforcement level was selected for the computation for other types of reinforcement protection. The FEM Macro used to model diffusion process has capabilities to automatically repeat with randomly selected input variables with respect to Monte Carlo simulation used for probabilistic analysis.

SBRA Application

A probabilistic approach was adopted to address the inherent variability of important input parameters used to describe the onset of corrosion. Diffusion coefficients, reinforcing steel depths, and chloride threshold histograms were developed based on computation from fundamental electrochemistry and lab and field data presented herein. Crack frequency distributions were estimated based on observed behaviour, but can be frequently made for in-situ data. The onset of corrosion depends on the chloride concentration level and this is significantly affected by the random proximity of the cracks to exposed steel from holidays. Probabilistic analysis can be used to adequately approximate this relationship.

The SBRA Module for ANSYS is a tool for managing the probabilistic Monte Carlo simulation process with random variables distributions characterized by frequency histograms according to [16]. The SBRA Module [13] runs the FEM Macro containing the diffusion process description. Random variable parameters in the FEM macro were automatically replaced by randomly generated variables throughout the Monte Carlo simulations.

The level of acceptable probability of corrosion initiation is not well defined yet. Tikalsky (2003) [17] discusses the target probabilities and states the probability that the structure will perform for 50 years at 3 out of 4 (respective P_d would be 25 %) and the probability that the structure will perform for 100 years is 1out of 2. Teplý [24] discusses the P_d for the corrosion imitation according to EN1990 and recommended value of the target probability for serviceability (irreversible state) as $P_d = 7\%$. Acceptable corrosion initiation likelihood $P_f <= 0.25$ is selected here because the corrosion initiation does not immediately threat safety ($P_d<7\times10^{-5}$) nor serviceability ($P_d < 0.07$).

Probabilistic 2-D Diffusion Analysis

Incorporation of reliable diffusion coefficients for HPC concrete in probabilistic model is a significant part as diffusion coefficient plays a major role in corrosion initiation. Histograms for diffusion coefficients for different high performance cementitious materials along with some impermeable and OPC mixtures are computed from two different approaches namely Nernst-Plank and Zhang-Gjorv method [25]. These two approaches involved with fundamentals of electrochemistry, where they compute steady state diffusion coefficients from chloride ion penetration test (CIPT) data. Chloride migration rate were computed from 5 minutes to 360 minutes CIPT and it was adjusted due to the Joule effect. Nernst-Plank and Zhang-Gjorv method were applied after the verification of steady state condition. These two methods provide reliable prediction of steady state diffusion coefficients to be incorporated in the corrosion initiation model. Four different types of simulation were run in ANSYS including High Performance Concrete (HPC) and Ordinary Portland Cement (OPC) concrete for comparison of probability of corrosion initiation.

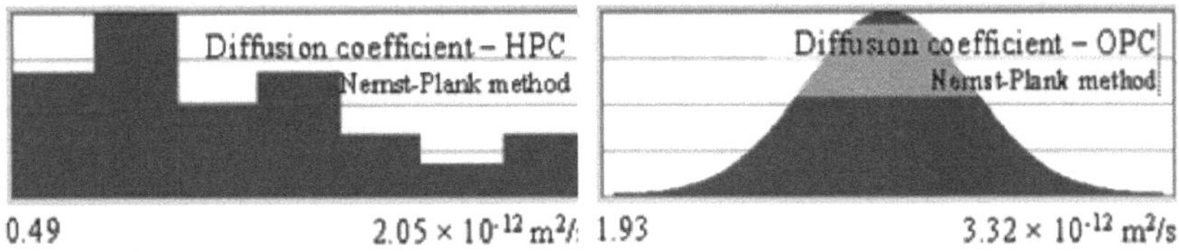

Figure 3. Histograms for diffusion coefficients for Nernst-Plank method HPC (left), OPC (right)

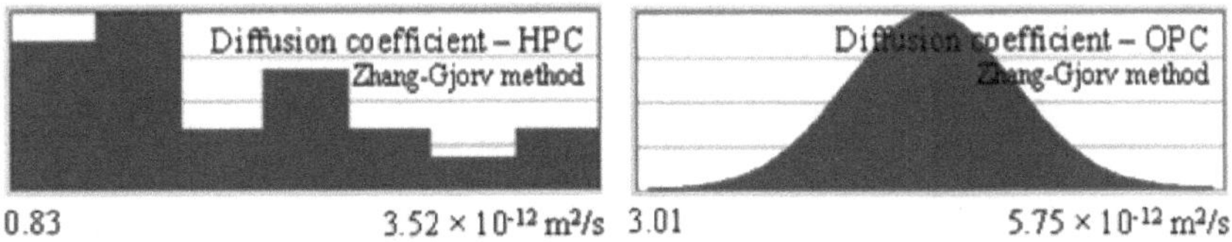

Figure 4. Histograms for diffusion coefficients for Zhang-Gjorv method HPC (left), OPC (right)

Histogram of reinforcing steel depth, presented in [26], is based on the measurement of chloride penetration and concrete cover from more than 200 samples taken from 40 bridge decks.

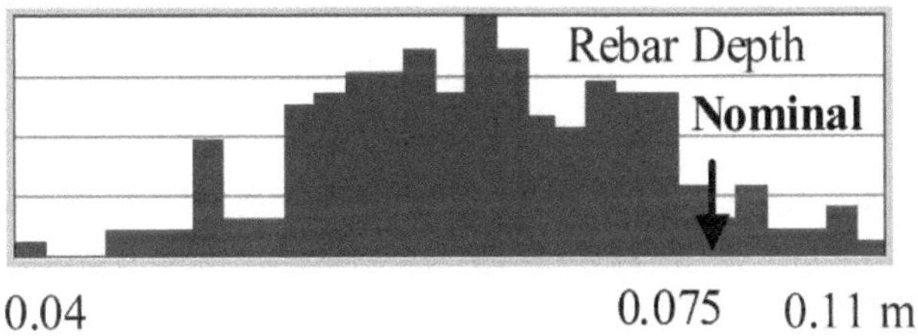

Figure 5. Histogram of Cover (depth of reinforcement) [26]

Use of alternative corrosion resistant reinforcement is already in use for the last 30 years in the USA as a measure of corrosion protection strategy. Two widely used reinforcements are epoxy coating rebar and galvanized rebar to provide long term performance against corrosion. In this SBRA model, performance of epoxy coating and galvanized rebar is compared on the basis of corrosion initiation time with black bar as control reinforcement. The chloride threshold distribution for black bar and galvanized steel is based on the data published in [19]. The chloride threshold distribution for epoxy-coated reinforcement is considered to be based on the black bar performance (Fig. 6 left) while the performance of the epoxy is assessed by the evaluation of the chloride ion concentration at the nearest holiday to the crack.

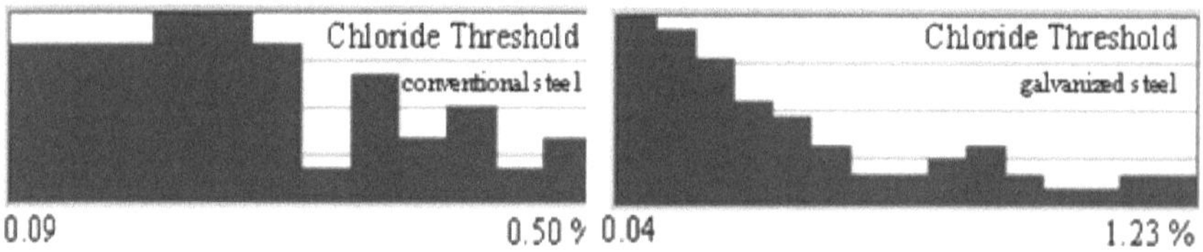

Figure 6. Histograms for chloride threshold for black bar and galvanized rebars [19]

The frequency of holidays in epoxy-coating rebar and surface chloride concentration is based on field data obtained from study performed in Virginia [18].

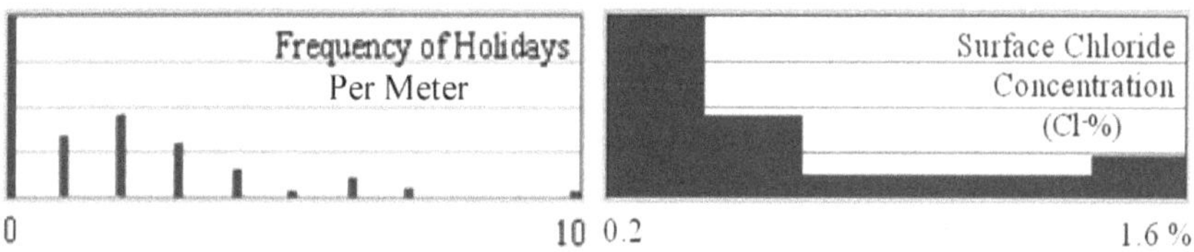

Figure 7. Histogram for frequency for holidays and surface chloride concentration [18]

Distribution for crack depth is based on the engineering estimation by exponential distribution. Minimal value is zero depth and maximal value is all the way through the deck thickness.

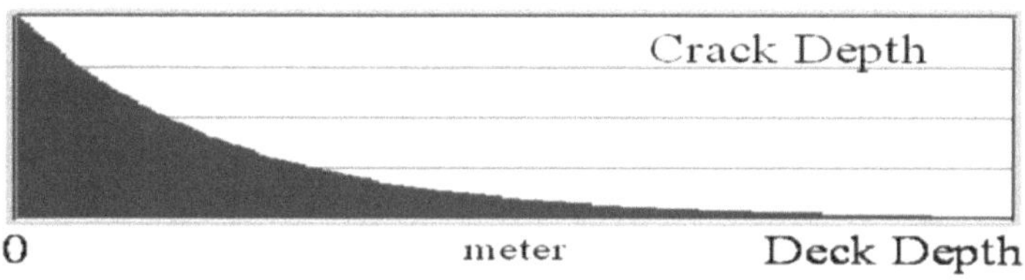

Figure 8. Distribution of crack depth

Crack spacing distribution is assumed to be correlated with deck thickness and width of model slab is equal to the random variable crack spacing. For this paper the worst case scenario is used with frequent cracks throughout all bridges. Spacing of the cracks is also estimated as a normal distribution with the mean value 0.7 m that is 3 times the thickness of the slab. Standard deviation is 0.15. The distribution is truncated within boundaries <0.25, 1.15> m.

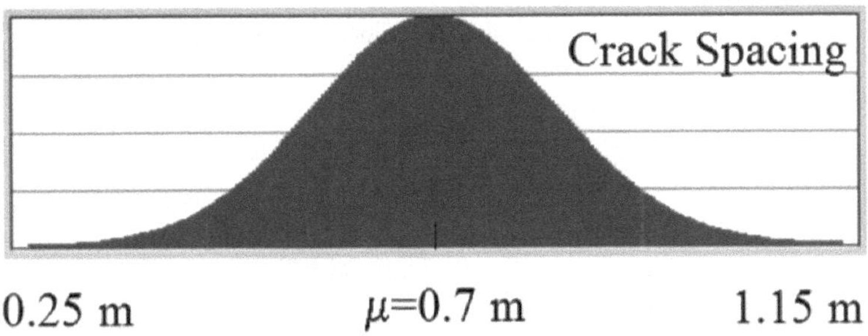

Figure 9. Distribution of crack spacing for steel girder bridges

Following Table I on the next page summarize deterministic and random input variables for specified analyses.

Precision of Monte Carlo simulation

The SBRA module governs the probabilistic analysis with 5,000 Monte Carlo simulation steps. The error of resulting probabilities obtained by the Monte Carlo simulation can be estimated using Eqn. (5) because the precision depends on the number of simulations.

$$[-\varepsilon;\varepsilon]=[-t\sigma;t\sigma]=\left[-t\sqrt{\frac{P_f(1-P_f)}{N}};t\sqrt{\frac{P_f(1-P_f)}{N}}\right] \tag{5}$$

where ε is confidence region, N is the number of simulations, P_f is desired precision and t represents the confidence level. If the desired precision is in order of percent (P_f=1/100), 5000 simulation is applied and t=1.6449 for 90% confidence level is applied than the resulting probabilities will be $P_f \pm 0.0023$ with 90% confidence.

Table I. Random and deterministic input values for ANSYS model

Parameter	Range	Distribution
HPC - Diffusion Coeff. D_c [$10^{-12}m^2/s$](Nernst-Plank)	0.49-2.05	Histogram on Fig. 3
OPC - Diffusion Coeff. D_c [$10^{-12}m^2/s$](Nernst-Plank)	1.93-3.32	Normal Distribution N(2.6,0.23)* Fig. 3
HPC - Diffusion Coeff. D_c [$10^{-12}m^2/s$](Zhang-Gjorv)	0.83-3.52	Histogram on Fig. 4
OPC - Diffusion Coeff. D_c [$10^{-12}m^2/s$](Zhang-Gjorv)	3.01-5.75	Normal Distribution N(4.38,0.46)* Fig.4
Rebar Depth (Cover) R_{ebd} [m]	0.04-0.11	Histogram on Fig. 5
Frequency of holidays M_{ashn} [m^{-1}]	0-10	Histogram on Fig. 7
Crack Spacing $Crck_s$ [m]	0.25-1.15	Normal Distribution N(0.7,0.15)* Fig.9
Crack Depth $Crck_{dpt}$ [m]	0-Depth	Exponential Fig.8
Relative Damage Area Position $Crack_i$	0-1	Uniform distribution
Surface Soluble Chloride Concentration C_0 [%]	0.2-1.6	Histogram on Fig. 7
Chloride Threshold (epoxy-coated bar) C_{th} [%]	0.09-0.50	Histogram on Fig. 6
Chloride Threshold (black bar) C_{thb} [%]	0.09-0.50	Histogram on Fig. 6
Chloride Threshold (galvanized bar) C_{thg} [%]	0.04-1.23	Histogram on Fig. 6
Depth of Slab D_{epth} [m]	0.23	Constant value
Life Span t [years]	100	Constant value

Results

The sample output for epoxy-coated reinforcement is indicated next on the example of HPC concrete with diffusion constant derived using the Nernst-Plank method. The chloride ion concentration at the rebar level at the exposed reinforcing steel areas of and performance function distributions are shown after 100 years in service. The ACI threshold C_{th}=0.2% is shown in Fig. 10 for illustration purposes only. The performance function, which is shown in Fig. 11, consists of random variable chloride concentration as well as random variable threshold based on the particular distribution.

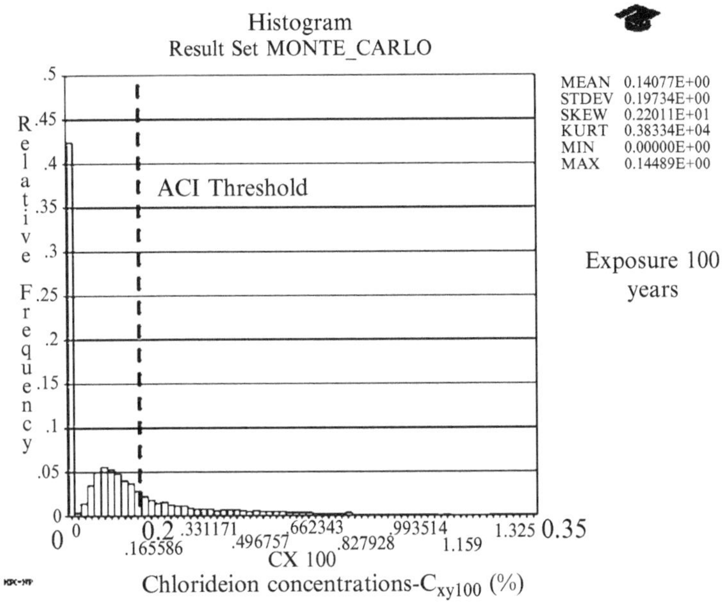

Figure 10. Chloride ion concentration $C_{xy,t}$ at the most exposed location of epoxy coated reinforcement

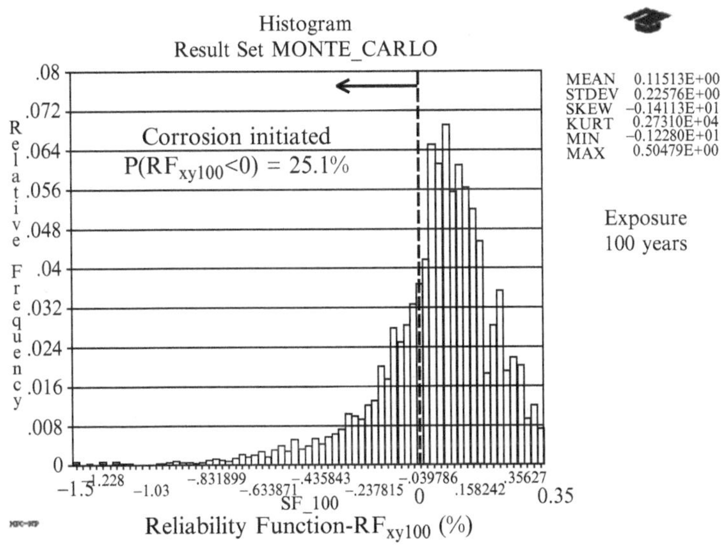

Figure 11. Performance function RF_t for epoxy-coated reinforcement in 100 years

Probability of corrosion initiation on the epoxy-coated reinforcement for service life 100 years is $P_{f,100}$=25.0% as can be seen in Fig. 11. The probabilities can be obtained throughout the whole service life and the corrosion initiation risk can be drawn as a function of time (see Fig. 12).

The probabilities of chloride threshold exceedance are shown in Figs. 12 and 13. For the Nernst-Plank method, black bar in HPC will reach the P_f=25% of corrosion initiation in 18 years, where as in OPC it will reach the P_f=25% of corrosion initiation in 8 years. Galvanized rebar and epoxy coating can delay probability of 25% of the bridges with corrosion initiation up to 95 years and 100 years respectively in HPC. Even in OPC or permeable mixtures they will reach the P_f=25% threshold in 38 years and 45 years respectively.

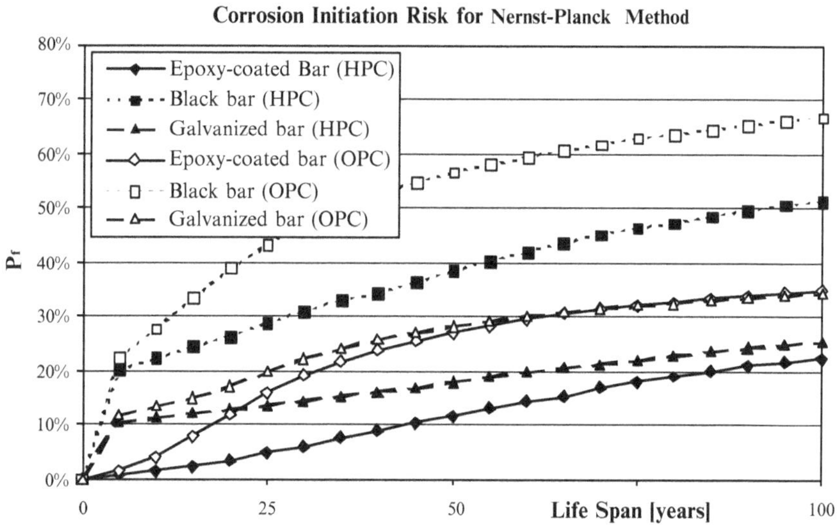

Figure 12. Performance of all the reinforcements (Nernst-Plank method)

For the Zhang-Gjorv method, performance of galvanized rebar and epoxy coating rebar is not as promising as compared to Nernst-Plank. Galvanized rebar will reach the P_f=25% of corrosion initiation in 22 years and 58 years for OPC and HPC, where as Epoxy coating rebar will reach the P_f=25% threshold in 28 and 65 years for OPC and HPC respectively as shown in Fig. 13.

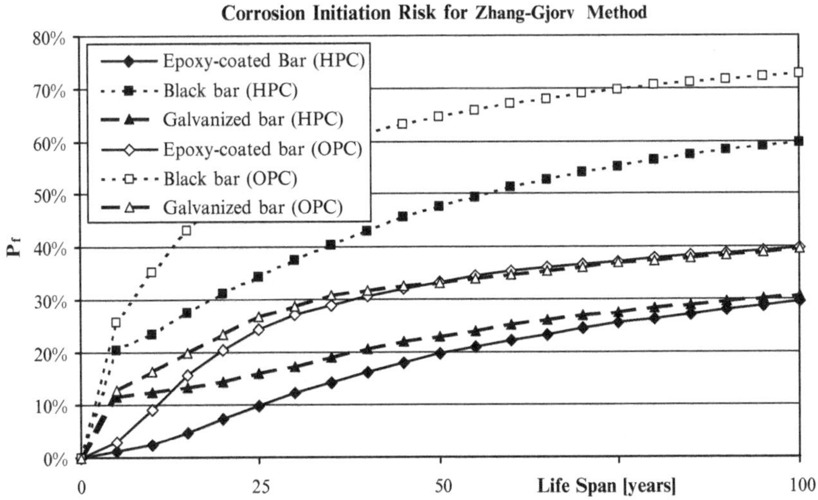

Figure 13. Performance of all the reinforcements (Zhang-Gjorv method)

Discussion

The performance of epoxy-coating and galvanized protection of steel reinforcement in concrete was better than the behaviour of uncoated reinforcement in simulated concrete bridge deck with crack exposed to deicing chemicals effect. The epoxy coating performs significantly better than the galvanized steel in first four decades where the effect of protective epoxy coating and bridge deck crack effect interaction is most significant.

The FEM in conjunction with a Simulation Based Reliability Assessment (SBRA) method was used to estimate the probability of corrosion initiation from chloride ingress of longitudinal epoxy-coated reinforcing steel bars throughout the life of a typical bridge deck.

Conclusions

1. This research demonstrates the necessity of the formulation of probabilistic corrosion initiation model of HPC materials with inclusion of reliable steady state diffusion coefficients and several corrosion resistant steel reinforcements.
2. HPC mixtures have large effect for reduction of diffusion coefficients by enhancing the corrosion initiation time. Both Nernst-Plank and Zhang-

Gjorv provide extended service life for HPC mixtures as compared to OPC mixtures.

3. Epoxy coating rebar and galvanized rebar can extend the corrosion free service life of severely cracked bridges over 100 years if embedded in HPC materials considering the applied probability. In OPC and other permeable concrete materials they can only delay the corrosion initiation maximum up to 45 years. This numerical model does not take into account of corrosion in epoxy coating or propagation stage of galvanized rebar. The propagation stage of corrosion plays an important role in service life prediction of concrete structures. Proper maintenance, installation and alternative corrosion protection strategy according to environmental condition should be implemented along with the use of corrosion resistant steel in HPC materials. This result is comparable in accordance with other researchers study about the long term suitability study for epoxy coated rebar and galvanized rebar [27].
4. The authors considers the Nernst-Plank method is the preferred method for computation of extended corrosion free service life as it provides lower values of diffusion coefficients. But, further research is needed to compute steady state diffusion coefficients from actual field core samples to compare with lab data.
5. Overall, the findings from the probabilistic corrosion initiation model can provide effective decision support in the design of new durable structures, as well as in the optimization of inspection and maintenance of existing structures in chloride laden environments.

References

[1] Zemajtis, J. (1998), Modeling the time to corrosion initiation for concretes with mineral admixtures and/or corrosion inhibitors in chloride laden environments, Dissertation at Virginia Polytechnic Institute and State University, Virginia, USA.

[2] Weyers, R.E., Pyc, W. and Sprinkel, M.M. (1998), Estimating the service life of epoxy-coated reinforcing steel, *ACI Mat. J.*, vol. 95, n. 5, pp. 546-557.

[3] Boddy, A., Bentz, E., Thomas, M. D. A. and Hooton, R.D. (1999), An overview and sensitivity study of a multi- mechanistic chloride transport model, *Cem. and Conc. Res.*, vol. 29, pp. 827-837.

[4] Alisa, M., Andrade, C., Gehlen, C., Rodriques, J. and Vogels, R. (1998), Modelling of degradation, *European Union – Brite Eurram*, CT95-0132, Project BE95-1347, Document BE95-1347/R0.

[5] Papadakis, V.G. (2000), Effect of supplementary cementing materials on concrete resistance against carbonation and chloride ingress, *Cem. and Conc. Res.*, vol. 30, n.2, pp. 291-299.

[6] Teplý, B., Novák, D., Keršner, Z. And Lawansuit, W. (1999), Deterioration of reinforced concrete: Probabilistic and sensitivity analyses, Acta Polytechnica, Prague.

[7] Bentz, E. and Thomas, M. D. A. (2001), Life-365 service life prediction model, Computer Program for Predicting the Service Life and Life-Cycle Costs of Reinforced Concrete Exposed to Chlorides.

[8] Wheeler, M. (2003), Parameters influencing the corrosion protection service life of epoxy coated reinforcing steel in Virginia bridge decks, Master thesis, Virginia Polytechnic Institute and State University, Blacksburg, Virginia, U.S.A.

[9] Lounis, Z. and Amleh, L. (2004), Reliability-based prediction of chloride ingress and reinforcement corrosion of aging concrete bridge decks, In: *Probabilistic Modelling of Deterioration Process in Concrete Structures,* 3rd IABMAS Workshop on Life-Cycle Cost Analysis and Design of Civil Infrastructure Systems/JCSS Workshop, March 24-26, Lausanne, Switzerland.

[10] Lounis, Z. (2003), Probabilistic modelling of chloride contamination and corrosion of concrete bridge structures, In: *Uncertainty Modelling and Analysis*, Proceedings of 4th International Symposium, pp. 447-451, September 21-24, Maryland, U.S.A.

[11] Daigle, L., Lounis, Z. and Cusson, D. (2004), Numerical prediction of early-age cracking and corrosion in high performance concrete bridges– case study, available online: http//www.tac-atc.ca/english/pdf/conf2004/Daigle.pdf.

[12] Tikalsky, P.J., Pustka, D. and Marek, P. (2005), Statistical variations in chloride diffusion in concrete. ACI Structural Journal, *ACI Str. J.*, vol. 102, n.3, pp. 481-486.

[13] Konečný, P., Tikalsky, P. J. and Tepke, D. G. (2007), Performance evaluation of concrete bridge deck affected by chloride ingress: Simulation-Based Reliability Assessment and Finite Element Modelling, *Trans. Res. Rec.*, vol. 2020, Transportation Research Board of the National Academies, ISSN: 0361-1981, Washington, DC, U.S.A.

[14] Shim, H. (2005), Design & analysis of corrosion free service life of concrete structures using Monte Carlo method, *KSCE J. of Civil Eng.*, vol.9, n.5, pp.377-384.

[15] Lounis Z., Zhang, J. and Daigle, L. (2004), Probabilistic study of chloride induced corrosion of carbon steel in concrete structures, In: *probabilistic mechanics and structural reliability,* 9th ASCE Joint specialty conference on, Albuquerque, New Mexico, July 26-28, pp.1-6.

[16] Marek, P., Guštar, M. and Anagnos, T. (1995), Simulation-based reliability assessment for structural engineers, CRC Press, Inc., Boca Raton, Florida.

[17] Tikalsky, P. (2003), Chapter 20 durability and performance-based design using SBRA, *Probabilistic assessment of structures using Monte Carlo simulation*, Basics, Exercises, ITAM Academy of Sciences Czech Republic, 2nd Edition, ISBN:80-86246-19-1.

[18] Pyc, W. (1998), Field performance of epoxy-coated reinforcing steel in Virginia bridge decks, Doctoral dissertation, Virginia Polytechnic Institute and State University, Blacksburg, Virginia, USA.
[19] Darwin, D., Browning, J., O'Reilly, M., Xing, L. and Ji, J., (2009), Critical chloride corrosion threshold of galvanized reinforcing bars, *ACI Mat. J.* vol.106, n.2, pp. 176-183.
[20] Tuutti, K. (1982),Corrosion of steel in concrete, CBI Research Report 4:82, Swedish Cement and Concrete Research Institute, Stockholm, Sweden.
[21] Hooton, R.D., Thomas, M.D.A. and Standish, K. (2001), Prediction of chloride penetration in concrete, Federal Highway Administration, Washington, D.C., No. FHWA-RD-00-142, pp. 405.
[22] Collepardi, M., Marcialis, A. and Turrizuani, R. (1972), Penetration of chloride ions into cement pastes and concretes, *J. of American Cer. Res. Soc.*, vol.55, n.10, pp. 534-535.
[23] Ansys 11.0 Release Documentation, 2009.
[24] Teplý, B., Keršner, Z., Rovnaník, P. and Chromá, M. (2005a), Durability vs. Reliability of RC structures, In: *Durability of Building Materials and Components,* proceedings of 10DBMC International Conference on, Lyon, France, April 17-20.
[25] Pratanu, G., Hammond, A. and Tikalsky, P. (2010), Prediction of equivalent steady state chloride diffusion coefficients, *ACI Mat. J.* (in press for publication).
[26] Sohanghpurwala and Scannell, W.T. (1994), Verification of effectiveness of epoxy-coated rebars, Final Report to Pennsylvania Department of Transportation, Project No. 94 005, pp. 97.
[27] Yeomans, S.R. (1994), Performance of black, galvanized, and epoxy-coated reinforcing steels in chloride-contaminated concrete, *Corrosion*, vol.50, n.1, pp. 72-81.

The Condition Assessment of Reinforced Concrete Lining of Exploration Galleries Exposed to Chemically Aggressive Environment and Its Further Durability Prediction

Amos Dufka

Brno University of Technology, Faculty of Civil Engineering, Institute of Technology of Building Materials and Components, Czech Republic

Abstract. The article describes the diagnostic procedures and methods used for monitoring of exploration gallery lining exposed to action of an aggressive bottom environment. It concerns the so-called control galleries, which became in the framework of tunnel building the part of the primary tunnel lining. The galleries pass through very complex terrain in light of morphological geology. The effect of this fact is that the galleries are in their different parts exposed to attacks of water, the aggressiveness of which towards reinforced concrete structures is significantly variable. The realised research showed that it is necessary to classify the water, which penetrates through the gallery lining in some localities as moderately aggressive towards reinforced concrete.

The objective was to assess the condition of reinforced concrete gallery lining and mainly the further durability prediction of such constructions in given environment.

Introduction

The objects under consideration are in our case the exploration galleries, which are parts of vehicular tunnel construction. They are two galleries 800m long (further the galleries are indicated I and II) which will be connected in future construction and their walls become parts of a primary lining of a tunnel tube.

The tunnelling of galleries was done mainly by mechanical driving technology, final treatment of the stope by classic manual driving. The stope area of one gallery is 13–14 m^2 (according to kilometre spacing). The pin piles were executed at the

C. Andrade and G. Mancini (eds.), *Modelling of Corroding Concrete Structures*, RILEM Bookseries 5, DOI 10.1007/978-94-007-0677-4_6, © RILEM 2011

beginning sections. The lining and the invert are made by shotcrete with guaranteed compression strength 20MPa. The lining of the exploration galleries was reinforced by the mesh reinforcement (Ø6 -100/100 mm) and supports (steel profiles K24). The thickness of the shotcrete is 100mm and was applied using the method of so called "dry guniting". The Portland cement CEMI 42,5R (according to ČSN EN 197-1) was used in content 430 kg per m^3 of concrete and the aggregate of fraction 0-4 a 4-8 mm as a filling. The mixture was modified by the alkali-free rapid cement agent.

The linings in some parts of the galeries are exposed to very intensive attack of groundwater. The attack of groundwater is in some localities so much intensive, that it causes relatively massive penetration of water through the lining of galleries. This fact is documented by following figures:

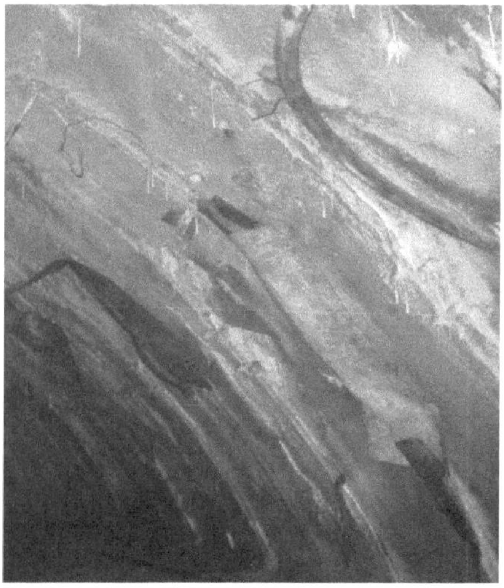

Figure 1: View of the locality where intensive water penetration through the lining from sprayed concrete takes place.

Figure 2: Detailed view of the place where massive penetration of water through the lining takes place. The new formations: on the surface of the lining are formed by calcium carbonate. These new formations are the result of successive "washing" of the concrete cement matrix .

The extent of attacking water aggressiveness in the locality of the galleries moves in a broad range considering the complex geomorphological conditions in the locality of galleries (see Table I). The subject of this paper are the diagnostic methods which have the aim of the galleries lining complex evaluation and especially the prediction of its further service life in connection with the realization of the proper tunnel body.

Specification of the Environment

The geological prospecting showed that the galleries are situated in localities where the earth is formed by tertiary and quaternary sediments. This fact determines the considerable differences in the chemical composition of water which attacks the lining of galleries. Samples were taken to determine and to evaluate correctly the effect of these waters by chemical analysis. The sampling was realized in the way to examine separately waters flowed through the tertiary and through the quaternary sediments. The results of the waters chemical analyses you will find in the following table:

Table I: The results of chemical analysis

Chemical composition	**Sample localization**					
	Gallery I			**Gallery II**		
	Kilometre spacing A - earth formed by tertiary sediments	**Kilometre spacing B** - earth formed by quaternary sediments	**Kilometre spacing C** - earth formed by quaternary sediments	**Kilometre spacing A** - earth formed by quaternary sediments	**Kilometre spacing B** - earth formed by tertiary sediments	**Kilometre spacing C** - earth formed by tertiary sediments
Alkalinity	7.7	7.4	7.7	7.2	7.3	7.4
Sulphates	1420 $mg.l^{-1}$	154 $mg.l^{-1}$	160 $mg.l^{-1}$	320 $mg.l^{-1}$	1490 $mg.l^{-1}$	480 $mg.l^{-1}$
Chlorids	160 $mg.l^{-1}$	151 $mg.l^{-1}$	240 $mg.l^{-1}$	--	--	--
Nitrates	131 $mg.l^{-1}$	107 $mg.l^{-1}$	162 $mg.l^{-1}$	--	--	--
Ammonia ion	0.3 $mg.l^{-1}$	--	--	8.4 $mg.l^{-1}$	--	17.1 $mg.l^{-1}$
Mineralization	1014 $mg.l^{-1}$	--	--	--	--	1396 $mg.l^{-1}$

With regards to the sulphate presence it is possible to classify the environmental aggressiveness in each gallery according to the Czech standard ČSN EN 206-1 “Concrete - Part1: Specification, performance, production and conformity” as is given in Table II:

Table II: Specification of the environment aggressiveness

Localization		Aggressiveness of the Environment
Gallery I	**Kilometre spacing A**	Semi aggressive chemical environment
	Kilometre spacing B	Without any sulphate aggressiveness
	Kilometre spacing C	Without any sulphate aggressiveness
Gallery II	**Kilometre spacing** A	Slightly aggressive chemical environment
	Kilometre spacing B	Semi aggressive chemical environment
	Kilometre spacing C	Slightly aggressive chemical environment

According to the outcomes obtained by the comparison of results of geological research and chemical sample analysis it is possible to state that the waters penetrating through the quarternary sediments practically do not threaten the concrete with the sulphate corrosion at all. It is different when the waters penetrate through the tertiary sediments. Such waters can classified according to the Czech standard ČSN 206-1 as waters with semi-sulphate aggressiveness.
As the consequence of the sulphate ions effect on concrete conditions can be achieved for the corrosion of so-called II and III type [7]. The corrosion of II type is characterized by interaction between the cementing compound and the sulphates compounds which results in the formation of substances which don't have binding properties. During the corrosion of III type reaction of sulphate ions with components of cement matrix take place. The product of these reactions are voluminous crystalline new formations (firstly ettringit, monosulphate or eventually gypsum). The primary sign of the III, type of corrosion is the accumulation and crystallisation of salts or reaction products, the action of aggressive media in capillary tubes and pores of the matter connected with the increase of their volume [XX]. The precipitated new phase and the grow of crystals can evoke considerable pressure on the walls of pores and in this way initiate the destruction of the concrete matter.

It is possible to summarize given facts into a conclusion that the construction durability of the exploration galleries is limited either by the chemical status of the environment or by the possibility of stray currents occurrence.

All upper mentioned facts needed to be taken into the consideration also when assessing the gallery lining condition, resp. when predicting their further durability.

Principle of Implemented Analysis

The objective of implemented analysis was to assess the shotcrete lining condition and to predict their further durability on the basis of acquired facts. The condition analysis of lining material was based on the determination of physico-mechanical and physicochemical parameters of the shotcrete. Further durability prediction was based mainly on comparison of the results obtained from the check tests done during the construction and after two years of exploitation. The shotcrete parameter determination was carried out on sample pieces resp. samples prepared from the core taken from the walls of assessed galleries. The following parameter assessments of shotcrete were made:

Physico-mechanical parameters:

- Volume mass assessment (according to ČSN EN 12390-7),
- Compression strength assessment (according to ČSN EN 12390-3),
- Elasticity modulus assessment (according to ČSN ISO 6784).

Physiochemical assessment:

- Chemical analysis (according to ČSN 720100 and related ČSN resp. ČSN EN),
- X-ray diffraction analysis (according to methodical procedure Matoušek, Drochytka [8]),
- Gap thermal analysis (according to methodical procedure Matoušek, Drochytka [8]),
- Metering of extract pH-value (according to methodical procedure Matoušek, Drochytka [8]).

The test pieces for the determination of the sprayed concrete strength parameters were prepared from core holes sampled from the lining of galleries. The test pieces were prepared considering the demands of quoted Standards and the character of linings in a way in which they represented de facto the whole thickness of the lining. The samples for the evaluation of the size and the degradation of concrete were sampled in borings to represent the concrete, which is in immediate contact with the earth and is exposed to relatively most intensive action of the penetrating water.

The next fixated aspect was the determination of the reinforcing steel corrosion size. This was made on the one hand by non-destructive testing with use of half-cell methods, on the other hand directly in search units.

The sampling of sprayed concrete and the realization of search units for the determination of the corrosion degree are presented in the following figures:

Figure 3: View of the locality in which intensive water penetration through the lining from sprayed concrete takes place.

Figure 4: Detailed view of the place in which massive water penetration through the lining takes place. The new formations on the surface of lining are formed by calcium carbonate. These new formations are formed as the result of gradual "washing" of the concrete cement matrix.

The objective of implemented analysis was to assess the shotcrete lining condition and to predict their further durability on the basis of acquired facts. The condition analysis of lining material was based on the determination of physico-mechanical and physicochemical parameters of the shotcrete. Further durability prediction was based mainly on comparison of the results obtained from the check tests done during the construction and after two years of exploitation.

Results of Executed Analysis

In the following text the information obtained from the monitoring of gallery linings are presented. For objective condition assessment of the shotcrete were chosen the parameters obtained from places with different water aggressiveness, i.e. the places in which the lining is attacked, either by waters without any aggressiveness or waters with semi or slight aggressiveness. The information obtained by tests resp. parameter analysis of the lining concrete is possible to summarize in the following way:

Summary of physico-mechanical characteristics of reinforced concrete:

- Lining thickness (i.e. the shotcrete thickness) is between 120 to 170 mm,
- According to the Czech standard ČSN EN 206-1 it was possible to classify the shotcrete as concrete grade C20/25,
- The elasticity modulus of shotcrete ranged from 24 400 to 26 200 $N.mm^{-2}$.
- It was proved that in places where the steel reinforcement is protected by adequately strong and compacrt layer of concrete no significant concrete corrosion takes place.

Results summary of physical and chemical analysis:

- In such places, where the walls are exposed to acting of semi-sulphate aggressive waters (i.e. the waters penetrating through the tertiary sediments) the concrete layers adjacent to soil are impaired. The degradation rate of concrete creating the surface of the walls after two years of exploitation is relatively low till this time. This statement was formulated on the base of complex composition analyses of assessed samples. The up to now low degree of degradation was confirmed firstly by the low content of corrosive new formations identified in samples taken from layers of linings closely adjoined to the earth (for instance the content of secondary ettringite was only in traces). The results of chemical analyses showed, that the content of sulphates or other aggressive substances in concrete is low. Important fact from the point of view of the structure service life is, that the concrete of the lining is practically not attacked by chlorides, i.e. the ability of concrete to passivate the reinforcement against corrosion is not negatively influenced by the effect of water. The concrete, which forms the "internal mass" of the lining, was not affected by the attack of corrosive substances.
- In places, where the walls are exposed to acting of slightly aggressive resp. without any sulphate aggressiveness, the concrete degradation caused by these waters was practically negligible.
- The rate of degradation caused by gradual disintegration (leaching) of cement matrix by water penetrating into the concrete structure after two years of gallery exploitation is low. This detection results again from the set of physico-chemical analyses. It is based mainly on the comparison of the binder quantity, determined in samples taken from concrete, which forms the "internal mass" of the lining (i.e. of concrete not exposed to significant action of external surroundings) and of the binder quantity determined in samples of concrete which is in immediate contact with the adjacent earth.

Shotcrete Durability Prediction

One of the most important aspects for durability prediction of shotcrete lining was comparison of results of compression strength tests applied during the construction and after two years of lining exploitation. According to the results of check tests carried out during the gallery construction (in a way according to ČSN EN 206-1) the shotcrete was classified into the concrete grade C16/20 (i.e. concrete grade given by the project documentation). After two years of exploitation it was even possible to grade the shotcrete C20/25. The fact, that there were obtained higher values of compression strength after two years of exploitation than right after the gallery finishing, is related especially to graduate growth of clinker mineral hydratation which is accompanied with the cement matrix "strengthening". It was proved by the physicochemical analysis that the shotcrete degradation is low. This fact totally corresponds to the results of compression strength tests of the concrete. With regards to ascertained facts there was created mathematical model which allowed to simulate the behaviour of reinforced concrete gallery linings and to predict their further durability. The mathematical model involved the results of executed tests (mainly the comparison of the data obtained during the gallery construction and parameters obtained after two year exploitation), the conditions of exploitation itself and also the empirical experience with similar types of structures. The mathematical model was compiled in the way that the development of strength values was expressed by the polynomial of third order. This process was based on the methods described in [7, 8] as the result of long time monitoring of similar type of structures.

With regard to results of mathematical simulations, it is possible to predict the durability in a following way:

Prediction of concrete durability – locations with penetrating non-aggressive or slightly aggressive waters

It is possible to express the development of shotcrete compression strength (in places where the gallery walls are exposed to penetrating waters without any or with a slight sulphate aggressiveness) in a way which is shown in Figure 5.

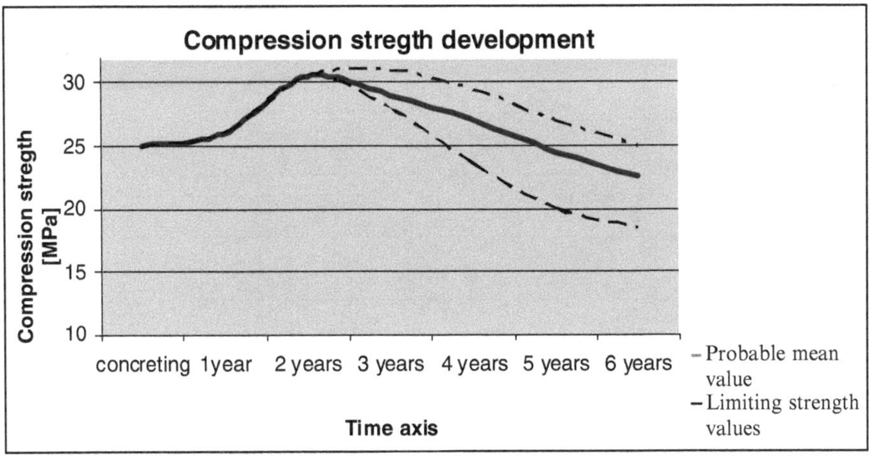

Figure 5: The compression strength development in concrete – waters without or with a slight sulphate aggressiveness

Remark: The expected values of concrete compression strength will range between the dash and dash-dot lines of boundary values.

Prediction of concrete durability – locations with penetrating semi aggressive waters

It is possible to express the development of shotcrete compression strength (in places where the gallery walls are exposed to penetrating waters with semi sulphate aggressiveness) in a way shown in Figure 6.

Remark: The expected values of concrete compression strength will range between the dash and dash-dot lines of boundary values.

Conclusion

The objective of the complex analyses described in this article is to assess the reinforced concrete lining of the exploration galleries exposed to impacting waters with higher chemical aggressiveness and to predict their further durability. Based on detailed and complex analysis was created the mathematical model which simulated the behaviour of reinforced concrete gallery linings depending on the aggressiveness degree of penetrating waters and the exploitation time.

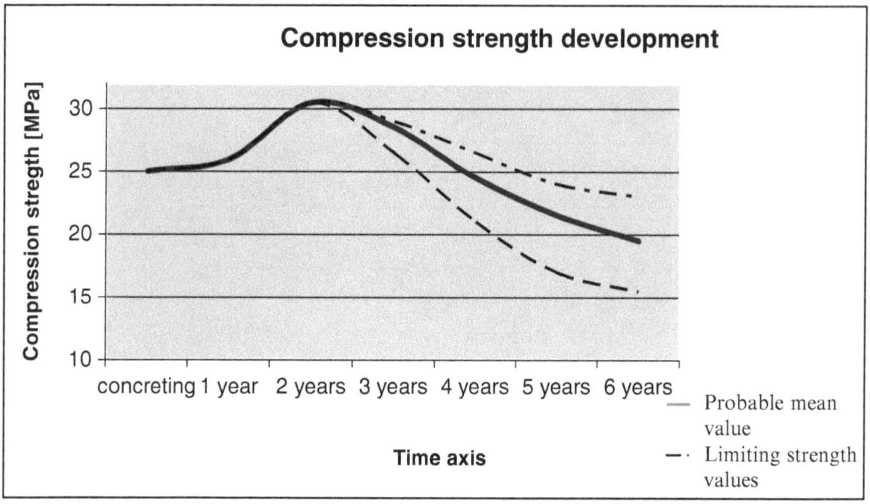

Figure 6: The compression strength development in concrete – waters with semi sulphate aggressiveness

As already mentioned the evaluated galleries are exploration galleries. It means that consequently in the frame of the proper tunnel building the lining of the galleries becomes to be the part of the primary lining. The mentioned mathematical model consequently enables to prognosticate the property changes of sprayed concrete (hereafter shotcrete) and to predict its parameters in the time of the proper tunnel primary lining realization. Considering the economical problems a significant prolongation of the building period took place and the time delay between the realization of exploration galleries and the start of the proper tunnel building is expected 5 till 6 years. We can expect with respect to results determined by the help of the mathematical model that in localities in which the lining of galleries is exposed to non-aggressive water or water with mild aggressiveness no decrease of the shotcrete strength values under the value of 20 MPa will take place. On the contrary in localities where the lining is attacked by waters having higher aggressiveness will in the time interval of 5 till 6 years the mechanical properties decrease of shotcrete be decidedly striking. These facts must be understandably respected in the building of the proper tunnel. It can be stated that in the realization of the primary tunnel lining will be necessary to redevelop the exposed places of the lining and eventually to replace it. The results of mathematical models compiled for specific application enabled in this case to obtain the technical basis which will be effectively respected in the technology of the proper tunnel building.

Acknowledgements

The work was solved with the support of GACR 103/08145 "The Development of Sprayed Concrete for Exposure in extreme Conditions" and VVZ MSM 0021630511 "Progressive Building Materials with Utilization of Secondary Raw Materials and their Influence on the Standard Life of Structures".

References

[1] Balueva, A.V., Zazovskij, A.F.: Elastic-hydrodynamic problem of inflow of fluid in cracks in porous media, AN Makhanika Tverdoge Tela, Moscow, 2007.
[2] Magureanu, C., Rosca, B, Heghes, B.: Influence of aggressive environmental effects on high strength concrete, 7th International Symposium on High Strength/High Performance Concrete, Washington, 2005.
[3] Hunkeler, F.: Corrosion in reinforced concrete: Procces and mechanisms. Corrosion in reinforced concrete structures, Cambridge, 2005.
[4] Melchers, R.E.: Modelling immersion corrosion in natural fresh and brackish waters, Corrosion Sience 48, 2006.
[5] Mainguy, M., Tognazzi, J.M., Torrenti, Addont E.: Modelling of leaching in pure cement and mortar, Cement and Concrete Research 30, 83-90, 2000.
[6] Collepardi, M.: Ordinary and long term durability of reinforced concrete structures, Fifth International Conference on Durability of Concrete, ACI Publication, 2000.
[7] Dobrý, O., Palek, L.: Corrosion of concrete in building practice, Praha 1998.
[8] Matoušek, M., Drochytka, R.: The atmospheric corrosion of concrete, IKAS Praha, 1998.
[9] Attiogbe, E.K. and Rizkalla, S.H.: Response of concrete to sulfuric acid attack, ACI Mater. J. 84(6), 481-48, 1998.
[10] Bickley, J.A, Hooton, R.D., Hoover, K.C.: Performance specifications for durable concrete, Concrete International, 51-57, 2006.
[11] Ho, D.W.: A performance specification for durable concrete, Construction and Building Materials 10, 375-379, 1996.

Systematic Laboratory Test on Structural Performance of Corroded Reinforced Concrete and Its Utilization in Practice

Takashi Yamamoto[1], Michiaki Oyado[2], Yasuhiro Mikata[3], Koichi Kobayashi[4] and Takumi Shimomura[5]

[1]Kyoto University, Kyoto, Japan
[2]Railway Technical Research Institute, Tokyo, Japan
[3]Osaka Institute of Technology, Osaka, Japan
[4]Gifu University, Gifu, Japan
[5]Nagaoka University of Technology, Niigata, Japan

Abstract. A great number of loading tests have been performed on load-carrying behaviour of RC members with the corroded reinforcing steel. However, it is difficult to find the quantitative relationship between the degree of corrosion and the loading capacity. This may be attributed to a scatter in the results of the loading test due to the non-uniformity in the reduced cross sectional area of the corroded reinforcement and some differences in the experimental technique. Such scatter of the results also leads to the difficulty in the selection of the target of the numerical analysis, which must be useful for the estimation of the structural performance. JSCE (Japan Society of Civil Engineers) Committee 331 conducted a systematic laboratory test on the structural performance of RC beam deteriorated by the corrosion of the reinforcement according to the common experimental procedure. The objectives of the common test are to grasp the essence of the scatter in the result of load-carrying capacity. In this study, the flexural behaviour of RC beam, which is degraded due to the chloride ion attack, was focused on without taking into consideration shear failure or bond failure of the tensile reinforcement. The results of the common test show that the large ratio of the corrosion weight loss was highly likely to induce the failure due to the fracture of the longitudinal reinforcement, because the longitudinal scatter in the cross-sectional area of corroded reinforcement increases as the corrosion loss increases.

C. Andrade and G. Mancini (eds.), *Modelling of Corroding Concrete Structures*,
RILEM Bookseries 5, DOI 10.1007/978-94-007-0677-4_7, © RILEM 2011

Introduction

Background

This paper introduces a systematic laboratory test on structural performance of reinforced concrete members with corroded reinforcing steels conducted by Technical Committee 331 in Japan Society of Civil Engineers (JSCE-331). It is important for an ideal service life design and management of concrete structures to evaluate structural performance of concrete structures, namely deflection, load-carrying capacity and ductility, taking into account material deterioration, such as reinforcement corrosion, frost damage and alkali-silica reaction. JSCE-331 committee chaired by Prof. Takumi Shimomura tried to establish scientific and technical platform to comprehensively solve these problems.

A reduction of reinforcement in its cross sectional area due to corrosion causes degradation in structural performances of RC member. In a performance based design concept, a mechanical performance of RC members should be predicted well during its whole life. Therefore, a great number of loading tests have been performed on load-carrying behaviour of RC members with the corroded reinforcing steel by many researchers and institutes up to now [1, 2, 3]. However, it seems to be difficult to find the obvious and quantitative relationship between the degree of corrosion and the loading capacity, owing to the scatter in the test results. Such scatter also leads to the difficulty in the selection of the target data of the numerical structural analysis, which must be useful for the estimation of the structural performance. The followings may be causes to bring the scatter in the test results.

Non-uniformity of corrosion in concrete. A chloride ion attack hardly produces an even reduction in cross sectional area of reinforcements in concrete due to corrosion. A mean value of the degree of corrosion, such as the average cross sectional area loss, is often used to arrange the results of the laboratory test. Is it proper to arrange the results with the only average value?

Differences in experimental way. There must be a lot of differences in the experimental way employed by each institute, such as a manufacture of test specimen, a simulating technique of corrosion, an equipment of the loading test and a measurement method of the corroded reinforcement. The technical quality of the experiment or experimenter may also influence the test results, because there are few standard methods in this kind of the experimental field.

Significance and objective of systematic laboratory test

Non-linear numerical structural analysis, such as a finite element analysis, is available to estimate the structural performance of RC member. Especially, it must be useful for the analysis of the corroded RC member, because the fine analytical

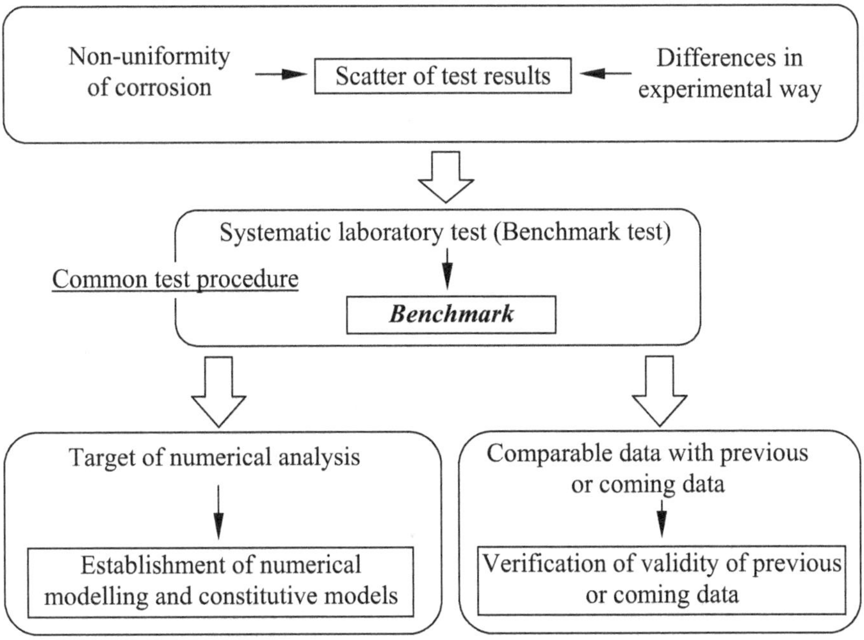

Figure 1. Scheme of systematic laboratory test.

modelling close to the practical corrosion distribution can be made [4]. However, the scatter of the loading test results as mentioned above leads to the difficulty in the selection of the experimental data, namely 'benchmark', to compare with the output data of the numerical structural analysis. This systematic laboratory test is called 'benchmark test' hereafter. If there is no fixed 'benchmark', the suitability of the numerical modelling or constitutive model for the corroded material is difficult to be confirmed.

A common experimental procedure may reduce the scatter of the test results, controlling the technical quality in the experiment. Therefore, JSCE-331 committee conducted a systematic laboratory test on the load-carrying behaviour of RC beam deteriorated by the corrosion of the reinforcement according to the common experimental procedure, including the design of the specimen, the simulating technique of the corrosion, the measuring method of the corrosion and the loading procedure. The flexural behaviour of RC beam, which is degraded due to the chloride ion attack, was focused on without taking into consideration shear failure or bond failure of the tensile reinforcement. The scheme of this systematic laboratory test is shown in Figure 1. Nine institutes participated in this test with the multiple RC beam specimens. This paper shows the common experimental method and the results of this systematic test in the first phase. The reliable and fixed 'benchmark' data will make re-evaluation of the same kind of previous experimental data possible. The validity of the coming results derived from the further test under another experimental factor such as frost damage or alkali-silica reaction can be also evaluated in comparison with the 'benchmark' data. The common test procedure and the test results are shown in the following sections.

Experimental Procedure

RC beam specimen

Figure 2 shows the shape and dimension of RC beam specimen. RC beam specimens consist of a rectangular cross-section with a design concrete strength of 24 N/mm^2. The longitudinal reinforcing steel bars of a deformed 13 mm with the cover thickness of 20 mm was embedded in the lower side of the cross-section of RC beam, giving a tensile reinforcement ratio of 0.014. The yield strength, f_{sy} of the sound longitudinal reinforcement used in all institutes was 360 N/mm^2 up to 369 N/mm^2.

Rectangular stirrups of a deformed 6 mm were provided with a spacing of 100 mm in the span of RC beam. The ratio of designed shear loading capacity to shear force at the designed flexural capacity of RC beam was around three, when the designed shear/flexural capacity were calculated according to Standard Specifications for Concrete Structures [5]. Those arrangements of the reinforcements ensure a flexural failure in RC beam specimens at its sound state, because this bench-mark test focuses on the influence of the corrosion of the longitudinal reinforcement on the flexural behaviour of RC beam. On the other hand, RC beam specimen had to be planned with the smallest possible dimension under the condition that RC beam showed the flexural behaviour as above mentioned. This is attributed that many research institutes can take part in the benchmark test with a large number of RC beam specimens in order to find the statistical result of the benchmark test.

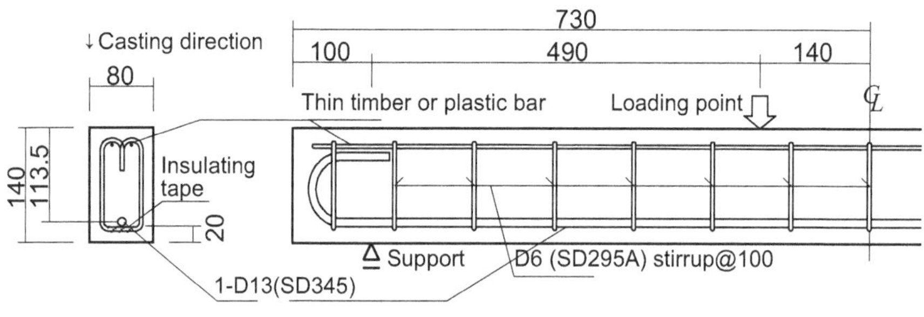

(The right and left side span of the beam is symmetrical.)

Figure 2. Shape and dimension of RC specimen for benchmark test (units in mm)

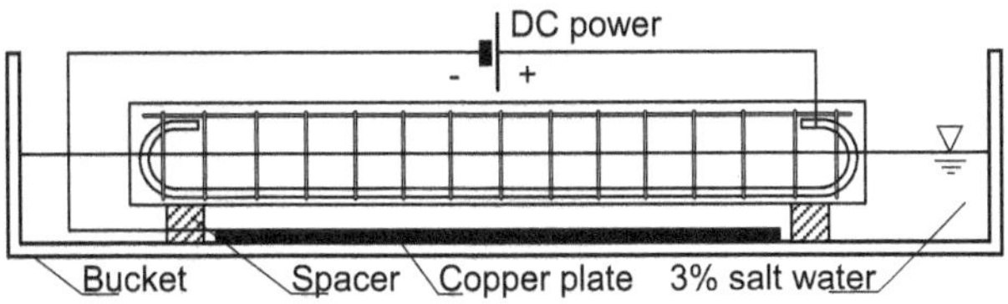

Figure 3. Electrolysis corrosion setup

A contact point on the bottom side of the stirrup with the longitudinal reinforcement was insulated using a vinyl tape so as not to pass an electric current through the contact point. A thin timber or plastic bar with an electric non-conductance was used for the longitudinal reinforcement to fix the spacing of the stirrup in the upper side of the cross-section.

Experimental parameters

The benchmark test was conducted at nine institutes marked as A to I. The amount of corrosion was indicated by an average ratio of corrosion weight loss, assuming that the full length of the longitudinal reinforcement had been uniformly corroded. Four levels of 0% (the sound state), 3%, 10% and 30% were planned. In the institute of H, two specimens with the corrosion ratio of 20% were additionally planned. The total number of 68 specimens was arranged.

Simulating method of corrosion

An electrolysis corrosion technique using a DC power supply was employed to simulate the corrosion of reinforcements. Figure 3 shows a setup of an electrolytic bath used in the electrolysis corrosion technique.

An integrated current, which leads to the target level of the ratio of corrosion weight loss, has to be estimated in the electrolysis corrosion method. Based on the relationship between the corrosion weight loss and the integrated current as shown in the literature [6], the average ratio of corrosion weight loss of 3%, 10%, 20% and 30% corresponded to the integrated current of 69.2 A×hr (5 days), 230.8 A×hr (16 days), 461.6 A×hr (32 days) and 692.4 A×hr (48 days), respectively.

Large corrosion weight loss of 30% will bring the degradation in bond strength between concrete and corroded reinforcement. Especially, electrolysis corrosion causes severe bond condition [7]. However, the flexural capacity will not decrease significantly unless the shear and bond failure mode are distinguished, and also the spalling of cover on a large scale occur [8]. Although it is difficult to find the ratio of the reduction caused by the degradation of the bond strength in the flexural capacity, the degradation of bond strength influences the stiffness and the flexural deformation more significantly than the flexural capacity [9].

Measurements of corrosion

The ratio of corrosion weight loss was calculated from the sound (original) weight of the reinforcement and the corrosion weight loss in the overall length. The sound (original) weight of the reinforcement was derived from measuring the reinforcement alone before casting the RC beam. The corrosion weight loss was measured by removing the rust of corroded longitudinal reinforcement, which was taken out of RC beam after the loading test, using a 10% di-ammonium hydrogen citrate solution at 60°C.

The diameter of the longitudinal reinforcement was measured using a slide calliper at 25 points separated by a longitudinal spacing of around 50 mm regardless of the location of the pitting corrosion, which located at the plane body of the reinforcement with no lateral rib. The minimum diameter was chosen as the measuring direction in the cross-section of the corroded reinforcement. The cross-sectional area was estimated from calculating the circular area of the measured minimum diameter.

Loading test procedure and measurements

Unidirectional static loads were applied to two symmetrical points so that flexural span was 280 mm, the overall span being 1260 mm, as shown in Figure 2. The loading test was performed up to the point at which the load dropped below 80% the maximum load in the post-peak region. The ratio of shear span length to effective depth, a/d is 4.3. Measured items in the loading tests include the applied load, the deflection at the centre span, the deflections below two loading points, the deflections at the two supports, and the strain at the extreme upper end of the cross-section at the centre span.

Results and Discussion

Measurement of corroded reinforcement and failure mode

Figure 4 shows the relationship between the target and measured corrosion loss ratio. The measured corrosion weight loss of up to 20% indicated good agreement with the target values. However, the target corrosion weight loss of 30% brought some smaller values in the measured corrosion weight loss ratio. This may be attributed that a leakage of the current to the stirrup and/or the hooked anchorage of the longitudinal reinforcement reduces the corrosion weight loss of the target region in longitudinal reinforcement between supports.

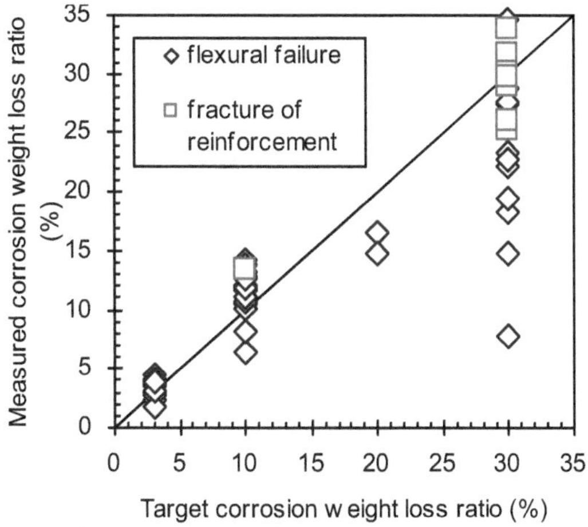

Figure 4. Relationship between the target and measured corrosion weight loss.

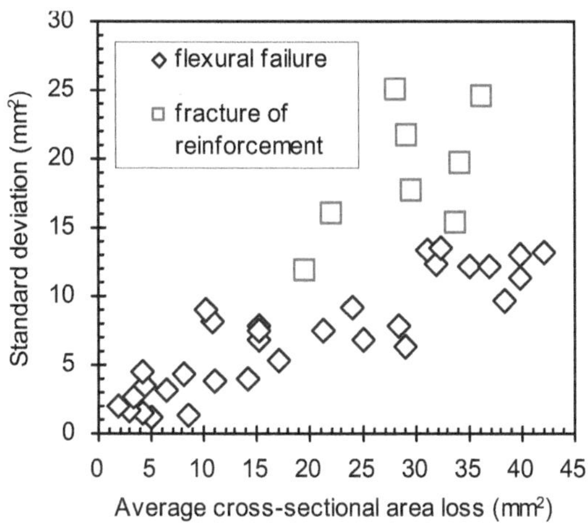

Figure 5. Relationship between the average cross-sectional area loss and its standard deviation of the corroded reinforcement.

Figure 5 shows the relationship between the average cross-sectional area loss and its standard deviation of the corroded reinforcement. The standard deviation of some corroded reinforcements with the large cross-sectional area loss increase significantly. This may show that the longitudinal scatter in the cross-sectional area of corroded reinforcement increases as the corrosion loss increases. Furthermore, the standard deviation above around 15 mm^2 brings the failure due to the fracture of the longitudinal reinforcement.

Nine specimens were failed with the fracture of the longitudinal reinforcement and the others resulted in the flexural compression failure of concrete after the yield of the longitudinal reinforcement. (There are only eight plotted points classified into the fracture of reinforcement in Figure 5, because the longitudinal diameter distribution of a few specimens using the calliper was not measured.) All specimens that failed with the fracture of the longitudinal reinforcement had the standard deviation of the cross-sectional area above around 15 mm^2 as shown in Figure 5. As mentioned above, the longitudinal scatter in the cross-sectional area of corroded reinforcement increases as the corrosion loss increases. Strain in the tensile reinforcement after yielding of the beam may have concentrated in a part of the reinforcement with a locally reduced cross-section, leading to the reinforcement fracture. This may lead the failure mode of the heavily corroded beams to the fracture of the longitudinal reinforcement. Hence, a large standard deviation (> 15 mm^2) and large average cross-sectional area loss (> 20 mm^2) was highly likely to induce the failure due to the fracture of the longitudinal reinforcement.

Load-deflection curve

Figure 6 shows the load-deflection curve of RC beam with the corrosion weight loss ratio of 0%, around 10%, 20% and 30%. Since the load-deflection curve of

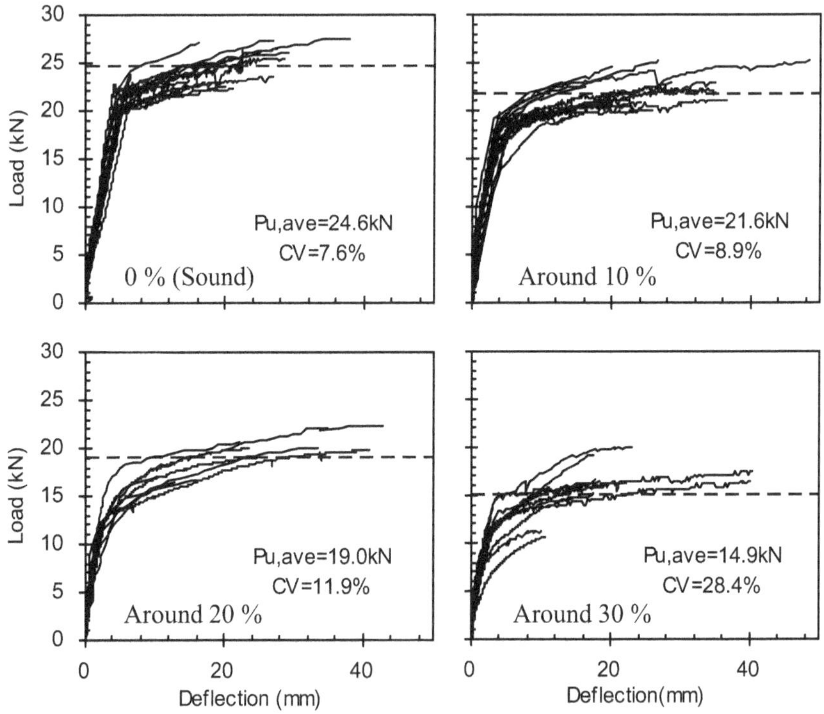

Figure 6. Load-deflection curves

around 3 % was very similar to that of 0 %, it was omitted from Figure 6. The ultimate load, $P_{u,ave}$, and the coefficient of variation, *CV*, are also indicated in these figures. With increasing ratio of the corrosion weight loss, the average of the ultimate load decreased and the coefficient of variation increased, while the ultimate deflection decreased. In particular, specimens with the corrosion weight loss ratio of around 30 % had the smallest flexural capacity and the largest scatter. This is because this group includes most of the specimens that failed with the fracture of the longitudinal reinforcement and their flexural capacity is pretty small. Therefore, the severe corrosion enhanced the reduction in the flexural capacity and its scatter. The difference of the load-deflection curve before the yield of the longitudinal reinforcement was negligible in the case of the corrosion weight loss up to around 20%. However, the stiffness of specimens with the corrosion weight loss ratio of around 30 % decreases. The extremely small section of the corroded reinforcement due to the localized corrosion brought the earlier yield.

Yield load

Figure 7 shows the relationship between the yield load ratio and the corrosion weight loss ratio. The non-uniformity in the cross-sectional area of the corroded longitudinal reinforcement sometimes brings the unclear yield load. In this benchmark test, the yield load of the corroded specimen was defined with the point in load-deflection curve, where the stiffness of the corroded RC beam begins to decrease. The yield load ratio is defined as the ratio of the yield load of the corroded specimen to that of the sound specimen. The dotted line indicates the average reduction in the yield load when the yield load of RC specimen decreases in proportion to the cross-sectional area loss of the longitudinal reinforcement due to the corrosion. The yield load of RC specimen was calculated using the RC sectional analysis. The longitudinally average cross-sectional area of the corroded rein-

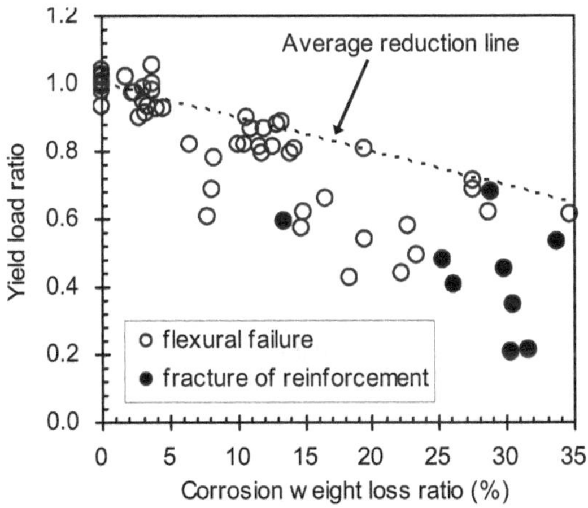

Figure 7. Relationship between yield load ratio and corrosion weight loss ratio

forcement corresponding to the corrosion weight loss was applied to the cross-sectional area of the longitudinal reinforcement used in the RC sectional analysis.

The yield load ratio decreased as the corrosion weight loss ratio increased. However, the reduction rate of the yield load with the increasing of the corrosion weight loss ratio is larger than the average reduction rate indicated by the dotted line. This may be attributed to the locally large reduction of the cross-sectional area in the flexural span with the uniform bending moment. Hence, the minimum cross-sectional area of the longitudinal reinforcement in the flexural span causes the early yielding.

Ultimate load

Figure 8 shows the relationship between the ultimate load ratio and the ratio of the corrosion weight loss. The ultimate load ratio is defined as the ratio of the flexural capacity of the corroded specimen to that of the sound specimen. The dotted line indicates the average reduction in the flexural capacity when the flexural capacity of RC specimen decreases in proportion to the cross-sectional area loss of the longitudinal reinforcement due to the corrosion. The flexural capacity was calculated in the same way as the yield load.

The ultimate load ratio decreased as the corrosion weight loss ratio increased. The reduction rate of the ultimate load with the increasing of the corrosion weight loss ratio shows good agreement with the average reduction rate indicated as the dotted line. The local reduction in the cross-sectional area of longitudinal reinforcement in some degree may not bring the significant reduction in the ultimate load, because the effect of the strain hardening of the reinforcement can compensate the tensile force reduced due to the cross-sectional loss of the reinforcement in the

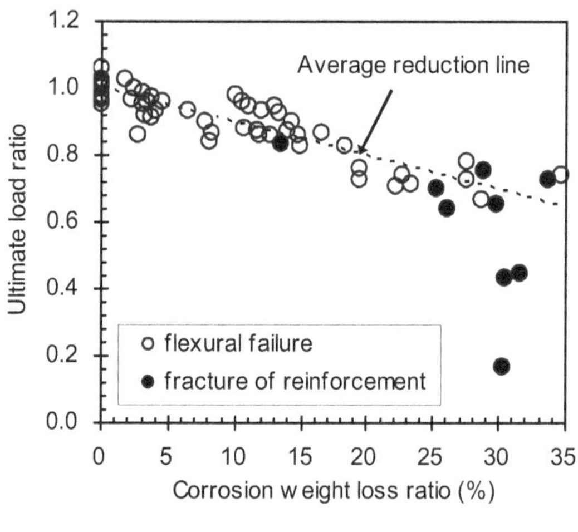

Figure 8. Relationship between ultimate load ratio and corrosion weight loss ratio

cross-section of RC beam, unless the fracture of the longitudinal reinforcement in the ultimate state. However, the specimens with the large ratio of the corrosion weight loss are likely to reveal the large scatter of the ultimate load ratio. In particular, specimens that failed with the fracture of the longitudinal reinforcement demonstrated this tendency and their ultimate flexural capacity were quite small.

Conclusions

The main results obtained in the present study are summarized as follows:

The longitudinal scatter in the cross-sectional area of corroded reinforcement increases as the corrosion loss increases. Consequently, the large ratio of the corrosion weight loss was highly likely to induce the failure due to the fracture of the longitudinal reinforcement, because the strain in the tensile reinforcement after yielding of the beam concentrated in a part of the reinforcement with the locally reduced cross-section.

The yield load ratio decreased as the corrosion weight loss ratio increased. However, the reduction rate of the yield load with the increasing of the corrosion weight loss ratio is larger than the average reduction rate on the assumption of the longitudinally uniform corrosion, because the minimum cross-sectional area of the longitudinal reinforcement in the flexural span causes the early yielding.

The ultimate load ratio decreased as the corrosion weight loss ratio increased. The specimens with the large ratio of the corrosion weight loss show the significant reduction and the large scatter in the flexural capacity, because of the failure due to the fracture of the longitudinal reinforcement with the large standard deviation in the cross-sectional area.

This systematic laboratory test was performed under the limited test conditions, such as the electrolysis corrosion technique and the flexural load-carrying behaviour. Since the corrosion distribution due to the electrolysis corrosion is quite different from that due to the chloride and carbonation induced corrosion, further investigation on the artificially simulating method of the corrosion should be needed. However, the results of this test were obtained in conformity with the common experimental procedure. Therefore, this systematic laboratory test contributes to the clarification of the effect of the longitudinal distribution of the cross-sectional area in the corroded reinforcement on the load-carrying behaviour and to the verification of the output data of the numerical simulation.

Moreover, although the sufficient accumulation and comparative analysis of loading test results in the practically corroded RC beam must be performed in the coming research, the result that the structural performance of RC member with the high corrosion levels had so much scatter in its capacity suggests an application of larger safety factor in the estimation of the capacity at maintenance phase than that at the design phase. Therefore, this systematic test must be necessary and helpful to produce an appropriate safety factor in the estimation of load-carrying capacity of the degraded RC member.

Acknowledgements

The authors wish to express their sincere gratitude to all the members of JSCE-331 committee for their eager contributions to this project.

References

[1] Okada, K., Kobayashi, K. and Miyagawa, T. (1988), Influence of Longitudinal Cracking due to Reinforcement Corrosion on Characteristics of Reinforced Concrete Members, ACI Structural Journal, Vol. 85, No. 2, pp. 134-140.

[2] Castel, A., Francois, R. and Arliguie, G. (2000), Mechanical behaviour of corroded reinforced concrete beams-Part 1: Experimental study of corroded beams, Materials and Structures, Vol. 33, pp. 539-544.

[3] Yamamoto, T. and Kobayashi, K. (2006), Report of Research Project on Structural Performance of Deteriorated Concrete Structures by JSCE-331 - Review of Experimental Study, Proceedings of the 1st International Workshop on Life Cycle Management of Coastal Concrete Structures, pp. 171-180, Niigata.

[4] Lee, H. S., Tomosawa, F. and Noguchi, T. (1998), FEM analysis for structural performance of deteriorated RC structures due to rebar corrosion, Proceedings of the 2nd International Conference on Concrete under Severe Conditions, pp. 327-336.

[5] Japan Society of Civil Engineers (2008), Standard Specification for Concrete Structures-2007, Design, Japan Society of Civil Engineers.

[6] Tamori, K., Maruyama, K., Odagawa, M. and Hashimoto, C. (1988), In: Characteristics of crack induced by corrosion of reinforcing steel embedded in concrete, Proceedings of the Japan Concrete Institute, Vol. 10, No. 2, pp. 505-510, Tokyo (in Japanese).

[7] Yuan, Y., Ji, Y. and Shah, S. P. (2007), Comparison of two accelerated corrosion techniques for concrete structures, ACI Structural Journal, Vol. 104, No. 3, pp. 344-347.

[8] Cairns, J. and Zhao, Z. (1993), Behaviour of concrete beams with reinforcement exposed, Proceedings Institution of Civil Engineers, Structures and Buildings, Vol. 99, pp. 141-154.

[9] Aoyama, T., Shimomura, T. and Maruyama, K. (1998), Flexural behaviour of RC members with corroded reinforcing bars due to salt attack, Proceedings of the Japan Concrete Institute, Vol. 20, No. 2, pp. 883-888 (in Japanese).

Comparison of Resistance to Chloride Penetration of Different Types of Concrete through Migration and Ponding Tests

Luca Bertolini, Federica Lollini and Elena Redaelli

Diparimento di Chimica, Materiali e Ingegneria Chimica, Politecnico di Milano, via Mancinelli 7, 20131 Milan, Italy

Abstract. Recently, performance-based methods using probabilistic approaches have been proposed to allow a quantitative evaluation of the service life of a structure with respect to reinforcement corrosion. In these models the resistance of concrete to chloride penetration is evaluated by means of accelerated tests that provide specific parameters, e.g. an apparent chloride diffusion coefficient. These parameters can be used to predict the chloride penetration in a structure only if they are corrected through coefficients that take into account the real environmental exposure conditions. However there is a lack of information about the relationship between long term performance of concrete and results of short term tests. Data on concretes of different compositions from accelerated tests should be compared to data from real structures or, at least, with results of medium term tests on specimens exposed to conditions more similar to that of real structures. In this work the resistance to chloride penetration of different types of concrete was measured by means of the rapid chloride migration test and 1-year ponding test. The role of concrete composition on the resistance to chloride penetration was assessed and the correlation between results of the two tests was investigated.

Introduction

Due to the increase of degradation of reinforced concrete (RC) structures, especially due to steel corrosion, durability has become a critical issue in their management [1]. To prevent steel corrosion, besides the European standards [2, 3], which however recommend simply deemed-to-satisfy rules, performance-based methods using probabilistic approaches have been proposed. They allow a quantitative evaluation of the service life of a structure with respect to

C. Andrade and G. Mancini (eds.), *Modelling of Corroding Concrete Structures*, RILEM Bookseries 5, DOI 10.1007/978-94-007-0677-4_8, © RILEM 2011

reinforcement corrosion. Some of them, like that proposed by the international federation for structural concrete (FIB) [4], are based on a probabilistic or semiprobabilistic approach similar to that used in the structural design: limit states that indicate the boundary between the desired and the adverse behaviour of the structure are defined. In these models the concrete behaviour, for instance its resistance to chloride penetration, is evaluated by means of an accelerated test, which provides an apparent chloride diffusion coefficient. However the test result cannot be directly used to predict the chloride penetration in a structure, but should be corrected through coefficients that take into account, for instance, the real environmental exposure conditions. Therefore there is a need to correlate data obtained from accelerated tests with data from real structures or, at least, with results of medium term tests on specimens exposed to conditions more similar to that of real structures.

Several accelerated tests were proposed in the literature to evaluate the concrete resistance to the transport of chlorides and different correlations were found between results of long and short term tests [5-8].

This paper reports the resistance to chloride penetration of different types of concrete. Concretes with different binders, water/binder ratios and binder dosages were subjected to the rapid chloride migration tests. Furthermore, reinforced concrete specimens made with the same compositions were subjected to ponding tests and chloride penetration was evaluated at different times by measuring chloride profiles. The paper compares the results of the two series of tests in relation to the assessment of the role of the concrete composition on the resistance to chloride penetration.

Experimental Procedure

In order to investigate the influence of concrete composition on the resistance to chloride penetration, 26 types of concrete were cast. Different types of binder were obtained by replacing, in a cement factory, a portland cement with different percentages of ground limestone (15%LI, 30%LI), coal fly ash (30%FA), a natural pozzolan (30%PZ), and ground granulated blast furnace slag (70%BF). Water/binder ratios ranging from 0.42 to 0.61 and binder dosages from 250 to 400 kg/m^3 were considered. Major details on concrete compositions, curing and characterization with regard to mechanical properties are given elsewhere [9-10].

Resistance to penetration of chloride ions was tested by means of the so-called rapid chloride migration test, according to NT-BUILT 492 standard, which led to the calculation of a non-steady state chloride diffusion coefficient (D_{RCM}) [11]. A plastic tube was mounted coaxially on a 50 mm thick concrete cylinder, cured 28 days, and a chloride-free solution was poured inside. The specimen, laid on an inclined plastic support, was placed in a container filled with 10% NaCl solution. A potential difference of 30 V was applied, the initial current was measured and, according to its value, the applied voltage was adjusted and the duration of the test, which varies between 6-96 hours, determined. At the end of the test, the specimen was split axially, and on its fracture surface a 0.1 M $AgNO_3$ solution was sprayed on the fracture surface. The average chloride penetration depth x_m (m) was measured and the chloride diffusion coefficient D_{RCM} (m^2/s) was calculated as:

$$D_{RCM} = \frac{RT}{zFE} \cdot \frac{x_m - \alpha\sqrt{x_m}}{t} \tag{1}$$

where R is the gas constant (J/K mol), T the average temperature in the anodic solution (K), z the absolute value of charge number (1), F Faraday's constant (96500 C/mol), t time (s); E =(U-2)/L (U is the applied voltage in V, L the thickness of the specimen in m) and is defined as:

$$\alpha = 2\sqrt{\frac{RT}{zFE}} erf^{-1}\left(1 - \frac{2c_d}{c_0}\right) \tag{2}$$

where c_d is the concentration at which the colour change is observed (assumed equal to 0.07 N) and c_0 the chloride concentration of the test solution (2 N) [11]. Ponding tests were carried out on reinforced concrete specimens with three rebars at different cover depth (Figure 1). This test was performed on specimens cured 7 days and for some types of concrete also on specimens cured 28 days.

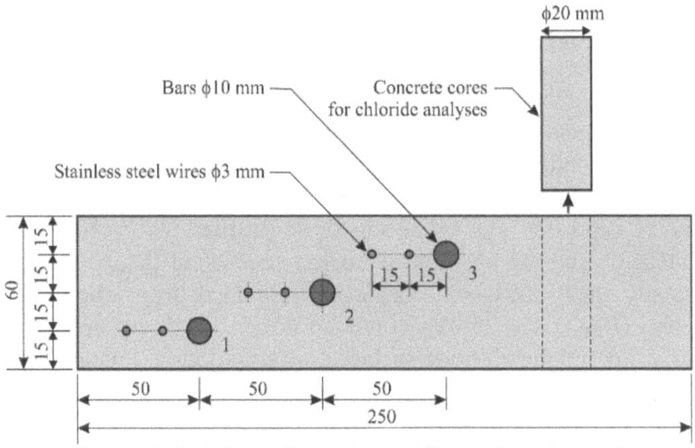

Figure 1. Reinforced specimen (dimensions in mm).

A pond was mounted on the upper surface of the specimens and a 35 g/l chloride solution was poured inside, in order to simulate seawater. The four lateral faces of the reinforcement specimens were masked with epoxy. Evaporation was allowed from the bottom face of specimens. These specimens were cast in order to collect data on the initiation (i.e. the critical chloride threshold) and propagation of corrosion (i.e. the corrosion rate) by means of periodically electrochemical measurement (results are discussed elsewhere [10]). Moreover, on these specimens after one year of exposure, a sample was cored and the chloride profile was measured, by analyzing the total chloride content (acid soluble) at intervals of 10 mm. The chloride profiles were fitted with the relationship:

$$C(x,t) = C_s\left[1 - erf\left(\frac{x}{2\sqrt{Dt}}\right)\right] \quad (3)$$

and the apparent diffusion coefficient (named D_{pond}) and the surface content C_s were calculated.

Results and Discussion

Table I summarises the composition of the different concrete mixes and the compressive strength evaluated on replicate specimens after 28 days of wet curing. Compressive strength after other curing times and other properties of mixes are given and discussed elsewhere [9-10]. It was found that a reduction in the compressive strength of concrete on 28-day cured specimens was observed replacing part of portland cement with mineral additions (Table I). For a given *w/b*, a decrease in the strength was observed as the binder content increased; this could be explained by the increase in the amount of hardened paste and, thus, in the total porosity of the concrete [9-10].

In order to obtain an estimation of the resistance to the penetration of chloride ions of each material accelerated chloride migration tests were performed. It is well known that results from accelerated tests allow to achieve a screening on the performance of different types of concrete with regard to the resistance to chloride penetration, however they cannot be used to predict the behaviour of the material under real environmental exposure conditions. In order to take into account more realistic exposure conditions, results of the ponding tests on reinforced specimens were also considered. On these specimens the chloride profiles were measured after 1 year of exposure and, by fitting these profiles with relationship (3), the surface content C_s and the apparent diffusion coefficient (D_{pond}) were calculated. The non-steady state diffusion coefficient obtained from the rapid chloride migration test (D_{RCM}) on 28-day cured specimens and the apparent diffusion coefficient (D_{pond}) and the chloride surface concentration (C_s) obtained after 1 year of ponding on specimens cured 7 and 28 days are also shown in Table I.

Figures 2 and 3 show, respectively, the diffusion coefficient D_{RCM}, measured on 28-day cured specimens with the rapid chloride migration test, and the diffusion coefficient, D_{pond}, evaluated after 1 year of ponding test on specimens cured 7 days, as a function of concrete composition. In Figures 2 and 3 it clearly appears, as expected, that an increase in the water/binder ratio led to an increase in the diffusion coefficient. For instance, for concrete with 15%LI, D_{RCM} increased from 15.1 10^{-12} to 19.8 10^{-12} m^2/s and D_{pond} from 8.2 10^{-12} to 12.5 10^{-12} m^2/s increasing the *w/b* from 0.46 to 0.61. For 28-day cured specimens exposed to ponding test only *w/b* ratios of 0.42 and 0.46 were available and it was not possible to accurately investigate the relationship between D_{pond} and *w/b*. Occasionally D_{RCM} and D_{pond} increased after decreasing *w/b* ratio (for instance for concrete with 30%FA and *w/b* of 0.42 and 0.46); this might be due to higher uncertainty of experimental results for more impervious concrete.

Table I. Composition and properties of the tested concretes (average values of two replicate samples): *w/b* = water/binder ratio; *b* = binder content; R_c = 28-day compressive strength; D_{RCM} = rapid chloride migration diffusion coefficient, C_s and D_{pond} = chloride surface concentration and ponding diffusion coefficient after 1 year of exposure

	w/b	*b*	R_c	D_{RCM}	C_s		D_{pond}	
	-	kg/m³	MPa	10^{-12} m²/s	% vs cem		10^{-12} m²/s	
Curing	-	-	28	28	7	28	7	28
OPC	0.61	300	59.2	12.5	5.1	-	6.1	-
	0.46	300	87	7.7	1.7	1.66	8.2	7.4
		350	70.7	7.7	1.54	0.89	3.9	9.0
	0.42	350	86.6	5.4	0.81	1.05	14.4	7.0
15%*LI*	0.61	250	54.5	18.9	3.77	-	12.5	-
		300	45.2	19.8	5.14	-	12.9	-
	0.46	300	47.8	15.1	1.62	2.57	14.7	6.9
		350	67.3	13.6	1.84	1.11	8.2	5.9
	0.42	350	83.3	11.3	1.00	0.84	6.8	10.6
		400	63.7	11.4	1.77	1.39	9.6	4.1
30%*LI*	0.61	300	36.7	38.3	3.41	-	27.8	-
	0.46	300	62	18.1	1.49	3.39	20.5	6.3
		350	57.4	23.5	2.25	1.47	15.8	16.0
	0.42	350	66.6	13.1	1.63	1.63	14.9	5.1
30%*FA*	0.61	300	43.9	11.1	3.10	-	2.4	-
	0.46	300	76.2	4.4	3.35	3.83	1.8	1.8
		350	74.6	3.6	4.65	5.15	1.3	1.0
	0.42	350	80.5	3.8	1.33	3.10	8.4	3.5
30%*PZ*	0.61	300	42.7	22.7	3.52	-	9.2	-
	0.46	300	69	13.2	4.11	2.86	4.4	4.9
		350	45	12.8	2.06	1.92	4.7	4.9
	0.42	350	72.6	8.8	2.21	1.93	3.2	4.1
70%*BF*	0.61	300	45.3	3.0	3.93	-	4.6	-
	0.46	300	55.1	2.2	2.99	3.50	3.1	2.0
		350	54.7	1.5	2.78	4.46	1.5	0.8
	0.42	350	82.6	1.4	2.92	2.63	1.6	3.7

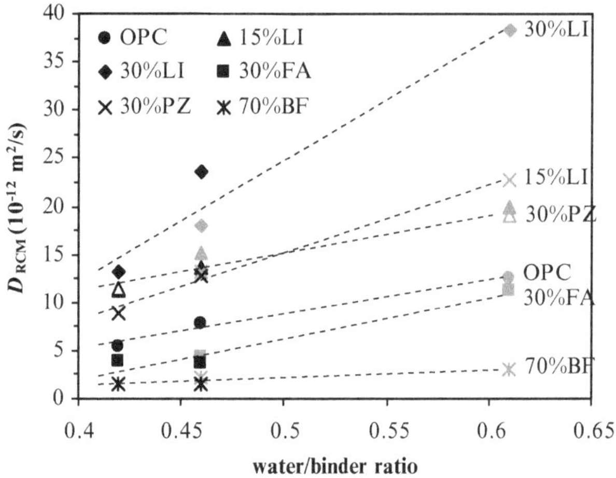

Figure 2. D_{RCM} coefficients obtained as an average value of two replicate specimens from the rapid chloride migration test as a function of *w/b*, mineral addition and binder content (250 kg/m^3 grey, 300 kg/m^3 filled grey, 350 kg/m^3 filled black, 400 kg/m^3 black)

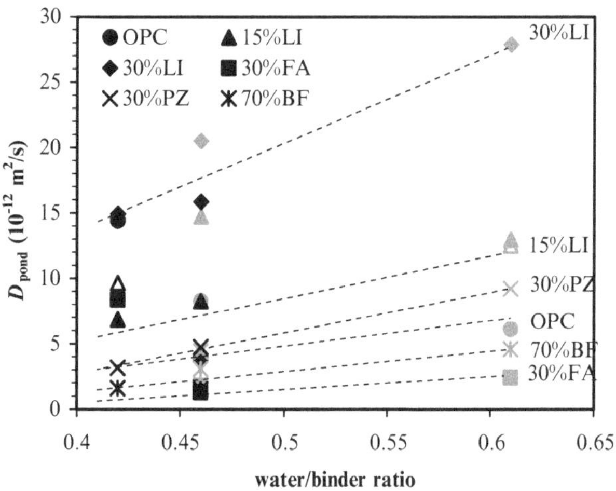

Figure 3. D_{pond} coefficient obtained from the ponding tests after 1 year of exposure as a function of *w/b*, mineral addition and binder content (250 kg/m^3 grey, 300 kg/m^3 full grey, 350 kg/m^3 full black, 400 kg/m^3 black) on specimens cured 7 days

The binder content did not show a significant influence on the resistance to chloride penetration, while a remarkable influence of the type of binder was observed.
The best performances were reached by the fly ash cement and the blast furnace slag cement. D_{RCM} of concrete with 30%FA was roughly halved with regard to OPC, for instance for concrete with *w/b* = 0.46 it decreased from about 8 10^{-12} to about 4 10^{-12} m^2/s. Similarly D_{pond} was more than halved and for concrete with *w/b* = 0.46 it decreased from about 8 10^{-12} m^2/s to about 1.8 10^{-12} m^2/s. With 70%BF both D_{RCM} and D_{pond} were significantly lower that with OPC (e.g. the D_{RCM} was lower than 2 10^{-12} m^2/s and D_{pond} than 3 10^{-12} m^2/s for *w/b* = 0.46). Conversely for concrete mixes with 15% of limestone the diffusion coefficient was more than doubled compared to OPC. For instance D_{RCM} and D_{pond} respectively increased from 12.5 10^{-12} to 19 10^{-12} m^2/s and from 6.1 10^{-12} to 12.9 10^{-12} m^2/s for concrete with *w/b* = 0.61. Both D_{RCM} and D_{pond} further increased with 30% of limestone and in concrete with *w/b* ratio of 0.61 values of D_{RCM} about 38 10^{-12} m^2/s and D_{pond} about 28 10^{-12} m^2/s were obtained. The rapid chloride migration test showed comparable performances in comparison performances for 30% natural pozzolan and 15% limestone, for instance for concrete with *w/b* = 0.46 D_{RCM} of 13.6 10^{-12} and 12.8 10^{-12} m^2/s were detected on concrete with respectively 15% limestone and 30% natural pozzolan. Instead the ponding test showed that improved resistance to chloride penetration were obtained with 30%PZ compared to 15%LI. For instance, considering concrete with *w/b* = 0.46, D_{pond} was about 8.2 10^{-12} m^2/s for concrete with 15%LI and it decreased to 4.7 10^{-12} m^2/s with 30%PZ. Therefore results obtained both with the rapid chloride migration and ponding tests are in agreement in showing the beneficial role of the partial replacement of the portland clinker with fly ash and ground granulated blast furnace slag in reducing the penetration of chloride and show the negative influence of the replacement with ground limestone.

Results shown in Figures 2 and 3 suggest that a linear relationship can be found between the *w/b* ratio and the diffusion coefficient [12]. Although it is clear that these relationships strongly depend on the type of binder. For 30% LI, the diffusion coefficients, both D_{RCM} and D_{pond}, show a higher dependence on the water/binder ratio, whereas for 70% BF this influence is much less evident. Moreover it can be expected that this relationship may depend on the type of test used to obtain the diffusion coefficient and the curing time of concrete, in fact different correlations were found taking into account results of rapid chloride migration and ponding tests.

Curing is a further parameter that has an influence on the resistance of concrete to chloride penetration. To investigate the role of curing, the relationship between the ponding diffusion coefficient obtained on specimens cured 7 days and 28 days is shown in Figure 4. Although there is a wide scatter of results, the diffusion coefficient slightly decreases when curing time increases. For instance for concrete obtained by replacing 15% of portland clinker with limestone, water/binder ratio of 0.46 and binder content of 350 kg/m^3 the ponding diffusion coefficient decreased from 8.2 10^{-12} to 5.9 10^{-12} m^2/s, increasing the curing time from 7 to 28 days. However few exceptions for concrete with the lower water/binder ratio were observed, probably due to the lower accuracy of chloride profiles measured on the more impervious concretes.; for instance $D_{pond,28}$ was higher than $D_{pond,7}$ for OPC with w/b = 0.46 and b = 350 kg/m^3, 15% LI, with w/b = 0.42 and b = 350 kg/m^3, 30% PZ with w/b = 0.42 and b = 350 kg/m^3 and, 70% BF with w/b = 0.42 and b= 350 kg/m^3.

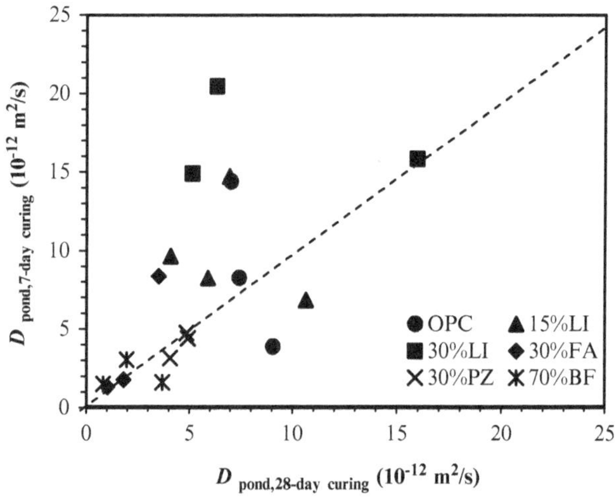

Figure 4. Relationship between D_{pond} coefficients obtained after 1 year of ponding on concretes cured 7 and 28 days

In order to predict the behaviour of the material under real environmental exposure conditions and to perform a service-life design, the resistance to chloride penetration evaluated by means of an accelerated test should be corrected with results of medium or long term tests. At this regard the comparison of results of rapid chloride migration tests and ponding tests could be useful to assess a relationship between short term and medium term results (Figure 5). In general it can be observed that on concretes cured both 7 (Figure 5*a*) and 28 days (Figure 5*b*), the diffusion coefficient is lower under ponding exposure than in accelerated conditions. For instance, for concrete with 15% LI and cured 28 days D_{pond} was roughly halved compared to D_{RCM}, in fact it decreased from 15.1 to 6.9 10^{-12} m^2/s for concrete with w/b of 0.46 and binder content of 300 kg/m^3. This is consistent with the well known dependence of the diffusion coefficient on time [4]. However the D_{pond} dependence on D_{RCM} seems to be rather complex and it could depend on several factors, for instance the concrete composition, the curing time, the

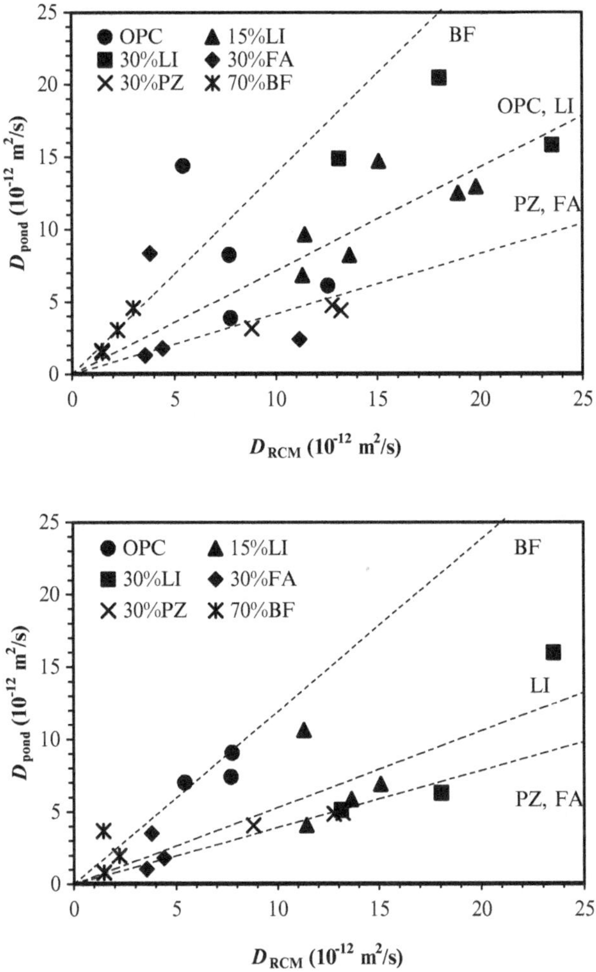

a)

b)

Figure 5. Relationship between D_{RCM} coefficients obtained from the rapid chloride migration test on specimens cured 28 days and D_{pond} coefficients obtained from the ponding test after 1 year of exposure on concretes cured 7 (*a*) and 28 days (*b*)

exposure conditions and the duration of exposure. Exposure conditions, which principally depend on temperature and environmental aggressiveness (i.e. the solution chloride concentration), were kept constant and equal during the tests. In particular both tests were carried out in the laboratory with a temperature of 20°C. As a result, the correlation found cannot depend on this factor. As far as the concrete composition is concerned, it clearly appears in Figure 5 that the D_{pond} dependence on D_{RCM} is influenced by this parameter. However it seems reasonable that the binder content does not influence it; in fact, as previously shown, the influence of binder dosage on the concrete chloride resistance is negligible. The water/binder ratio and the type of cement appear to be the most significant parameters which affect the correlation.

In fact, as it can be observed in Figure 5*a* for concretes cured 7 days, it seems not possible to determine a unique correlation valid for all the types of binder. Despite the wide scatter of results, different correlations could be detected for concrete with portland cement and limestone cement (both 15%LI and 30%LI), for concrete with pozzolan cement (both 30%FA and 30%PZ) and for ground granulated blast furnace slag. With regard to D_{RCM}, D_{pond} was halved for OPC, 15% and 30%LI, whilst it was more than halved for 30%PZ and 30%FA. It seems that D_{pond} was slightly higher than D_{RCM} for concrete with 70%BF, however for this binder very low chloride contents were measured after 1 year of ponding exposure and the determination of D_{pond} might be affected by measuring errors. Moreover the detected correlations were influenced by the curing of the specimens. In fact, a further reduction of D_{pond} compared to D_{RCM}, especially for concrete with limestone cement can be observed in Figure 5*b*. The correlation between D_{pond} and D_{RCM} could be also affected by the duration of the ponding test. In fact as proposed in the literature [4], the diffusion coefficient decreases in time, especially due to the hydration of the binder. Hence results of medium term tests (i.e. the ponding test), which duration is limited in time (i.e. 1 year), although they allow a preliminary indication on the relationship with accelerated test results, they might be not fully representative of the long term performance of concrete. Further studies are needed in order to evaluate the dependence of D_{pond} on the time and thus assess the role of this parameter on the correlation between D_{RCM} and D_{pond}.

Conclusions

The influence of several parameters on the resistance of concrete to chloride penetration was studied by means of the accelerated rapid chloride migration and medium term ponding tests. The effect of mineral additions in blended cement was studied by replacing part of the portland clinker with ground limestone, coal fly ash, natural pozzolan, ground granulated blast furnace slag. The beneficial effect related to the decrease in the water/binder ratio and the increase in curing time was assessed, while a negligible influence of the dosage of binder was observed. Both accelerated and medium term tests showed the beneficial role in the resistance to chloride penetration of fly ash and ground granulated blast furnace slag, while limestone led to a remarkable decrease in the chloride diffusion coefficient. The relationship between the rapid chloride migration coefficient and the ponding diffusion coefficient was investigated and it was observed that the diffusion coefficient measured under accelerated condition was higher than under ponding exposure however it seems not possible to determine a unique correlation valid for all the types of binder.

Acknowledgements

This research was financed by the Italian Ministry of University and Research (MIUR), Holcim Italia S.p.A. and Sismic.

References

[1] Bertolini, L. Elsener, B. Pedeferri, P. and Polder, R. (2004), *Corrosion of steel in concrete - Prevention, diagnosis, repair*, Wiley-VCH, Weinheim.
[2] EN 206-1 (2001), *Concrete - part 1. Specification, performance, production and conformity.*
[3] EN 1992-1-1 (2002), *Eurocode 2: Design of concrete structures – Part 1: General rules and rules for buildings.*
[4] FIB (2006), *Model code for service life design*, Bulletin N° 34.
[5] Andrade, C. Castellote, C. Alonso, C. and Gonzales, C. (2000), Mater. Struct, vol. 33, n. 225, pp. 21-28.
[6] Castellote, C. Andrade, C. and Alonso, C. (2001), Mater. Struct, vol. 34, n. 240, pp. 323-331.
[7] Yang, C. C. (2005), Mater. Struct, vol. 38, n. 277, pp. 313-320.
[8] Chiang, C.T. and Yang, C.C. (2007), Mater. Chem. Phys, vol. 106, n. 2-3, pp. 240-246.
[9] Bertolini, L. Lollini, F. and Redaelli, E. (2007). In: *Integral Service Life modelling of Concrete Structures*, Proceedings of the International RILEM Workshop, pp. 71-78, Ferreira, R.M. Gulikers, J. and Andrade, C. (Eds), RILEM Publications S.A.R.L., Bagneux.
[10] Bertolini, L. Lollini, F. and Redaelli, E. (2008). In: *Life-Cycle Civil Engineering*, Proceedings of the 1st International Symposium on Life-Cycle Civil Engineering, pp. 113-118, Biondini, F. and Frangopol, D.M. (Eds), Taylor and Francis Group, Raton, USA.
[11] NT BUILD 492 (1999), *Concrete, mortar and cement-based repair materials: chloride migration coefficient from non-steady state migration experiment*, NORDTEST.
[12] Polder, R.B. van der Wegen, G. Boutz, M. (2006). In: *Performances Based Evaluation and Indicators for Concrete Durability*, Proceeding of International RILEM Workshop, Baroghel-Bouny, V. Andrade, C. Torrent R. and Scrivener K. (Eds.), RILEM Publications S.A.R.L., Bagneux.

High Strength Steels Fracture Toughness Variation by the Media

J. Sánchez, J. Fullea and C. Andrade

Eduardo Torroja of Construction Science Institute (IETcc-CISDEM-UPM-CSIC), Serrano Galvache, 4, 28033 Madrid, Spain, javiersm@ietcc.csic.es

Abstract. The stress corrosion cracking process is at this moment an unknown mechanism of deterioration. It is a process that implies the joint action of the media, the presence of corrosion or a surface defect and of stress in the metal. Prestressing tendons can suffer SCC jointly with hydrogen embrittlement which dramatically changes not only the type of fracture (from ductile to brittle) but also the kinetics of the process leading to unexpected collapses. The metal should be resistant to this type of process which can be characterized by its toughness and therefore by its damage tolerance.

This research shows that the Fracture Toughness change when the steel corrodes, questioning the idea that is an intrinsic characteristic of the material. The reduction in the fracture toughness of steel wires when they are in contact to aggressive media involve that the material fractures with a lower crack depth for the same stress level. That means that the material becomes less damage tolerant, which implies that it is necessary to detect defects of smaller size, as for example, small notch, pits or superficial cracks. In the paper some results of the percentage of decrease of the toughness of prestressing wires suffering corrosion are presented.

Introduction

Concrete has an alkaline pore solution (pH > 12.6) that guarantees the passivation of steel reinforcement in addition to be a physical barrier against the penetration of environmental aggressives. This protection can be maintained indefinitely until an aggressive element in enough concentration reaches the bar. The most common causes of corrosion are the carbonation of the concrete cover, which produces a reduction of the pH of pore solution, and the penetration of chlorides, which induces pitting corrosion.

C. Andrade and G. Mancini (eds.), *Modelling of Corroding Concrete Structures*,
RILEM Bookseries 5, DOI 10.1007/978-94-007-0677-4_9, © RILEM 2011

A particular case of corrosion of the steel embedded in concrete is the Stress Corrosion Cracking (SCC), which can appear in prestressed structures. The SCC is produced by the simultaneous action and synergy of a mechanical tension and a corrosive media. Nucleated at the steel surface, the result is the appearance of microscopic cracks that are penetrating and inducing the brittle failure of the wire, due to a triaxial stress condition.

The Fracture Toughness (K_{IC}) is one of the most important parameters in Fracture Mechanics. Prestressed wires present high fracture toughness and, until now, this parameter has been considered as a characteristic of the materials. The fracture toughness is one of the fracture criteria [1]. This parameter is based on the knowledge of the stress ranges and displacement in the surroundings of the crack, that is to say, is based on the Stress Intensity Factor (K_I). Therefore, the fracture takes place when the stress intensity factor reaches the limit condition: $K_I=K_{IC}$.

There are a standard to measure the fracture toughness: ASTM E 399-90 (1997): "Standard Test for Method Plane-Strain Fracture Toughness of Metallic Materials". This standard provides details on the geometry of the specimens (Single Edge, Compact, Arc, etc.) and the minimum thickness based on the fracture toughness of the material and its elastic limit. This indicates that the fracture toughness varies with the thickness, decreases as increases the thickness until reaching a constant value from a big thickness. In addition, the fracture toughness depends on the rate of the test and the temperature.

The present work shows that the Fracture Toughness (K_{IC}) of steel varies when it remains in the media susceptible to the corrosion. That is to say, during the process of Stress Corrosion Cracking (SCC) the fracture toughness diminishes. Until now, the fracture toughness has been considered like a constant of the material. The reduction in the fracture toughness implies that the material, for a same tensional level, fractures with a defect much smaller. That is to say, the material becomes less tolerant to the damage, which implies that it is necessary to detect defects, like for example, small notches, superficial pits or cracks.

In order to support this statement it is shown some SCC results of high strength steel in carbonated solutions. In these tests, instead of generating the crack by fatigue, it is generated by means of controlled electrochemical and mechanical conditions. After that, it is possible to estimate the fracture toughness in a simple test. The obtained results show decreases around 30-40% of the fracture toughness with respect to the fatigue method value.

Experimental Method

Materials

A steel of eutectic composition have been tested in two conditions: cold drawn steel (1510 MPa Yield Strength) and the modified parent pearlitic steel (1300 MPa Yield Strength). The chemical composition of both is therefore the same and it is shown in Table I:

Table I. Chemical composition of parent pearlitic steel (%w)

C	Si	Mn	P	Cr	Ni	S
0.8	0.2	0.7	≤0.02	0.20	0.074	≤0.03

Parent pearlitic steel has treated thermally to a temperature about 250 ºC during 15 minutes [2-3]. The purpose of this treatment is to increase the yield strength from 950 MPa of the raw material to 1273 MPa. The value of the fracture toughness for this material is K_{IC} = 58 MPa $m^{0.5}$ [2]. And, the fracture toughness value for cold drawn steel is 107.9 MPa $m^{0.5}$ [4].

The samples were mechanized to a diameter of 2.5 mm and a length of 13.2 mm. In this case it is not possible to obtain standardized geometries, and then it is necessary to test cylindrical samples.

Methodology

A set of tests were carried out to localize the generation of single pit and avoid depassivation in the rest of the surface. The more realistic conditions are based in the generation of a crack by electrochemical dissolution from a pit [3,5-6], which may represent better the reality than to generate the crack by fatigue. After several trials, epoxy coating was used in order to avoid depassivation by generation of various pits. A notch artificially made to leave the steel surface in contact with the solution was used to reproduce a single pit.

In the test method the mechanical and electrochemical parameters are combined and it is made up of the following stages:

1. Fixed potential test in the media: The specimen is immersed in a solution of sodium bicarbonate at constant temperature. A fixed potential is applied, during around 100 hours, simultaneously a data logger registers the current. The specimen is strained to 80% of its yield strength. The objective of this stage is to generate an anodic zone and control the crack growth. After this stage, the specimens are removed from the solution and dried.
2. Slow strain rate test in air: SSRT is performed in air at a rate of $3*10^{-7}$ s^{-1} in order to determine the fracture toughness. It is possible to obtain the fracture

toughness using the fracture mechanics calculations from the stress and crack size data.

3. Scanning Electron Microscope analysis (SEM): In order to examine the fracture surface is used a scanning electron microscope. From this fractographic analysis is possible to evaluate the size of the crack in the fracture surface and the existence or not of brittle zones. In addition it is possible to determine the reduction of area, the different zones of surface of fracture and formed oxides.

Results

Once selected the right method of concentration of the damage [3, 6], some tests were carried out in order to know the influence of the applied potential on the crack propagation at constant load. The potential resulting more sensitive to induce SCC is -275 $mV_{Ag/AgCl}$ in the used bicarbonate solution.

Figure 1 shows two different behaviours. It is possible to see an example of a test for a material without defects and it is shown the curve corresponding to a material that has a crack generated in bicarbonate solution. The fracture for first is completely ductile (Figure 2) with the formation of micro-voids, whereas for the second case is completely brittle (Figure 3a). This type of fracture is characterized by a small area reduction and the fracture takes place in the same plane of the crack. This mechanical fracture is characterized by the appearance of brittle zones called cleavage (Figure 3b).

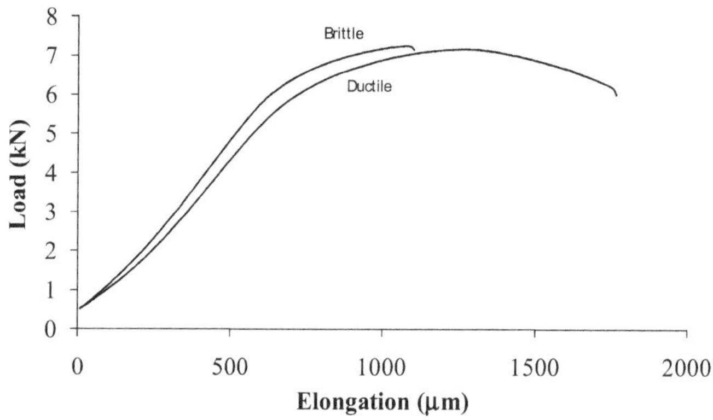

Figure 1. Ductile behaviour versus brittle behaviour

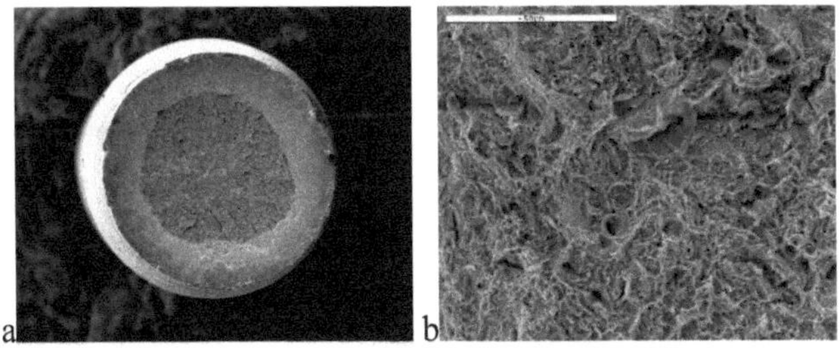

Figure 2. Ductile surface of fracture: a) general view, b) micro-voids

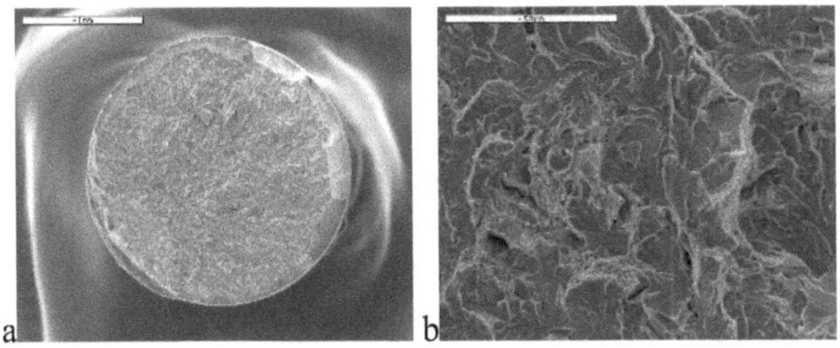

Figure 3. Brittle surface of fracture: a) general view, b) cleavage detail

Discussion

Due to limited size of the samples, prestressed steels cannot be prepared to obtain standardized specimens for testing fracture toughness of the material (ASTM E399-78) and therefore other approaches are necessary. For the case of a cylindrical geometry of the material, the calculation of the stress intensity factor and the criterion of fracture can be calculated through Astiz, Valiente and Elices's equation [1, 7] or through Levan and Royer's equation [8]. The above mentioned authors have assumed that cracks along the whole perimeter of the specimen are formed and the superficial cracks have semi-ellipse shape (Figure 4). Equations 1 and 2 give the expression corresponding to the stress intensity factor for a superficial crack with semi-elliptical form according Astiz and Levan's work.

$$\frac{K_I}{\sigma\sqrt{\pi a}} = \sum_i \sum_j C_{ij} \left(\frac{a}{2R}\right)^i \left(\frac{a}{b}\right)^j \quad (1)$$

$$\frac{K_I}{\sigma\sqrt{\pi a}} = \sum_i \sum_j \sum_k C_{ijk} \left(\frac{a}{R}\right)^i a^j \left(\frac{s}{Sm}\right)^k \quad (2)$$

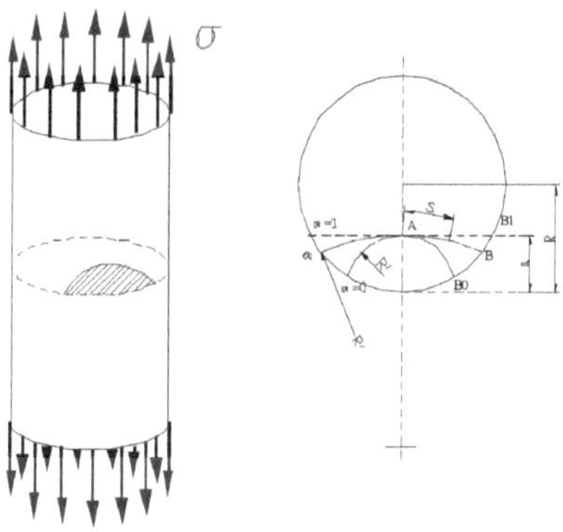

Figure 4. Superficial crack in a wire

However, the two previous equations have two main limitations to be applied in this study. Firstly, both are only applicable to linear elastic materials. And secondly, the geometry of the crack obtained by stress corrosion cracking must be for form factors = a/R ratios smaller than about 0.5 which is smaller than those observed in our experimentation [3, 6].

Both equations are represented in Figure 5. Until a depth of cracks an around 0.5 times the radius R, it is observed quite good accordance between the predictions of both equations. It is worth noting that, for greater depths of crack, as it is the case of those observed by us in prestressing wires, the value of stress intensity factor calculated from the equation of Levan [8] is much higher than that one calculated by the equation of Astiz [7].

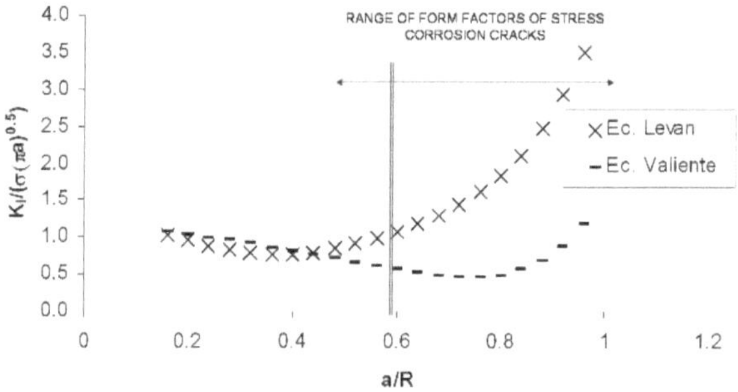

Figure 5. Comparison stress intensity form factors for specimens of cylindrical shape according Astiz and Levan et al.

For so acute cracks in cylindrical specimens it has not been found a rupture criterion that allows considering in a simple manner. Then, present paper shows previous calculations about Stress Intensity Factor for a bar with a crack geometry similar to that generated by stress corrosion, from solving the Integral J obtained by Finite Element calculation [9].

Results are given next for the different simulated crack sizes, always for the same value of semi-axis b and the same characteristics of the material, those corresponding to the steel of 6 sinking passages and to those of the parent steel. The crack length "a" varies between 200 μm and 1200 μm. Figure 6 shows the values of the Integral J for the cold drawn steel calculated in function of the stress. An abrupt change in the slope for the load level is observed where the plastic deformation of the surface of fracture begins. The same happens for parent steel (Figure 7). Once the values of the Integral J are obtained for each crack size, the stress intensity factor can be estimated according Irwin equation [10]. Figures 8 and 9 depict the values of KI so calculated, for the Cold drawn and Parent steels.

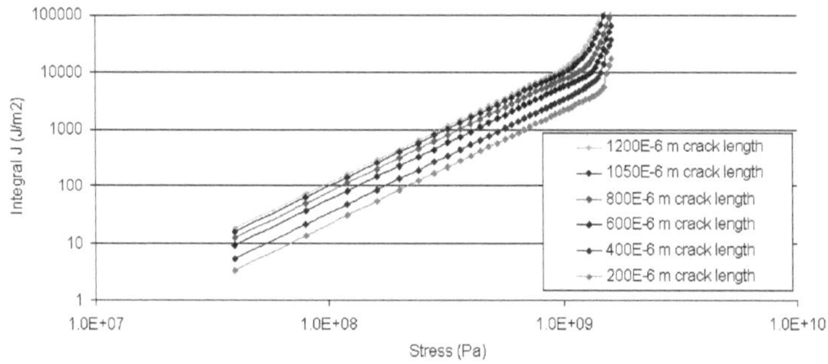

Figure 6. Integral J calculated for cold drawn steel in function of the stress

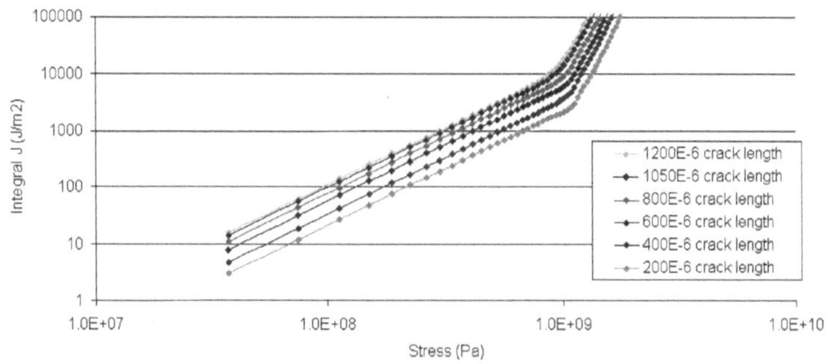

Figure 7. Integral J calculated for parent pearlitic steel in function of the stress

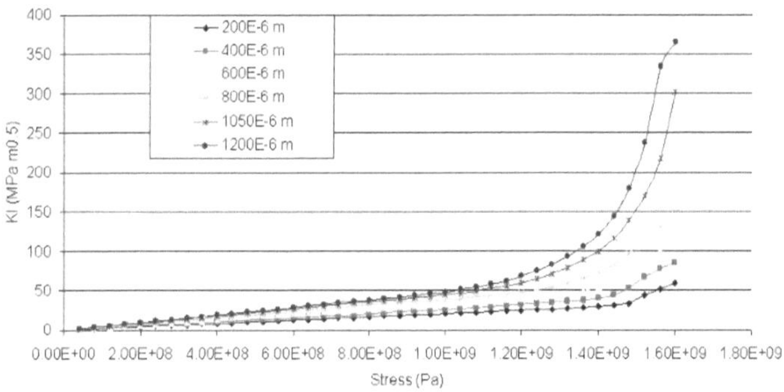

Figure 8. Values of K_I for the Cold drawn steel calculated from the Integral J

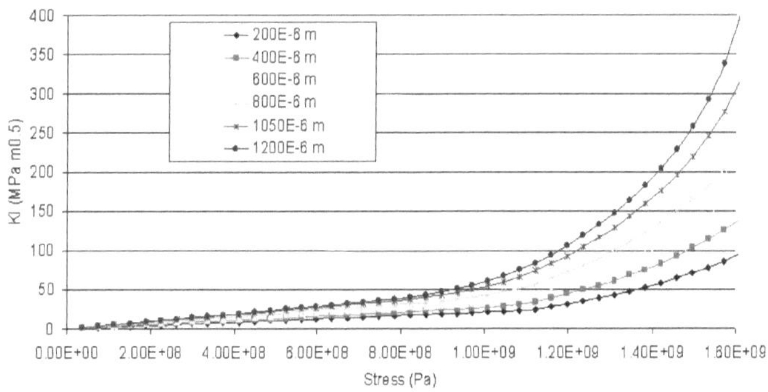

Figure 9. Values of K_I for the Parent steel calculated from the Integral J

As made in to previous studies [3, 5, 6] it is possible to calculate the fracture toughness of high strength steels from the crack geometry and the fracture parameters as shown in Figure 3 and according previously defined methodology. Figures 10 and 11 show the values of the fracture toughness for cold drawn and parent steel. The fracture toughness of the cold drawn steel is higher than parent steel one, although it can be distinguished a large variation of the fracture toughness in both materials. In all cases it is achieved very smaller values, around 50 MPam0.5 or less than the nominal ones. This means that the damage tolerance is reduced dramatically by the media, because when the crack grows up by stress corrosion cracking, the fracture toughness can be reduced around 40% by the hydrogen effect [11-13]. This reduction indicates the need to reconsider the crack depth needed to develop a brittle failure in the case of corroding high strength steels and therefore, to reduce the expected damage tolerance of these steels when they develop cracks by stress corrosion.

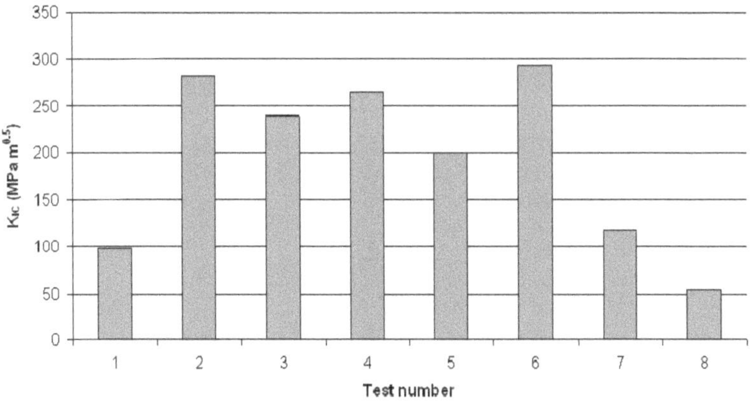

Figure 10. Fracture toughness of cold drawn steel before SCC test

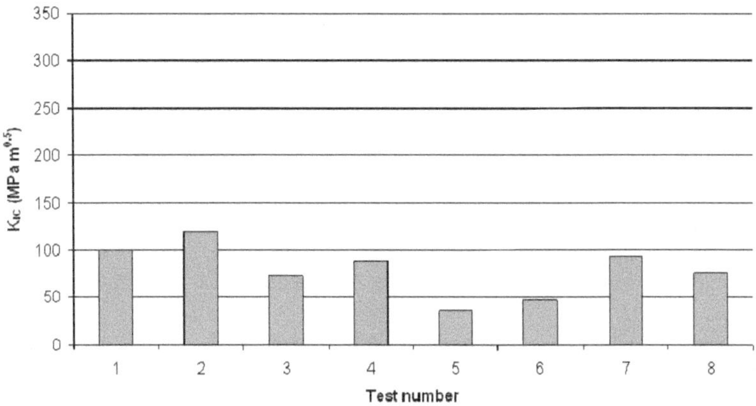

Figure 11. Fracture toughness of parent steel before SCC test

Conclusions

The fracture toughness of metals has been considered a material characteristic with fixed values that are normally determined by means of fatigue tests. Present results obtained from steels specimens coated with epoxy resin, were the crack is induced by corrosion on a notch made in the coating, have indicated that the toughness is significantly reduced when the material is immersed in media inducing stress corrosion cracking processes. Although these tests should be extended to other media in order to know how much this conclusion regarding prestressing steels, it is an indication of a need to review the damage tolerance of prestressed structures in some contaminated atmospheres.

Acknowledgments

The authors wish to thank Ministerio de Fomento of Spain for the funding the accomplishment of the project "Not destructive methods and strategies for the control of the corrosion in pretested steels", the INGENIO 2010-CONSOLIDER Project on "Safety and Durability of Structures: SEDUREC" and, specially, to Prof. Gustavo Guinea (UPM).

References

[1] Valiente A., Elices M. (1998), Eng. Fail. Anal. 5, 219.
[2] Parkins R. N., Elices M., Sanchez-Galvez V., Caballero L. (1982), Corrosion Science 22, 379.
[3] Sanchez J., Fullea J., Andrade C., Alonso C. (2007), Corrosion Science 49, 4069.
[4] Toribio J. (2004), Materials Science and Engineering a-Structural Materials Properties Microstructure and Processing 387, 227.
[5] Sanchez J., Andrade C., Fullea J. (2009), Corrosion 65, 368.
[6] Sanchez J. (2007), PhD Thesis. Complutense of Madrid.
[7] Astiz M. A. (1986), International Journal of Fracture 31, 105.
[8] Levan A., Royer J. (1993), International Journal of Fracture 61, 71.
[9] ABAQUS (2005).
[10] Irwin G. R. (1957), Journal of Applied Mechanics 24, 361.
[11] Sanchez J., Fullea J., Andrade C., de Andres P. L. (2010), Physical Review B 81, 132102.
[12] Sanchez J., Fullea J., Andrade C., de Andres P. L. (2009) Diffusion in Materials - Dimat2008 289-292, 203.
[13] Sanchez J., Fullea J., Andrade C., de Andres P. L. (2008) Physical Review B 78, 014113.

High pH Corrosion of Prestressing Steel in Segregated Grout

Luca Bertolini and Maddalena Carsana

Dipartimento di Chimica, Materiali e Ingegneria Chimica "G. Natta",
Politecnico di Milano, Italy

Abstract. Protection of prestressing tendons in post-tensioned structures can be ensured by the injection of a grout in the duct; since contact with hardened cement paste provides the conditions for passivity of steel. If ducts are fully injected corrosion protection may be guaranteed for a very long time thanks to the alkalinity of the grout, unless incomplete filling of the duct (and thus the lack of grout in contact with steel) or to penetration of chlorides through defects of sheath or to the heads of prestress occur. Recently, however, cases were reported where severe corrosion conditions took place in correspondence of areas where the grout was segregated. This phenomenon occurred mainly in the highest parts of external prestressing tendons with polyethylene ducts, where a whitish segregated grout was found, characterized by an alkaline pH and high content of sulphate ions. The corrosion attack encountered in these areas appeared much localized and, in some case, led to the failure of prestressing tendons; however, the mechanism of this type of corrosion has not yet been clarified. In order to investigate the corrosion of prestressing strands in segregated grout, corrosion tests were made in the laboratory. The results of tests suggest a complex mechanism which induces initiation of corrosion in the interstices among wires where the availability of oxygen is low. Subsequently propagation of localised attack is favoured by a macrocell that generates between the small depassivated areas and the surrounding passive areas.

Introduction

Prestressing steel, as traditional reinforcing steel, is passive in contact with alkaline cement materials [1]. For this reason, the protection of prestressing strands in post-tensioned structures can be ensured by the injection in the duct of a cement grout mixed with appropriate admixtures to increase its stability. The contact with the

C. Andrade and G. Mancini (eds.), *Modelling of Corroding Concrete Structures*,
RILEM Bookseries 5, DOI 10.1007/978-94-007-0677-4_10, © RILEM 2011

hardened cement paste provides the conditions for passivity of the steel. It is generally assumed that corrosion protection may be guaranteed by the alkalinity of the grout for a long time if the strands are fully embedded in it.
Indeed, cases of corrosion on prestressing tendons in post-tensioned structures are generally due to incomplete filling of the ducts (i.e., lack of grout in contact with steel), the penetration of chlorides through defects of the sheath or the prestressing heads or hydrogen embrittlement [2]. Recently, however, in some cases severe corrosion attacks have been reported on prestressing tendons, in areas where the grout was segregated [3]. This phenomenon occurred mainly in the highest parts of external prestressing tendons with sheaths made of polyethylene, where a whitish unhardened paste was found. Such whitish paste was characterized by an alkaline pH and high content of sulphate ions. Corrosion attacks found in these areas were very deep and, in some cases, led to the failure of prestressing tendons [3]. The consequences of this type of corrosion, which occurs soon after the filling of the ducts, may be quite severe, since it may lead to the failure of the ducts in rather short time (e.g. few years). To avoid segregation, new procedures for injection and a specific test (called the *inclined tube*) have been developed for checking the stability of cement grout during injection and at later stages [4]. Nevertheless, the mechanism of corrosion of steel in contact with the segregated grout has not yet been elucidated.
After a case of failure due to this phenomenon was experienced on a bridge, a laboratory study was carried out in order to reproduce this type of corrosion. This paper summarises the results of the study and proposes a possible corrosion mechanism for the initiation and propagation of localized attacks.

Case Study

The failure of an external prestressing cable of a bridge was observed after less than two years from the construction. The cable had 27 strands, a high density polyethylene duct and it was injected with a cementitious grout made with water/cement ratio of 0.32 and a commercial admixture specific for prestressing grouts. The failure was induced by the presence of deep localised attacks that resembled the form of pitting attacks, as shown in Figure 1.
After the failure of one cable, all the cables of the bridges were thoroughly inspected. A portion of the duct was removed in the vicinity of both ends of each cable, in the inclined parts were they rise to reach the prestressing heads. Although in most of the cables a conventional hardened cement paste, grey in colour, was observed, in some of the apertures the grout had a different aspect. Specifically, in some cases a whitish paste was observed which usually appeared as a plastic (unhardened) paste. After drying out, it became incoherent and dusty. This withish paste was identified as the result of segregation of the injected grout. Sometimes the segregated grout had a light grey colour with small black spots and was hardened, although weak; this type of grout was observed in intermediate position between the conventional grey hardened paste and the whitish paste.

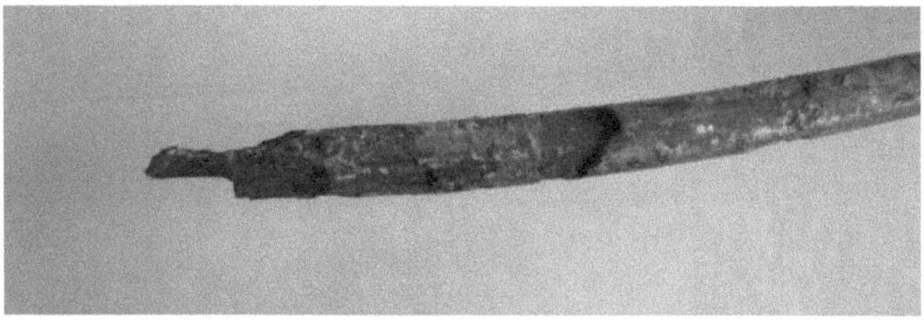

Figure 1. Example of penetrating corrosion attacks observed on a wire of the failed cable

Associated to the presence of the whitish segregated grout, heavy corrosion attacks were observed on the embedded strands (Figure 2). Cables showing corrosion attacks were removed and wires of the tendons were inspected along the length of the cable. It was confirmed that corrosion attacks only took place in the presence of the segregated grout, which was usually found only in the upper part of the cables, near their ends. Nevertheless, extremely localized and deep attack could be observed in these areas (Figure 3), which also led to failure of some wires (Figure 4). Corrosion rates of the order of several millimetres per year could be estimated, taking into account that a penetration depth of several millimetres (Figures 1 and 3) took place in less than 2 years.

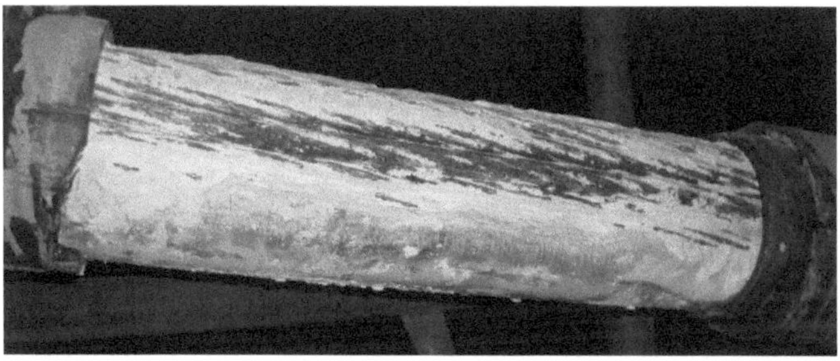

Figure 2. Example of whitish segregated grout embedding corroding strands

Microstructural analysis of broken wires, both of the cable failed under service and cables removed afterwards, showed that the failure was ductile and no cracks were observed which could be attributed to hydrogen induced stress corrosion cracking.

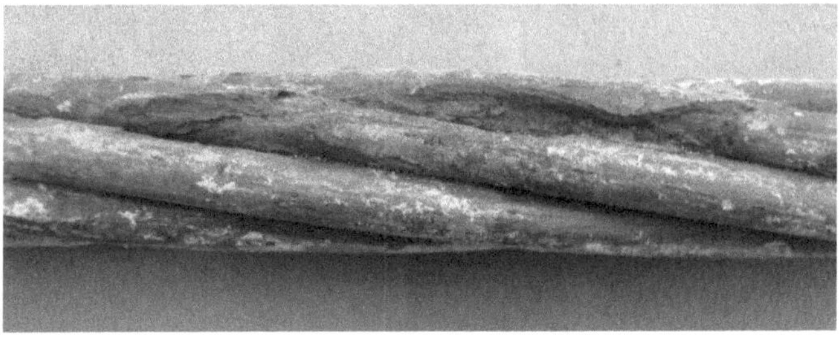

Figure 3. Example of corrosion attacks on a prestressing strand in contact with the whitish segregated grout

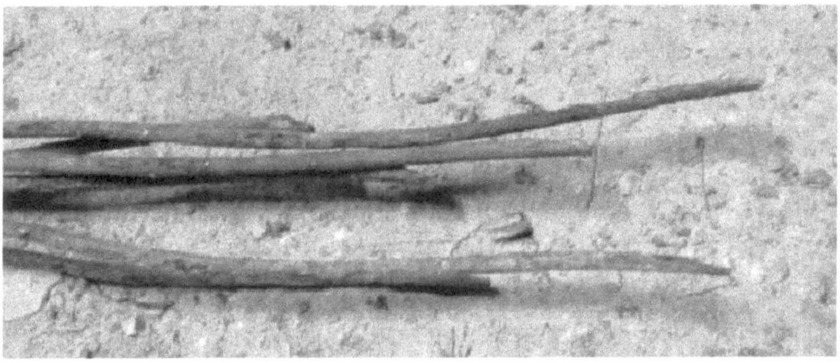

Figure 4. Example of failed wires in a prestressing strand in contact with the whitish segregated grout

Materials and Experimental Methods

In order to investigate on the causes of the corrosion attacks observed on the prestressing cables, chemical analyses were carried out on the segregated grout and corrosion tests were performed on the prestressing steel.
Different samples of segregated grout were collected from the ducts of prestressing cables of the bridge. These samples have been characterized by means of plasma emission spectrometry (*ICP-OES*), X-ray diffraction analysis (*XRD*), thermal analyses (*TGA* and *DTA*), density and water absorption measurements and scanning electron microscopy (*SEM*). Only ICP-OES analyses will be described in this paper.
Corrosion tests were carried out on pickled steel wires taken from a non corroding portion of a cable removed from the bridge. Specific cells were developed which allowed testing of one or more steel wires both in deaerated alkaline solutions and in water-soaked whitish segregated grout collected from the bridge. Sodium hydroxide solutions were used to obtain nominal pH variables between 13.8 and 14.3, in order to simulate chemical composition of the environment in contact with the strands; also 74 g/L of Na_2SO_4, corresponding to the addition of 5% by mass of

sulphate ions ($SO_4^{=}$), were added to some solutions. Some wires were also tested by fixing a plastic O'ring on their surface, in order to produce a crevice with limited access of oxygen.
The electrochemical behaviour of steel in the alkaline solution and in the segregated media was studied by measuring the corrosion potential and the corrosion rate (evaluated by means of polarization resistance method). Anodic polarization tests were also carried out on specimens immersed in the solutions, both by applying potentiodynamic scanning or potentiostatic steps of 1 hour.
At the end of the exposure period, corrosion attacks on the steel surface were observed and mass loss was detected.

Results and Discussion

The segregation of grout in prestressing ducts produces non-uniformity of composition of the filling material. Table I shows the results of the chemical composition measured by *ICP-OES* analysis on samples of grout both with and without segregation. From this type of analysis, the most important difference between the two types of grout arises from the content of the most soluble compounds, i.e. alkalis and sulphate ions.
Samples of non segregated cement grout (*NS*, with appearance and consistence of common hardened cement paste), show values of alkalis content in the ranges of 0.29-0.99% and 0.27-0.63% by dry mass, respectively for potassium oxide (K_2O) and sodium oxide (Na_2O). Sulphate content (expressed as SO_3) is in the range 1.75-4.15%. Conversely, in the case of segregated grout, which appears white and with plastic consistence when in wet conditions or as an incoherent compound when dry, the content of soluble compounds is much higher and more variable. The alkalis content is 0.66-4.00 and 0.47-2.40% (by dry mass) respectively for K_2O and Na_2O, while the sulphate content is between 3.58 and 7.88%. Thermal analyses also showed a lower content of calcium hydroxide in the segregated mortar. This is compatible with the higher content of alkalis. In fact, solubility of calcium hydroxide decreases when the concentration of alkalis increases, and $Ca(OH)_2$ precipitates as portlandite.
From chemical-physical characterization as well as microstructural analyses (not reported here), it was concluded that the whitish grout is a product of the segregation of the filling grout, which occurred before setting, i.e. during injection or soon after it. In fact, this material has a high concentration of the chemical compounds that are in solution in the initial stage of hydration of cement paste, specifically sulphates (due to the presence of gypsum as setting regulator) and alkalis (released in the early stage of hydration of clinker).
As a consequence, segregated grout in moist conditions is characterized by a rather high pH value. Considering the chemical composition of the whitish grout and the amount of water required to give to it a plastic consistence, a pH value of about 14.2 could be estimated for the liquid phase. Chemical analyses of the segregated grout showed that it contains a negligible amount of chloride.

Table I. Chemical composition (% vs dry mass) of grout, measured by *ICP-OES* analysis (ranges of variability).

Type of grout	*Non segregated (NS)*	*Segregated (S)*
Al_2O_3	1.61 - 2.47	1.18 - 2.02
CaO	39.6 - 47.2	32.4 - 47.0
K_2O	0.29 - 0.99	0.66 - 4.00
Na_2O	0.27 - 0.63	0.47 - 2.40
MgO	1.28 - 1.95	0.88 - 1.48
Fe_2O_3	1.13 - 2.27	1.02 - 1.93
SO_3	1.75 - 4.15	3.58 - 7.88
Residue	42.5 - 50.3	46.2 - 51.2

Due to the high pH and the absence of chloride ions, the usual causes of steel corrosion in concrete cannot be responsible of the corrosion attacks observed in the prestressing strands of the bridge. As a matter of fact, corrosion cannot be ascribed to carbonation, because of the alkaline nature of the segregated grout as well as the extremely localized morphology of the corrosion attacks (furthermore, carbonation is hindered inside the plastic ducts since penetration of carbon dioxide is prevented). On the other hand, the absence of appreciable amounts of chloride ions in the grout does not allow that pitting corrosion could take place. Even the presence of stray current inside the plastic ducts has to be definitely excluded. Therefore, as far as the cause of the corrosion attacks observed in the prestressing strands are concerned, different hypothesis have to be considered.
Based on the experiences of Pourbaix [5], corrosion is possible in the presence of high pH and in the absence of oxygen (resulting in a low corrosion potential for steel). Figure 5 shows a simplified Pourbaix diagram that shows possible corrosion of steel at pH values above about 13 and potential values below about -700 mV vs NHE. In his lectures on corrosion [6], Pourbaix describes a test during which he observed corrosion of steel in a degassed caustic soda (40 g/L) solution with associated hydrogen gas evolution, which ceased immediately soon after oxygen ingress was allowed in the solution.

In order to assess if corrosion can be attributed to the above mentioned mechanism, initially a series of corrosion tests was carried out on steel wires immersed in NaOH solutions of pH ranging from 13.8 to 14.3, which were deaerated both with nitrogen and helium. Also 5% by mass of sulphate ions were added to some of the solutions. However, in spite of a careful control of the flux of the deaerating gas through the cell, it was not possible to drive the potential to values lower than -800 mV vs SCE (i.e. about -550 mV vs NHE) and induce corrosion on the steel wires.

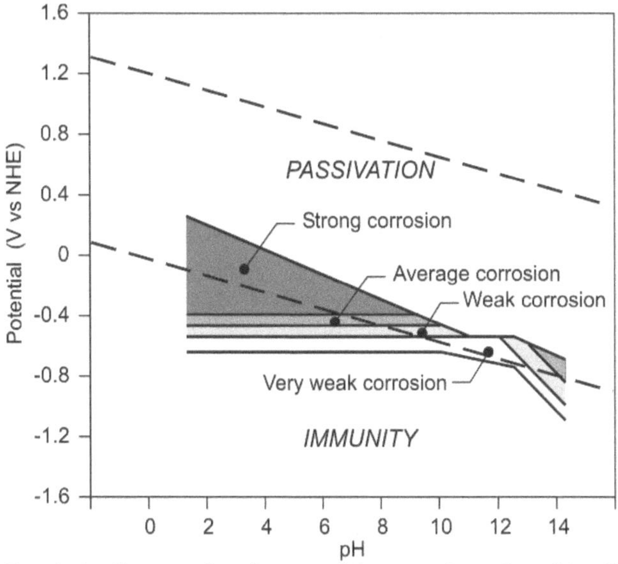

Figure 5. Pourbaix diagram showing possible corrosion of steel in alkaline and oxygen-free solutions [5]

Some uniform corrosion attacks could only be produced on steel wires that were subjected to a cathodic polarization to potential values of -1 V vs SCE (i.e. -750 mV vs NHE). Similarly steel wires embedded in the wet segregated grout did not show any corrosion even in sealed conditions. No localized corrosion was observed on the steel surface at the end of the tests in solution and in the moist segregated grout.

Nevertheless, localised attacks were observed in some cases in the presence of crevices that hindered the access of oxygen to the steel surface. This led to further tests and the proposal of a possible mechanism for the corrosion attack, which will be explained below.

Corrosion in the segregated grout can only take place if the grout is wet, so that alkali ions are dissolved and form an alkaline solution of pH close to or higher than 14. Furthermore, corrosion can only take place if the steel potential is low due to lack of oxygen. From the results of preliminary tests in deaerated alkaline solutions, it may be deduced that even extremely low quantities of oxygen are sufficient to promote passivity of steel at pH around 14. It is unlikely that inside the ducts of prestressing steel oxygen may be completely depleted. As a matter of facts, even if the segregated grout was fully saturated by water, some oxygen is dissolved in the mixing water and, until steel tendons are passive, oxygen is consumed very slowly by the cathodic process of oxygen reduction. Furthermore, even in the case of complete lack of oxygen, on the basis of Pourbaix diagram a generalized loss of the passive film would be expected, leading to uniform

corrosion attack, while corrosion attacks on the prestressing wires were highly localized (Figures 1, 3 and 4).
Results of tests on steel wires embedded in the segregated grout collected from the cables of the bridge provided further details useful to understand the phenomenon. These tests were carried out by embedding prestressing steel wires in samples of segregated grout, inside a sealed cell. The grout was dried, grounded and then mixed with distilled water in a amount enough to obtain such a consistence that it could be formed by hand (this is the consistence that was usually observed of the segregated grout immediately after opening the plastic ducts). Figure 6 shows an example of the results of monitoring of the corrosion potential and corrosion rate (average value on the exposed surface of the steel). It can be observed that in three days time the corrosion potential of steel decreased to values of about -500 mV vs SCE. Correspondingly, an increase in the corrosion rate from 10 mA/m^2 (i.e. about 10 μm/year) to 80 mA/m^2. After 50 days tests were interrupted and the steel wire was removed from the cell (the grout had to be broken, since it lost its plasticity and set during the test).

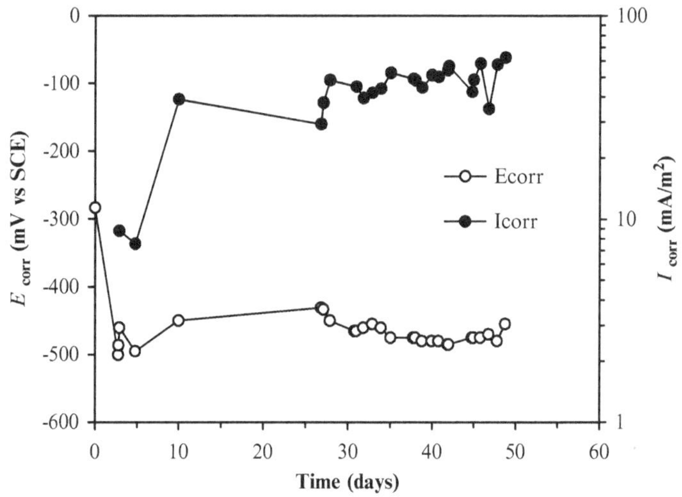

Figure 6. Corrosion potential and corrosion rate of a steel wire embedded in the segregated grout inside a sealed cell

The surface of the wire that had been in contact with the segregated grout was covered by iron oxides and adherent residues of the grout. After pickling the penetration of the corrosion attack was rather modest and it was uniformly distributed along the exposed surface. Nevertheless, a deep localised attack was observed in the upper part of the steel wire which was in the vicinity of the lid of the cell. The reduction in the cross section took place in the interstice produced by a teflon ribbon that was used to insulate the end of the steel wire. This result suggested that initiation of localized corrosion may be favoured in shielded areas (with a mechanism that could be similar to that of the well known *crevice*

corrosion [7]). Indeed, tests in deaerated solutions showed that, in the absence of interstices, steel is passive, as shown by the anodic polarization curves of Figure 7. The cathodic polarization curve shown in the same figure (measured on an inert platinum electrode) shows the presence of a small oxygen limiting current and, thus, the presence of traces of oxygen, which are enough to promote a passive behaviour on steel. However, a significant role of the pH was observed. By increasing the pH of the solution the anodic polarization curve shift rightwards, i.e. the passive current density increases. This may be assumed as an index of a higher reactivity of the material.
Assuming that corrosion conditions of steel are defined by the intersection of anodic and cathodic polarization curves, Figure 7 shows that as pH increases above 14 steel moves to a lower corrosion potential and a higher current density, i.e. towards conditions that promote corrosion initiation. A slight increase in the anodic current density was observed in the presence of sulphate ions; nevertheless the increase in the pH of the solution from 14 to 14.3 showed a more remarkable effect that the addition of 5% by mass of sulphate ions.

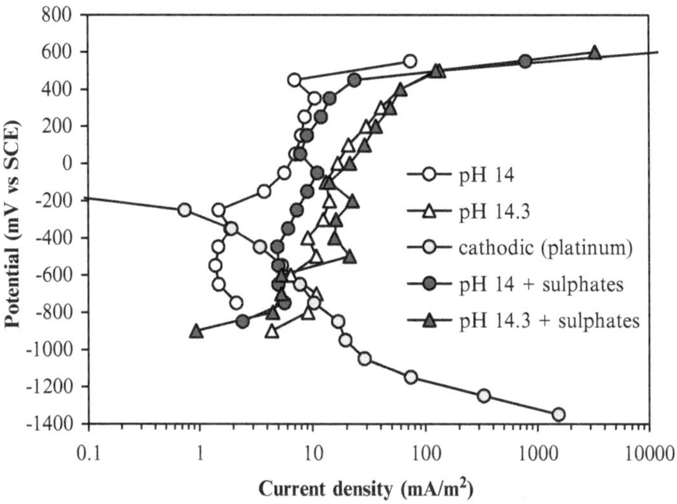

Figure 7. Anodic polarizzation curve of steel wires in different alkaline solutions and cathodic curve of an inert platinum electrode (potentiostatic steps of 1 hour)

Tests were also carried out on steel wires with one or more than one plastic ring on their surface, in order to produce interstices where oxygen ingress is hindered. Results of these tests confirmed that corrosion initiation is promoted inside interstices, where the low amount of oxygen present in the deaerated solution is quickly depleted and low potential values may be reached (Figure 8). Therefore, conditions similar to those that promote the well know crevice corrosion on many passive materials are established. Interstices are produced in real structures between wires of a single strand or between adjacent strands.

Once corrosion has initiated in localized spots between wires, different electrochemical conditions are produced inside the corroding area compared to the rest of the steel surface, as depicted in Figure 9. Corrosion tests showed that, in order to initiate corrosion in alkaline environment, a potential around -1 V vs SCE has to be reached.

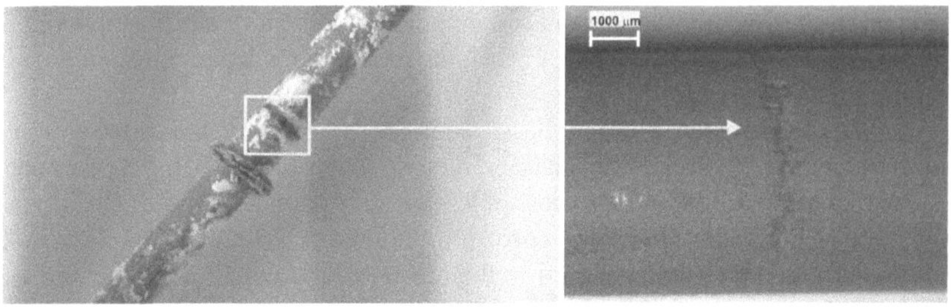

Figure 8. Corrosion attack on a steel wire embedded in the segregated grout, in an interstice produced by a plastic ring: (*a*) before and (*b*) after pickling

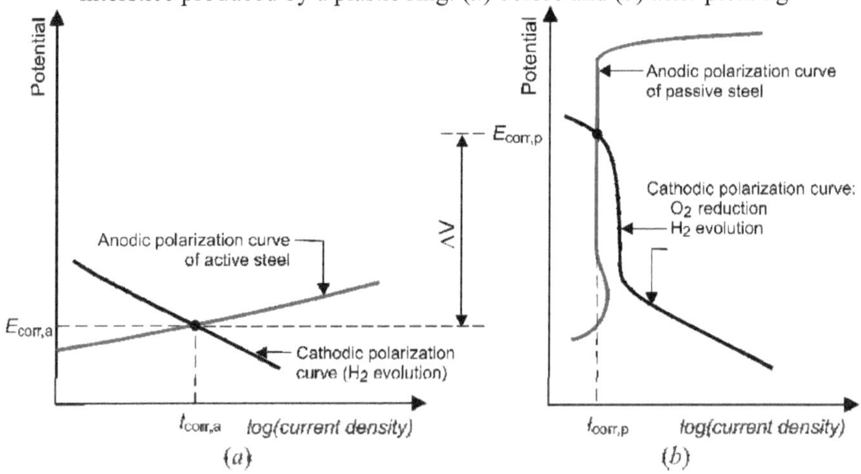

Figure 9. Schematic drawing of the polarization curves describing the electrochemical behaviour of active steel in alkaline environment without oxygen (*a*) and passive steel in alkaline environment with traces of oxygen (*b*)

Conversely, in areas where even a very small amount of oxygen is available (i.e. during deaerated tests), potential of steel in contact with highly alkaline solutions is at least 500 mV higher. The potential difference between the two areas promotes a macrocell that induces the circulation of an anodic current in the active area (which stimulates corrosion propagation) and a cathodic current in the passive areas (Figure 10). The increase in the corrosion rate in the depassivated area depends on the potential difference ΔV (Figure 9), but also on the geometry and characteristics of the segregated grout.

As far as geometry is concerned, the lower is the surface of the anodic area (i.e. the lower is the anode to cathode area ratio) the higher is the rate of penetration of the corrosion attack. In the case of strands inside prestressing cables, the surface of passive steel in contact with the segregated grout is usually quite large, since a large number of strands is present.

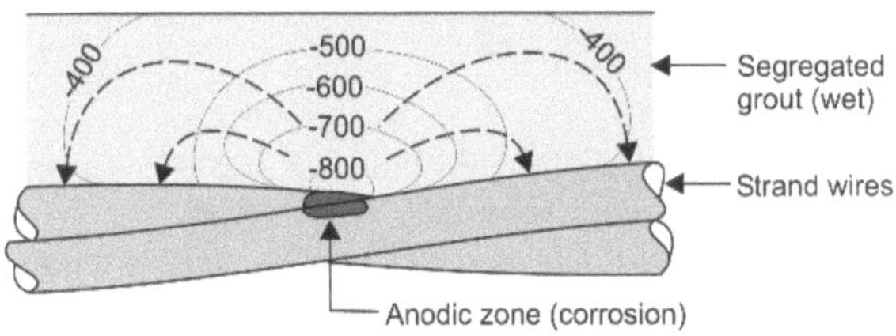

Figure 10. Schematic drawing of the possible macrocell between small anodic areas were corrosion has initiated and surrounding passive steel

In fact, if corrosion initiates in few points at the contact between wires, the small area where corrosion can initiate (i.e. the anodic area) is faced to a much larger surface of the passive wires (a cathodic area of at least one or two orders of magnitude higher can be expected). Concerning the properties of the segregated grout its electrical resistivity is the key factor, which is influenced by its composition and microstructure and by its moisture content. In wet conditions, the segregated grout has a rather low resistivity (values of the order of 0.5 Ω m were measured during laboratory tests) due to both the high ionic content (Na^+, K^+, OH^-, $SO_4^=$) and the low hydration (i.e. high porosity). Thus, several factors contribute to produce an extremely high penetration rate of the localized attack in the strands embedded in the segregated mortar and this can explain the reason why the attacks observed in the bridge were so severe.

Conclusions

Segregation of injection cement grout for prestressing cables may lead to the formation of a whitish phase with plastic consistence, characterized by high content of alkalis and sulphate ions. Steel in contact with this segregated material is subjected to deep localized corrosion attack that propagates with an extremely high penetration rate. Results of corrosion tests showed that, to induce the initiation of this type of attack, conditions of low availability of oxygen are essential. These may be produced inside the ducts especially in the interstices among wires of the strands. The high rate of penetration of the attack is favoured by a macrocell that generates between the small depassivated areas and the surrounding passive areas.

Acknowledgements

The authors are grateful to F. Brunella, N. Goffi, M. Ormellese and F. Traisci for their contribution to the tests.

References

[1] Bertolini L., Elsener B., Pedeferri P., Polder R. (2004), *Corrosion of steel in concrete - Prevention, diagnosis, repair*, Wiley-VCH.
[2] FIB Bulletin No. 33 (2005), *Durability of post-tensioning tendons*.
[3] Godart B. (2001), *Status of durability of post-tensioned tendons in France*, in Proc. of FIB Workshop *Durability of post-tensioning tendons*, Ghent, 15-16 November 2001, pp. 25-42.
[4] Lecinq B. (2004), *Recent research in France for the improvement of cement grouting*, in Proc. of 2nd Workshop on "Durability of post-tensioning tendons", 11-12 October 2004, ETH, Zürich.
[5] Pourbaix M. (1974), *Atlas of electrochemical corrosion*, NACE.
[6] Pourbaix M. (1973), *Lectures on electrochemical corrosion*, Plenum Press.
[7] Shreir, L.L., Jarman, R.A., Burnstein G.T. (1995), *Corrosion*, 3rd Edition, Butterworth-Heinemann.

Loading Test of RC Beam Bridge Built 80 Years Ago in Japanese Coastal Area

Yasushi Tanaka, Takumi Shimomura and Takayuki Yamaguchi

Nagaoka University of Technology, Japan

Abstract. Many concrete structures have to be now repaired or replaced due to salt attack in Japanese coastal area. However, their residual strength is not well considered in practice. This is because only few studies have so far been made on the residual strength of deteriorated concrete structures. Objective of this research is to study about the load carrying capacity and the failure behaviour of corroded real RC bridge.

Firstly, we sawed the reinforced concrete bridge to take out beam specimens. The bridge was seriously corroded because it was built about 80 years ago in Japanese coastal area. Two beam specimens were carried into experimental room and loaded until failure. Load carrying capacities of these two specimens were quite different. Maximum load of the relatively sound beam was about 96% of the estimated capacity of non damaged beam. On the other hand, the relatively deteriorated beam had only 48% of load carrying capacity. To examine the cause of the reduction in loading strength, layout of reinforcement and corrosion rate of longitudinal bars in concrete were carefully inspected. In the inspection, several lap splices were found in tensile area while the difference in corrosion rate is much smaller than the difference in strength. Thus, the cause of reduction in loading capacity is considered as bonding failure at lap splice.

Finite Element analysis was conducted to simulate test results. It was confirmed that FE analysis can estimate the structural behaviour of test beams if the bond strength of each lap splice is obtained first.

Introduction

The number of bridges deteriorated by salt attack is recently increasing in Japanese coastal area. Many of them are being repaired or replaced without evaluation of residual strength although residual strength is essential information to confirm the structural safety, to predict the degrading rate, and to make reasonable maintenance

C. Andrade and G. Mancini (eds.), *Modelling of Corroding Concrete Structures*, RILEM Bookseries 5, DOI 10.1007/978-94-007-0677-4_11, © RILEM 2011

plan. Therefore, the methodology to evaluate residual strength of deteriorated reinforced concrete (RC) structure should be established [1].

Many lab tests have been conducted regarding the structural performance of corroded RC members during past few decades [2]. The results of these researches should be verified with real structures to be practically applied to the maintenance of existing structures. However, only few attempts have been done about the residual strength of corroded real structures.

Objective of this research is to study about the load carrying capacity and the failure behaviour of corroded real RC bridge. In this research, corroded RC beams sawn from a real RC bridge were loaded until failure. Visual inspection and the measurement of corrosion rate were also conducted and compared with each other. Then, test results were simulated by means of FE analysis taking corrosion of rebar into consideration.

Outline of Tested Bridge

Two beam specimens were sawn from Nougawa bridge which was built in 1930 at Japanese coastal area. The distance from shoreline is about 100 m. Figures 1 and 2 show elevation and cross sectional view of Nougawa bridge, respectively.

Superstructure system is three span continuous RC beam bridge (bridge length: 131.8 m, span: 12@10.8 m) while substructure system is plain concrete bridge pier. The bridge was repaired by patching with mortar in 1959 (30 years from construction) due to the degradation caused by chloride attack. In 1996, restriction in service was applied due to the re-degradation. In 2006, the service was suspended. As shown in Figure 2, beam specimens were sawn from outer and inner bridge beam respectively in 2009 (79 years from construction).

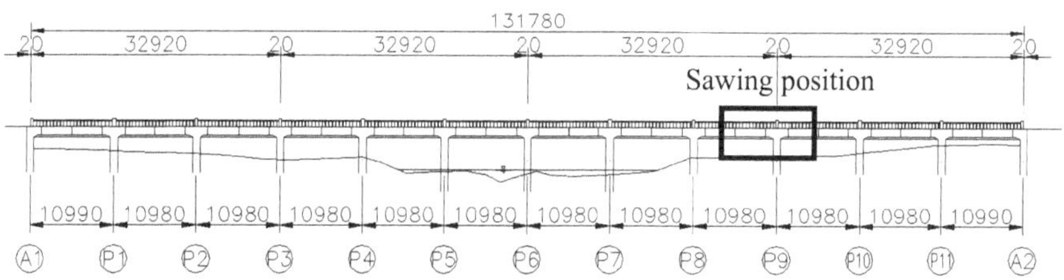

Figure 1. Elevation view of Nougawa bridge
(view from ocean side)

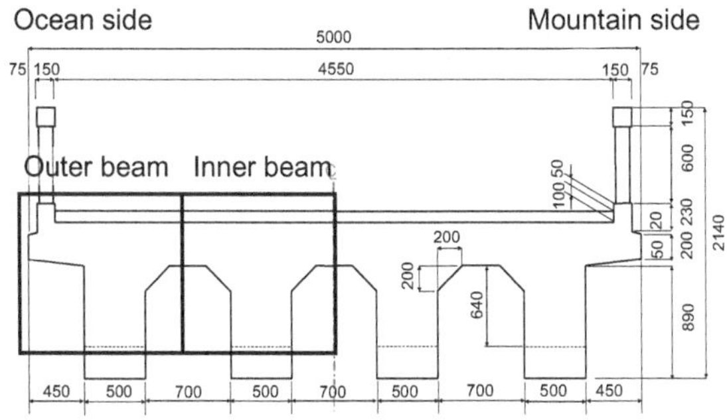

Figure 2. Cross sectional view of Nougawa bridge

Degrading Condition

Visual inspection

Figures 3 and 4 show the appearance of outer beam and inner beam, respectively. In the one side of outer beam, cover concrete was spalled entirely until longitudinal rebars were exposed. Ocean side rebars suffered from corrosion severely. Stirrups were corroded or ruptured within this range. On the other hand, inner beam looked relatively sound. There were longitudinal cracks along rebars, but spalling of concrete was not observed in inner beam.

Figure 3. Appearance of outer beam

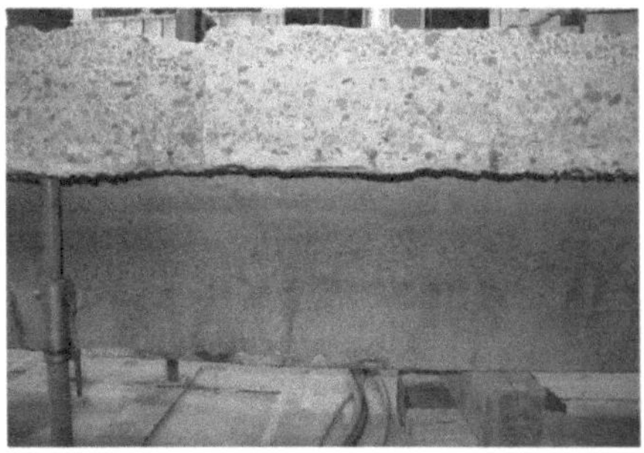

Figure 4. Appearance of inner beam

Cross section views in the middle of beam are shown in Figure 5. Each beam contained 8 round bars. The diameter of rebar was 25 mm. Four rebars were exposed in outer beam while no rebar was exposed in inner beam.

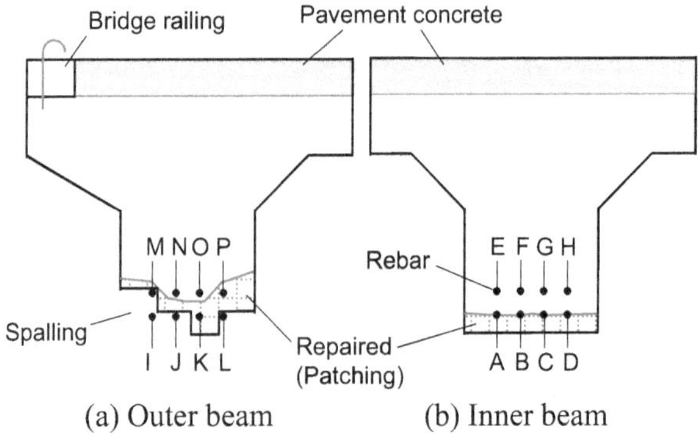

Figure 5. Degradation in the middle of specimen

Chloride content

Chloride content was analyzed between P9 and P10 in Figure 1. The results of total chloride content are summarized in Figure 6. Concrete cores were drilled from undersurface. Lateral cover depth was about 77 mm and bottom cover depth was about 47 mm as shown in Figure 7. Total chloride content at rebar location was higher than 1.2 kg/m^3 which is the threshold of total chloride content to avoid corrosion adopted in JSCE standard [3]. Figure 6 indicates that the chloride content at ocean side is not necessarily higher than others. Carbonation depth was almost zero everywhere.

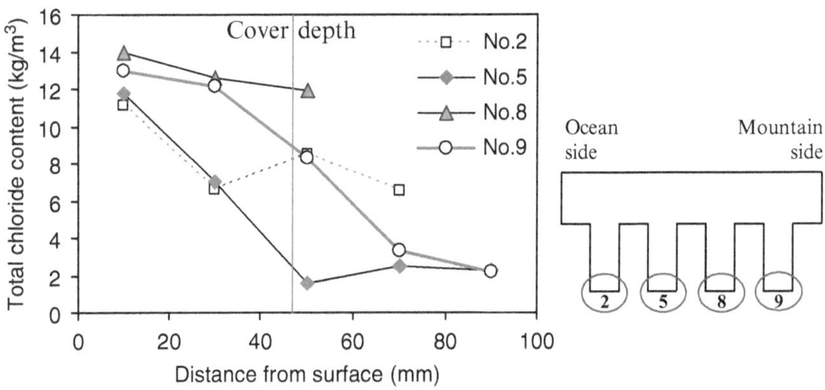

Figure 6. Total chloride content (P9-P10)

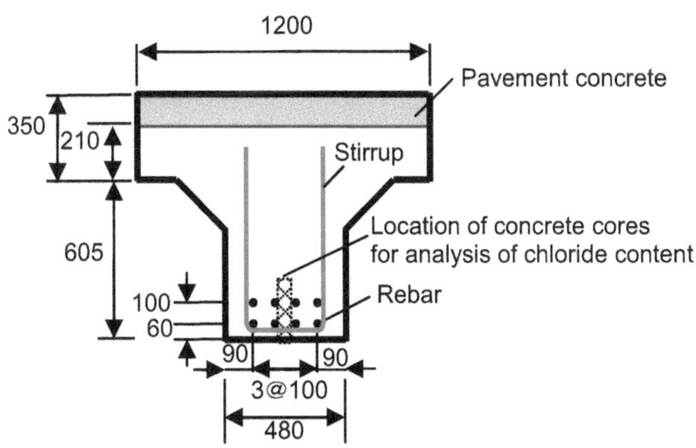

Figure 7. Cross section of specimen

Loading Test

Test method

Figure 8 indicates test setup. 8.9 m length of sawn beam specimen was simply supported with 8.0 m of loading span. Concentrated load was applied at two points to make 2.0 m of constant moment span. Figure 9 shows moment diagram of Nougawa bridge when design load is applied. Design load used here was established by Japanese Department of Interior in 1926. Positive flexural moment acts most part of specimen. The location of maximum flexural moment is within the constant moment span in loading test. Because anchorage failure may occur due to corrosion crack and spalling of cover concrete, the anchorage of rebar was strengthened by jacketing with concrete and by welding rebar and steel bearing plate.

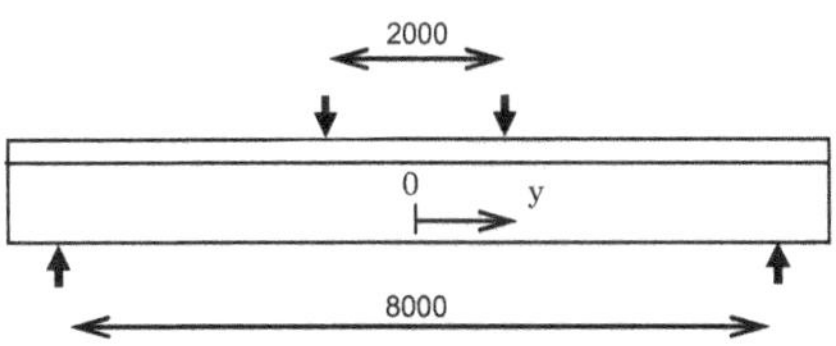

Figure 8. Loading conditions

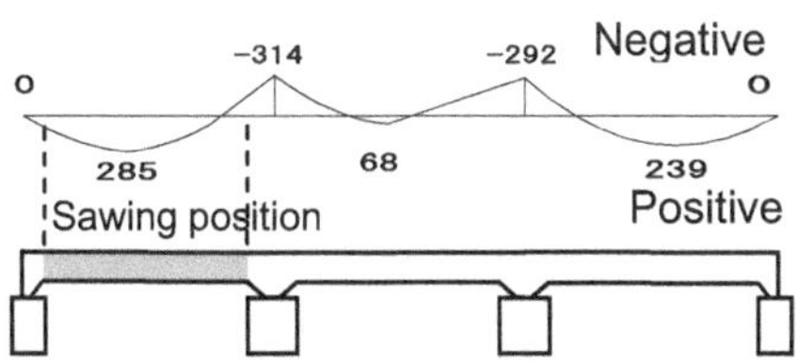

Figure 9. Moment diagram when design load is applied (Unit: kN·m)

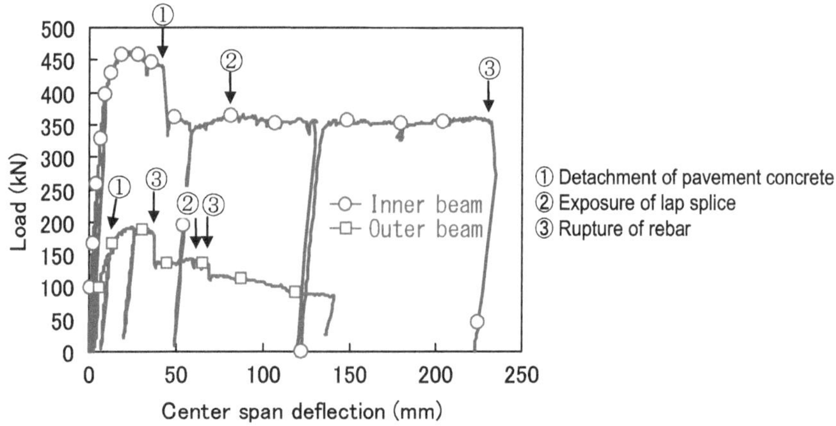

Figure 10. Load–center span deflection relationship

Test results

Figure 10 illustrates load–center span deflection relationship. In inner beam, concrete pavement was unified with T shaped RC beam until peak load. When center deflection reached about 40 mm, concrete pavement was detached suddenly resulting in load reduction by 20%. Longitudinal rebar (rebar A) was ruptured near the loading point when center deflection was 232 mm.

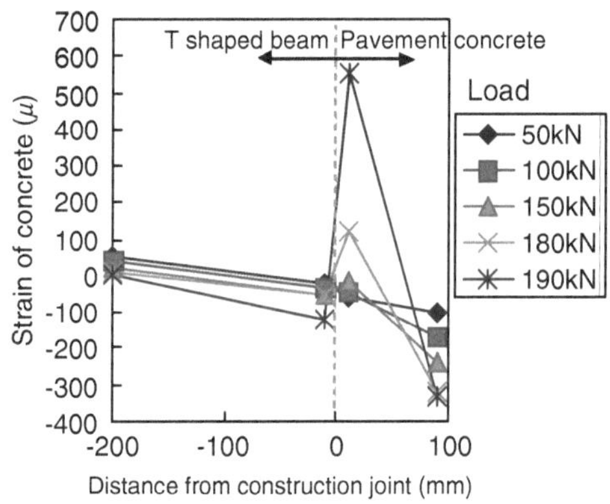

Figure 11. Strain distribution of outer beam (0.8 m from center, Mountain side)

On the other hand, concrete pavement was gradually detached before peak load in outer beam. Figure 11 shows concrete strain distribution of outer beam. Concrete strain got discontinuous at the boundary between concrete pavement and RC beam when load was higher than 150 kN. Measured concrete strain distribution corresponds to that of built up beam at maximum load (190 kN). Therefore, it is clear that concrete pavement was detached before maximum load. Longitudinal rebar (rebar O) was ruptured at 37 mm of center deflection in outer beam. Simultaneously, load was reduced by 25% (43 kN). Then, another rebar (rebar M) was ruptured when center deflection was 68 mm. However, the reduction in load was small (18 kN).

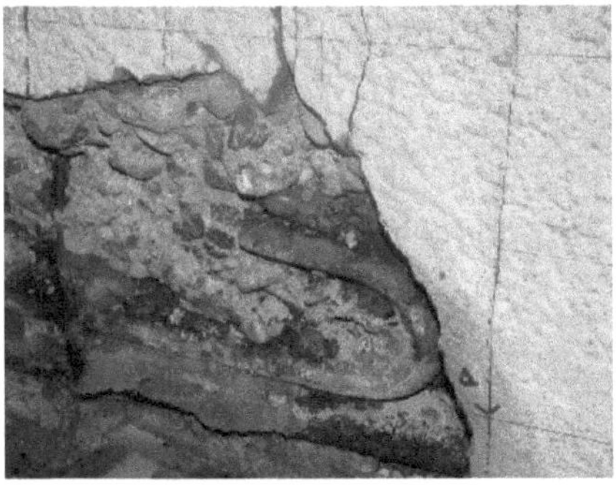

Figure 12. Exposed lap splice (Inner beam)

Concrete cover was spalled during loading test. Accordingly, lap splices were exposed as shown in Figures 12 and 13. Despite lap splices were exposed as a result of spalling, applied load was not reduced. Therefore, it is considered that these lap splices were already not effective from the beginning of the loading test.

In case of outer beam, a few rebars had been bent slightly before the loading test. These exposed rebars might be bent during sawing process or carrying process by accident. Because it was afraid that exposed rebars could not transfer tensile force due to the initial deformation, strain of exposed rebar was measured during the loading test of outer beam. Figure 14 illustrates center span deflection versus strain of exposed rebar of outer beam.

These measured strains were compared with numerical result. In the numerical analysis, exposed rebars were expressed as external cable element. Steel strain where strain gauge was attached was estimated by taking the local and average corrosion rate into account. In most cases, measured strain corresponded to analyzed strain. However, measured strain of rebar M which located at ocean side was very small compared with predicted strain. Initial deformation or pitting corrosion may be the cause of discrepancy.

Rebar I which also located at ocean side decreased its tensile force when center deflection exceeded 13 mm. Near the lap splice in rebar I, concrete cover was spalled. Therefore, anchorage failure tended to taken place in rebar I.

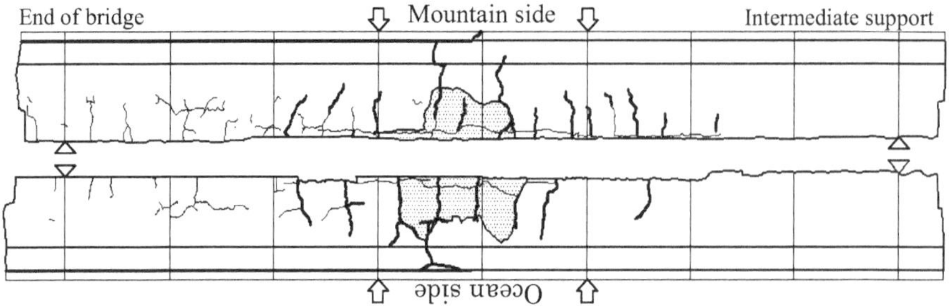

(a) Inner beam (Maximu load: 412kN)

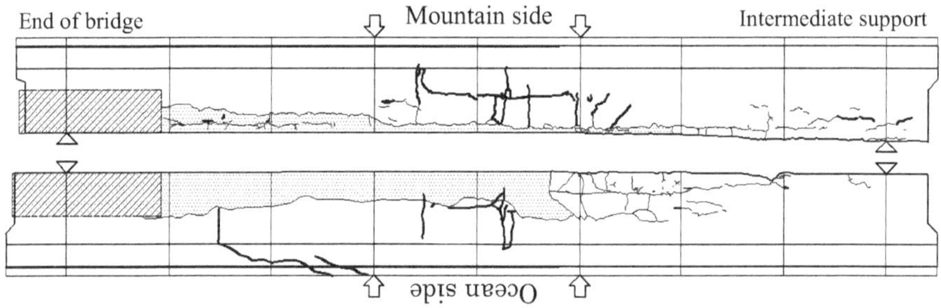

(b) Outer beam (Maximum load: 190kN)

Figure 13. Crack patterns observed after loading test (thin line: crack occurred in service; bold line: crack occurred during loading; hatching: spalling)

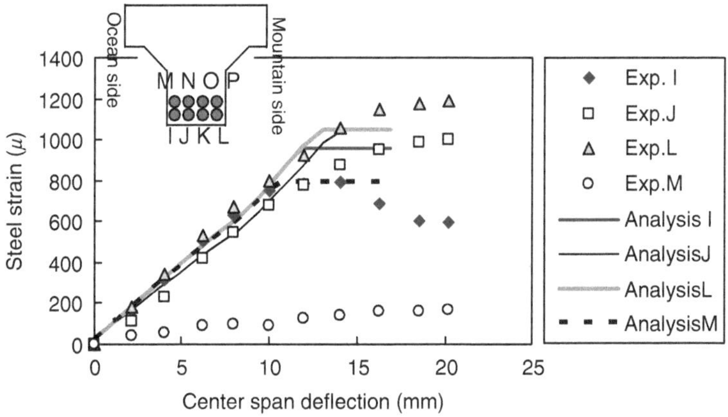

Figure 14. Strain of rebar versus center span deflection (outer beam)

Destructive Test after Loading

Testing of concrete cores and rebars

Compressive test was conducted using concrete cores taken from beam specimens after the loading test. Concrete cores were taken from the web of T shaped RC beam and pavement concrete. Test results are summarized in Table I. Concrete strength of pavement concrete was lower than that of T shaped beam.

Table I. Results of compressive test of concrete

	T shaped RC beam	Pavement concrete
Compressive strength	26 N/mm^2	13 N/mm^2
Elastic modulus	25,000 N/mm^2	22,000 N/mm^2
Max. aggregate size	50 mm	25 mm
Number of specimens	3	6
Coefficient of variance	5%	36%

Rebars were taken from RC beam and their tensile properties were tested. Table II shows test results of rebars. Though the standard of yielding stress of the original rebars are not known, these rebars corresponds to SR 235 class according to present classification.

Table II. Results of tensile test of rebar

	Longitudinal bar	Stirrup
Yielding stress	262 N/mm^2	270 N/mm^2
Elastic modulus	195,000 N/mm^2	198,000 N/mm^2
Bar diameter	25.4 mm	9.5 mm

Inspection of bar arrangement

Because neither drawing for design nor specification was preserved, bar arrangement was investigated during the demolishing work. Figure 15 illustrates the location of lap splices found by the demolition. There were 6 rebars which contain lap splice out of 8 rebars. These lap splices were dispersed in pairs to three positions. Because these rebars were round bars, the ends of rebars were bent to form semi circular hooks.

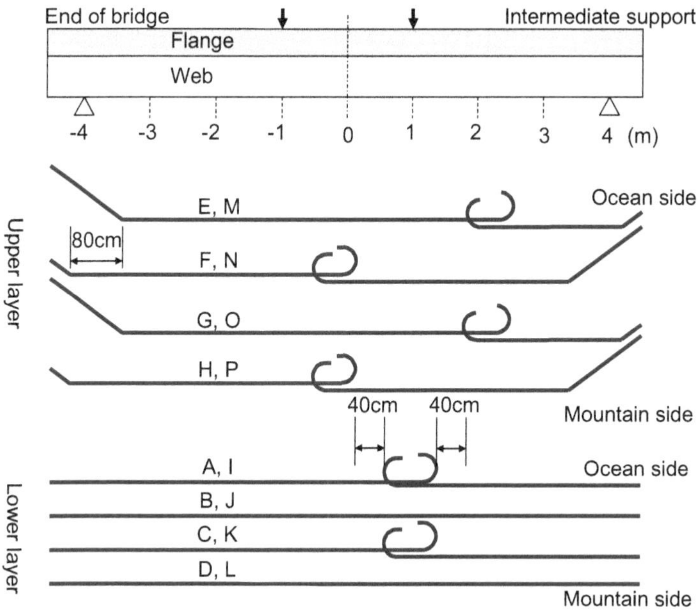

Figure 15. Location of lap splices

Figure 16 shows the arrangement of stirrups. Spacing of stirrups was small near support according to reacted shear force. Spacing of stirrup was about 480 mm in the middle, about 150 mm near the outer support and about 100 mm near the intermediate support.

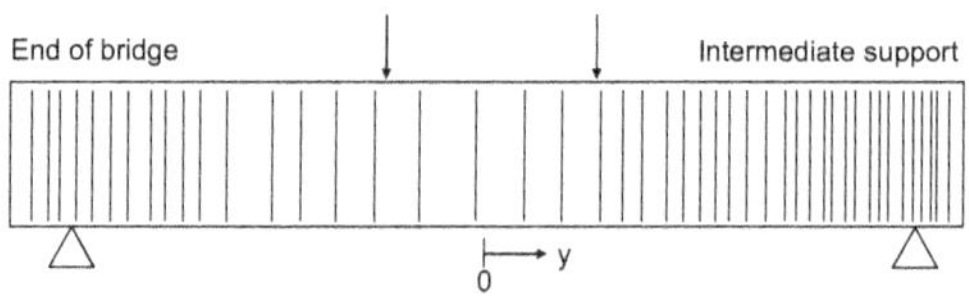

Figure 16. Arrangement of stirrups

Corrosion rate

Longitudinal rebars taken from beam specimens were immersed into 10% diammonium hydrogen citrate solution. After removing rust, the diameter of rebar was measured every 5 cm in orthogonal directions while the weight was measured every 10 cm. Corrosion rate was calculated by assuming the original diameter was 25.4 mm (1 inch) while specific weight was assumed as 7.8 g/cm^3. Figure 17 shows the distribution of the ratio of weight loss of inner and outer beam. Average percentage of weight loss of inner beam was 14 % while that of outer beam was 26%.

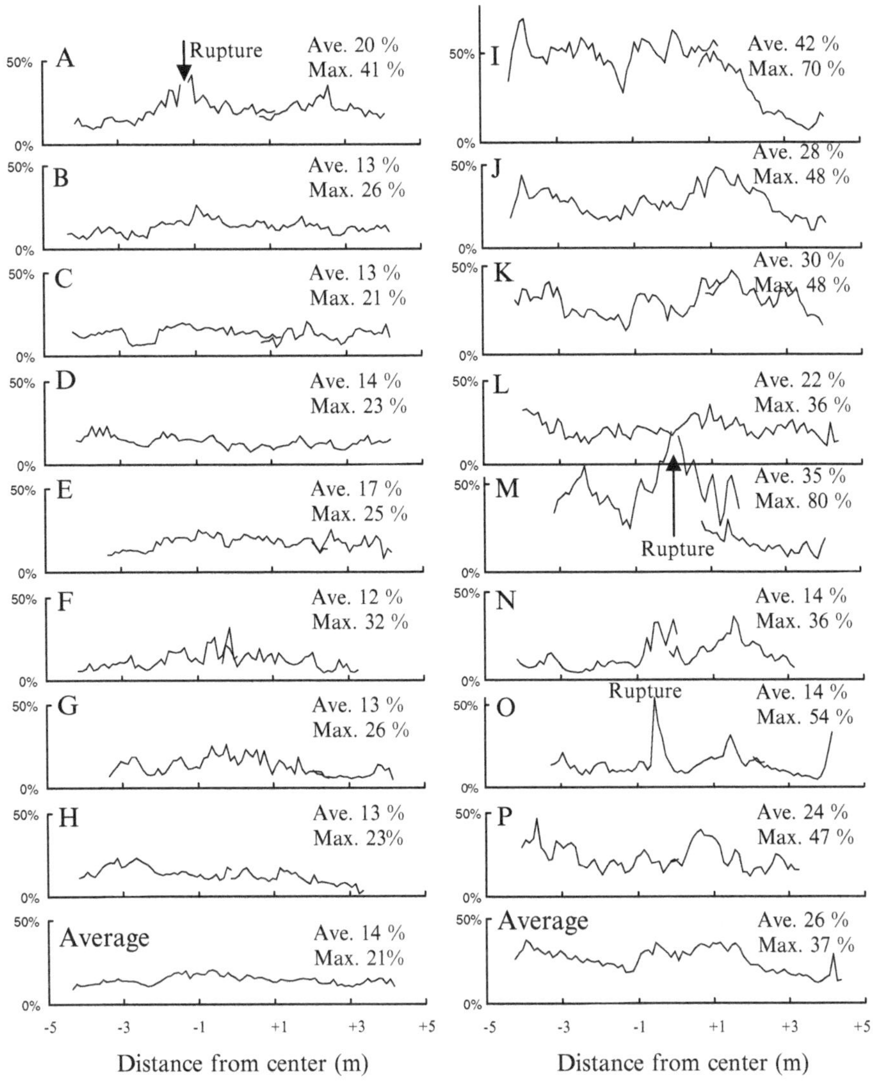

(a) Inner beam) (b) Outer beam

Figure 17. Distribution of the ratio of weight loss

Corrosion rate where concrete cover was spalled were larger than other portions. The influence of flexural crack on corrosion rate was not obvious. Rebars in ocean side were corroded heavily than other side in both specimens. The corrosion rate of individual rebar has considerable spatial variation. However, cross-sectional average has moderate variation by cancelling their variations each other. The reduction rate in cross sectional area was compared with the ratio of weight loss in Figure 18. The measured reduction rate in cross sectional area of rebars tends to be smaller than the ratio of weight loss although ideally they coincide with each other. The reason of the discrepancy will be attributable to the estimation process. As a first step to obtain the reduction rate in cross sectional area, the diameters of corroded rebar are measured in orthogonal directions. Then cross sectional area is calculated by assuming ellipse. As illustrated in Figure 19, this assumption is not adequate for corroded round bar because the measurement of external diameter cannot consider pitting corrosion. On the other hand, weighing method can include the influence of pitting corrosion. That will be the reason of the discrepancy shown in Figure 18.

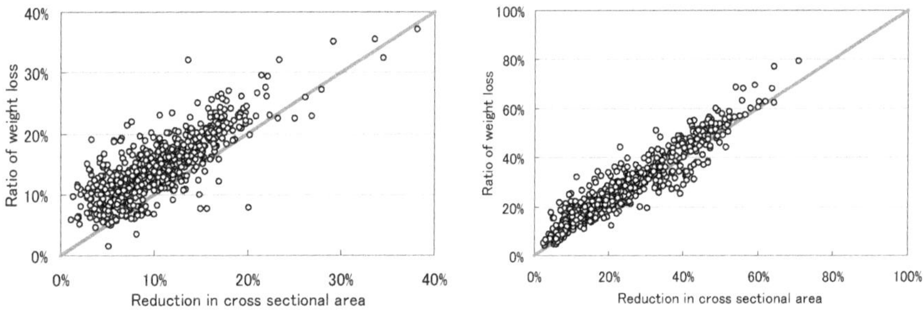

(a) Inner beam (b) Outer beam

Figure 18. Comparison between the reduction in cross sectional area and the ratio of weight loss

In case of outer beam, the reduction rate in load carrying capacity of the beam was much higher than estimated from the average corrosion rate. Figure 20 summarizes damage state observed after the loading test. Several lap splices were exposed during the loading test. That will be principal cause of the reduction in strength.

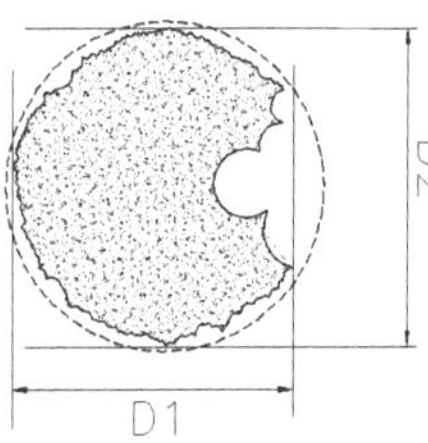

Figure 19. Cross section of corroded rebar and external diameter

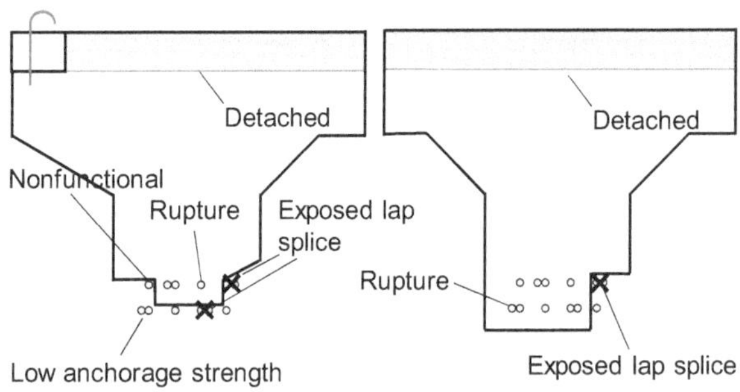

(a) Outer beam (b) Inner beam

Figure 20. Damage state after loading

Finite Element Analysis

Analytical method

Outline. FE analysis has possibility to reflect the information obtained from visual inspection or non destructive test on the verification of structural performance of corroded RC structures. However, only few studies have been made on FE analysis of corroded RC structures. Several problems should be overcome so that FE analysis is applied to real structures deteriorated by chloride attack. In this study, above experiment is simulated with three dimensional FE analysis to examine how we can deal with deteriorations due to corrosion of rebars in FE analysis.

Commercial FE software, ATENA ver.4.1.4 [4] was used in this study. Stress-strain model of concrete in CEB-FIB Model code [5] was used. It is assumed that the corrosion of rebars does not affect the material properties of concrete. Rebars were expressed as discrete line element as shown in Figure 21. Reduction in cross sectional area due to corrosion was represented by reducing the area of rebar element. Spatial variation of corrosion in longitudinal direction wais reflected in stress-strain relationship. Bond deterioration was expressed by introducing bond element between discrete rebar element and concrete element.

Distribution of corrosion rate. There are several ways to express the spatial variation of corrosion rate in FE analysis. To set up measured corrosion rate in each rebar element as it is should be most direct and reliable way. However, since this method requires numerous corrosion data and numerous analytical elements, it is difficult to be adopted to large scale structure. To reduce the number of required data in FE analysis, the influence of spatial variability of corrosion rate was considered in terms of stress strain relationship of rebar. In this study, tension test of corroded rebar has not been conducted because rebars were cut every 10 cm to estimate corrosion rate. Alternatively, numerical experiment was conducted by use of numerous measurement of corrosion rate.

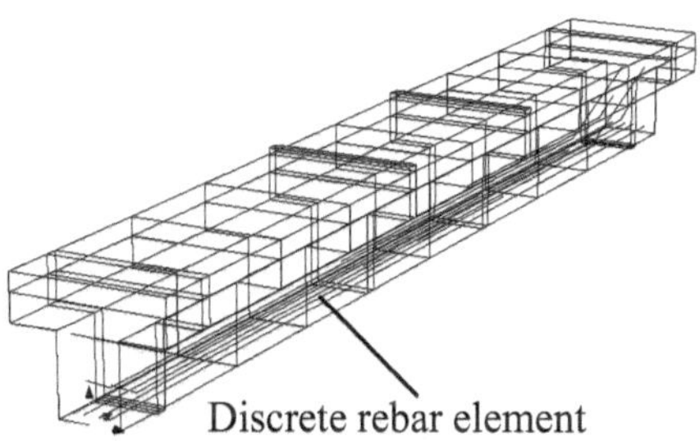

Figure 21. Three dimensional FE mesh

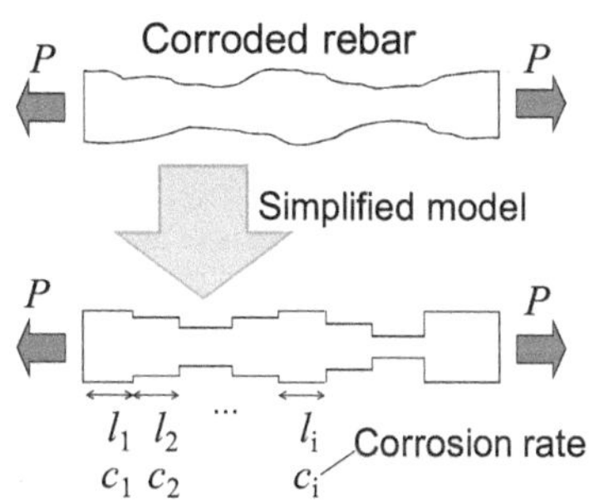

Figure 22. Simplified model for numerical experiment of corroded rebar

It is known that stress-strain relationship of corroded rebar is different from original stress-strain relationship. This is mainly caused by spatial variability of corrosion. It is reported that average stress strain curve can be estimated by taking local corrosion into consideration as shown in Figure 22 while original stress-strain relationship can be used for local constitutive law [6]. In numerical experiment, average stress–average strain relationship was estimated from following equations.

$$\varepsilon_m = \frac{1}{l}\int f^{-1}(\sigma_l)\,dl = \frac{1}{l}\sum_{1}^{n} f^{-1}(\sigma_i)\cdot \Delta l \tag{1}$$

$$\sigma_l = \sigma_m/(1-c) \tag{2}$$

where ε_m is average strain, $f^{-1}(\sigma)$ is inverse function of stress-strain relationship of rebar and σ_l is tensile stress in local cross section. σ_m is average stress which

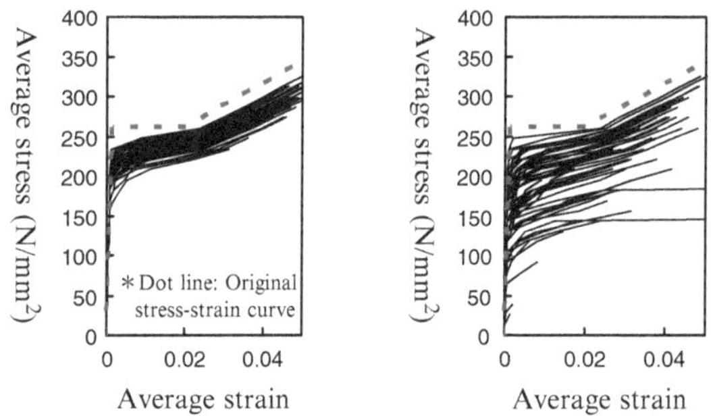

(a) Inner beam (b) Outer beam

Figure 23. Average stress–average strain relationship of 1 m length rebar

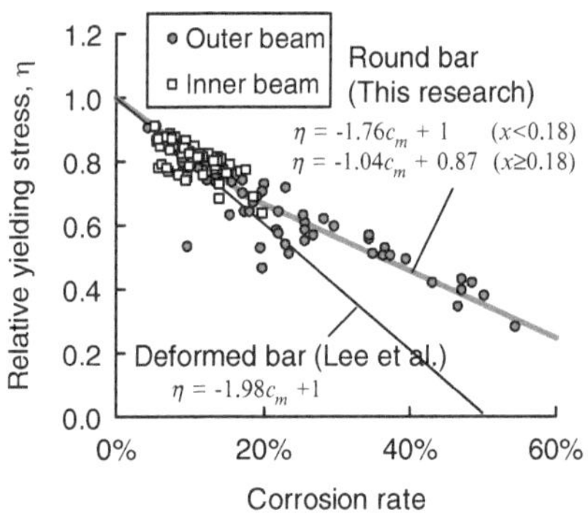

Figure 24. Corrosion rate versus relative yielding stress

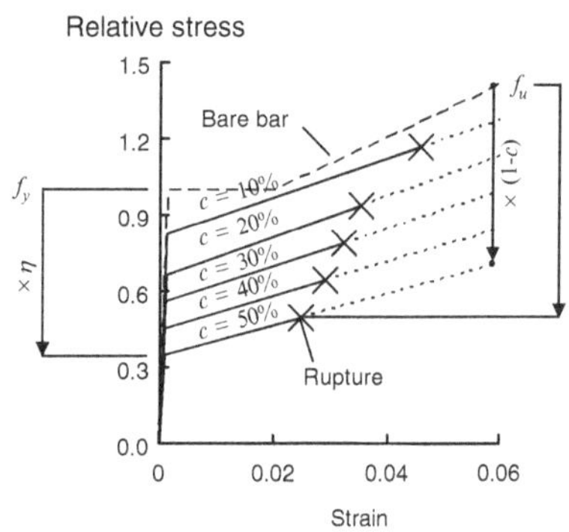

Figure 25. Bi-linear model for corroded rebar

equals tensile force divided by the original cross sectional area. c represents local corrosion rate. Calculated average stress–average strain relationship of 1 m length rebar is illustrated in Figure 23. In Figure 24, average corrosion rate is compared with relative yield stress. Relative yield stress η is defined as the ratio between the stress where yielding starts locally and original yielding stress. Bi-linear curve is obtained as a result of regression analysis of numerical experiment.

$$\begin{aligned} \eta &= -1.76c_m + 1 \qquad c_m < 0.18 \\ &= -1.04c_m + 0.87 \quad c_m \geq 0.18 \end{aligned} \tag{3}$$

The first equation in Eqn. (3) corresponds to Lee model [7] which was made from the regression analysis of deformed bar. The reduction rate in the second equation in Eqn. (3) is moderate compared with the first one. Spalling of concrete may lead to homogenization in corrosion. Consequently, the stress–strain relationship of corroded rebar is expressed by bi-linear model shown in Figure 25.

Bond-slip relationship. Bond stress–slip model followed CEB–FIP model (see Figure 26).

$$\begin{aligned} \tau &= \tau_{\max}\left(\frac{s}{0.01}\right)^{1/2} \qquad 0 \leq s \leq 0.01 \\ &= \tau_{\max} \qquad\qquad\qquad 0.01 \leq s \end{aligned} \tag{4}$$

where s is slip (mm). τ_{max} is bond strength obtained from following equation.

$$\tau_{\max} = \alpha\sqrt{f'_c} \tag{5}$$

α is the coefficient for bond condition. α equals 0.10 for good bond condition, while 0.05 for poor bond condition. Unbond condition was assumed where cover concrete had been spalled before loading.

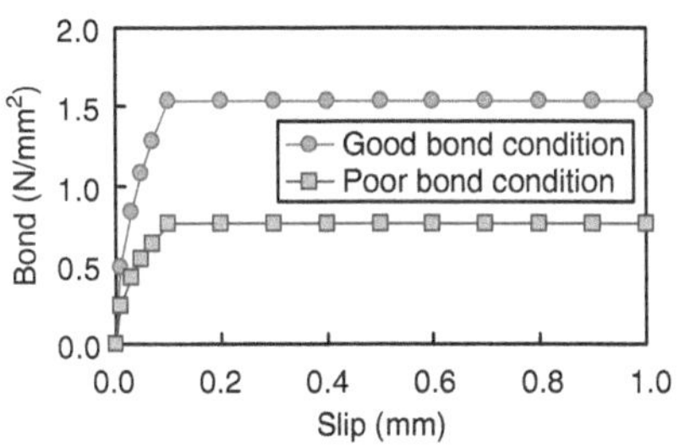

Figure 26. Bond–slip model based on CEB-FIP model code

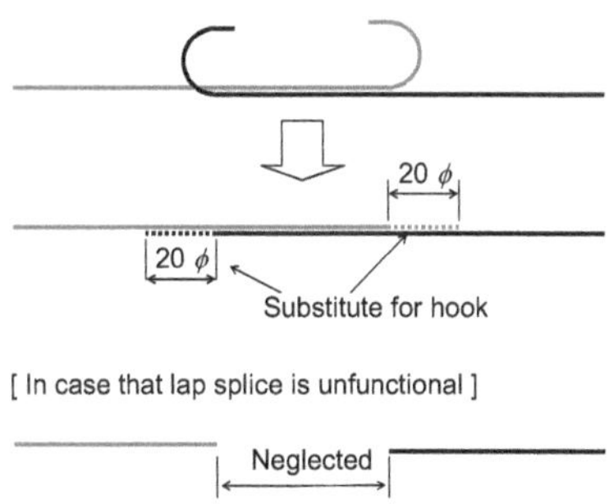

Figure 27. Expression of semi circular hook as line element

As shown in Figure 15, beam specimen contained several lap splices at the middle of loading span. Flexural strength of beam specimen was subjected to the strength of these lap splices. The strength of lap splice depends on the semi circular hook in

case of round bar. Therefore, anchoring strength of semi circular hook should be taken into account in FE analysis. In FE analysis, semi circular hook was replaced with 20 ϕ length of straight rebar for simplification.
In case that lap splice was not effective during the loading test, rebars were shortened as shown in Figure 27.

Analytical results

Inner beam. In Figure 28, numerical results of inner beam are compared with experiment in load–center span deflection relationship. Perfect bond was assumed between pavement concrete and T shaped RC beam. Good bond condition was adopted in the bond–slip model. The effects of pavement concrete and lap splice H were studied through the analysis as follows.

Before pavement concrete is detached (center span deflection is less than 40 mm), the numerical result agrees with the experimental one by assuming that lap splice H works effectively. After pavement concrete is detached, the numerical result corresponds to the experimental one when lap splice H is neglected. Lap splice H may have lost its anchorage performance simultaneously with the detachment of pavement concrete.

The influence of bond strength is studied in Figure 29. Analysis was conducted assuming perfect bond condition, good bond condition and poor bond condition. Stiffness of RC beam did not change regardless of bond condition. However, maximum load was subjected to the bond strength. This is because transferable force of lap splice is less than yielding force of rebar when bond – slip model is applied. Therefore, bond – slip model is essential for the analysis of corroded RC member which contains lap splices.

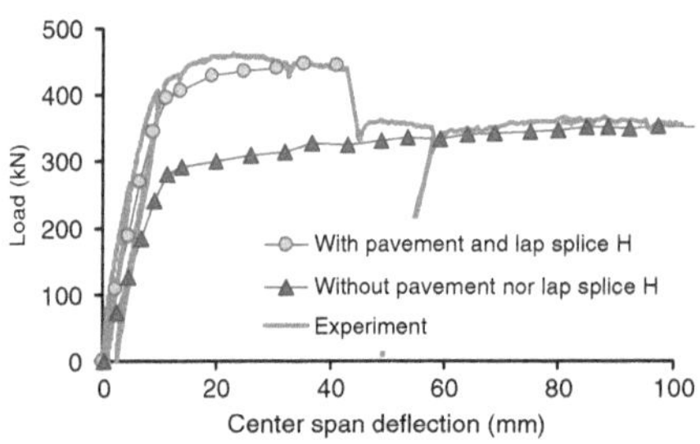

Figure 28. Influence of pavement concrete and lap splice H on analytical result (inner beam)

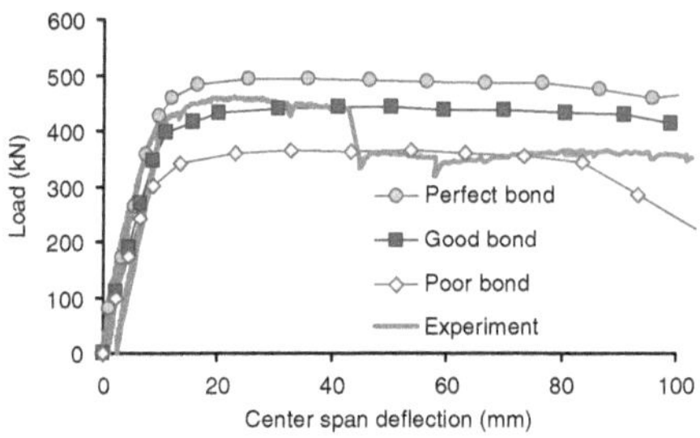

Figure 29. Influence of bond condition on analytical result (inner beam)

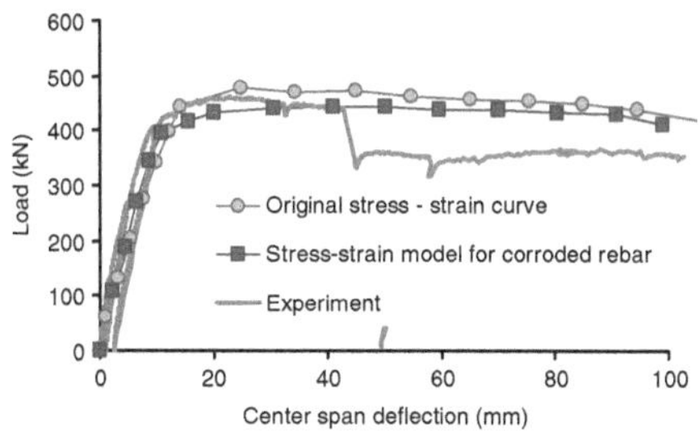

Figure 30. Influence of rebar model on analytical result (inner beam)

The influence of stress-strain relationship of rebar is studied in Figure 30. Analysis was conducted with use of original stress- strain law and bi-linear model illustrated in Figure 25. Bi-linear model proposed in this paper leads to less yielding load because spatial variability is considered. It is difficult to judge which model is better for analysis of corroded RC structures because the difference in load–deflection curve is small. In conclusion, structural performance of inner beam can be estimated if corrosion rate and bond condition are obtained preliminary.

Outer beam. Figure 31 shows load–center span deflection relationship of outer beam. As shown in Figure 20(a), lap splices of rebar K and rebar P were exposed during the loading test. Accordingly, these lap splices were neglected in FE analysis. Moreover, rebar M was also neglected in analysis because rebar M did not work efficiently as shown in Figure 14. Analytical result without pavement concrete is closer to the experimental result than the analytical result with pavement concrete. This result coincides with the consequence made from the distribution of concrete strain in Figure 11, i.e. pavement concrete had been detached before the maximum load. Despite rupture was considered in stress–strain law of rebar, numerical analysis could not simulate the rupture of rebar around 37 mm of center span deflection where rupture occurred in the experiment. Steel strain in the analysis was about half of rupture strain when center deflection was 37 mm. Further study is required in order to predict the rupture of rebar adequately in analysis.

Sound condition of lap splice and perfect bond condition are assumed in Figure 32 to estimate the potential strength of outer beam. Numerical result is much higher than the experimental one because the main cause of small strength in outer beam is not corrosion itself but the degradation of bond and lap splice.

As a result, bond condition of every rebar and the degree of degradation of every lap splice should be obtained as well as corrosion rate of rebar in order that structural performance is estimated accurately with FE analysis. However, it is almost impossible to judge the effectiveness of individual lap splice with high accuracy in advance. Therefore, we should in general make a conservative consideration in actual cases.

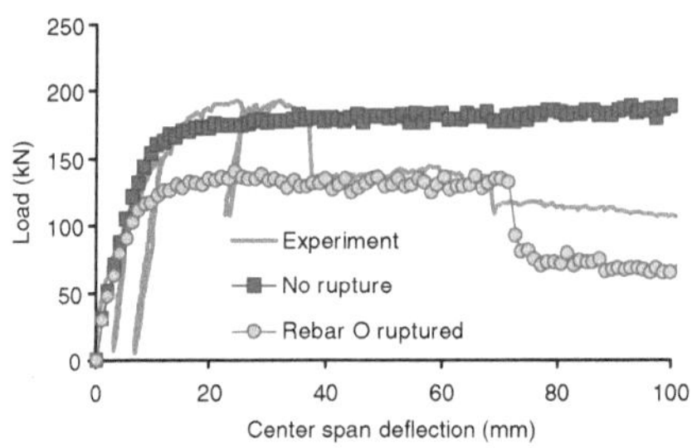

Figure 31. Comparison between analysis and experiment (outer beam)

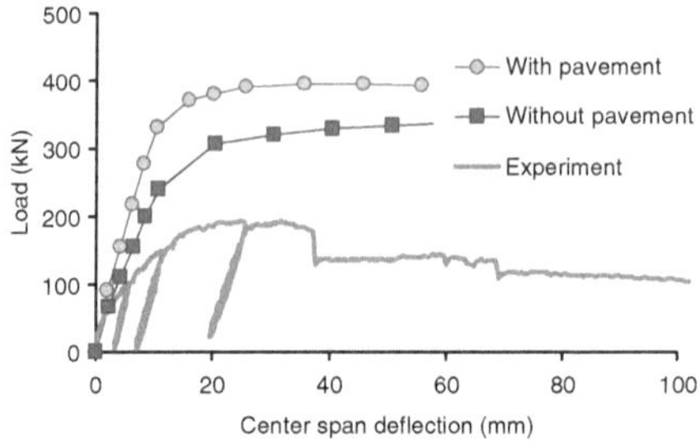

Figure 32. Analytical result when perfect bond is assumed (outer beam)

Corrosion rate and relative strength. Figure 33 shows average corrosion rate versus relative flexural strength of RC beam. Relative flexural strength is defined as measured flexural capacity divided by original flexural capacity of sound beam. Original flexural capacity is obtained from FE analysis. The reduction rate of flexural strength of inner beam corresponds to that estimated from average corrosion rate. Same tendency had been obtained from laboratory test [1, 2].

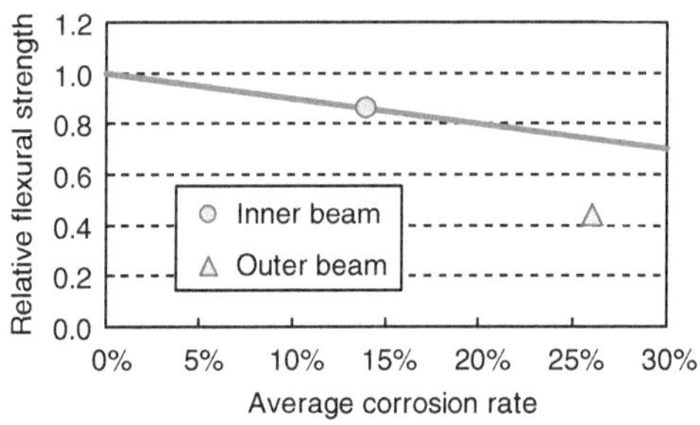

Figure 33. Average corrosion rate versus relative flexural strength

On the other hand, relative flexural strength of outer beam is lower than that estimated from average corrosion rate. Deterioration of lap splices and bond of rebars caused significant reduction in load carrying capacity of outer beam.

Conclusions

RC beam specimens taken from a real RC bridge which was degraded due to chloride attack were tested in the laboratory until failure. Influence of corrosion of rebar, bond strength and anchorage strength of lap splice were studied coupling with FE analysis. The following conclusions are obtained in this study:

1. If cover concrete is damaged where lap splice is arranged, the anchorage strength is significantly reduced. Consequently, residual strength of RC structure is also reduced.
2. Residual strength of RC member which is deteriorated by chloride attack can be estimated with FE analysis by considering corrosion rate of rebars, remaining anchorage strength of each lap splice and bond condition of rebars.
3. In case that several rebars are arranged and that bond strength and anchorage strength is not degraded, reduction in strength corresponds to the average corrosion rate. The influence of pitting corrosion is compensated by other rebars.

Acknowledgements

Authors would like to express their thanks to Itoigawa City Office, Nagaoka National College of Technology and Study Group of Concrete Maintenance in the Niigata prefecture for their cooperation.

References

[1] JSCE-331 (2006), *Report of Technical committee on structural performance of deteriorated concrete structures.*, JSCE, Tokyo (in Japanese)
[2] JSCE-331 (2009), *Report of Technical committee on structural performance of deteriorated concrete structures 2.*, JSCE, Tokyo (in Japanese)
[3] JSCE (2002), *Standard specifications for concrete structures-2002 "Materials and construction"*, JSCE, Tokyo
[4] Cervenka, V., Jendele, L. and Cervenka, J. (2007), *ATENA program documentation*, Cervenka Consulting, Prague
[5] CEB (1993), *CEB-FIP Model code 1990*, Thomas Telford, London
[6] Oyado, M., Kanakubo, T. Yamamoto, Y. and Iijima, T. (2007). In: *J. Materials, Conc. Struct. and Pavements*, JSCE, Vol.63, No.1, pp.143-155
[7] Lee, H., Noguchi, T. and Tomosawa, F. (1998). In: *J. Struct. Constr. Eng.*, AIJ, No. 506, pp. 43-50

Observations on the Morphology of Oxide Formation due to Reinforcement Corrosion

C. Andrade, F. Tavares, L. Toro and J. Fullea

Center for the Safety and Durability of Materials and Structures, CISDEM-UPM-CSIC, Madrid, Spain
andrade@ietcc.csic.es

Abstract. The oxide formed during corrosion has an expansive character which induces the cracking of concrete cover. The properties of this oxide layer are focussing the interest due the need to introduce its mechanical characteristics in the models to calculate crack width. In a previous paper it was suggested that the oxide behave as "water" and in consequence it has to be modelled so, which fitted very well in the experiments and model associated an that time. The attribution to the water was based in the observation that the rust has not mechanical consistency and was like a suspension. However, much recent papers insist in attributing other much higher young modulus to the rust. Present paper tries to insist in the original concept by illustrating the formation of the rust in solution and in mortar. The oxides are of mixed colours evolving in function of the time to contact to open air and they have not mechanical consistency, being apparently hydrophilic by retaining large amount of water.

Introduction

Concrete Reinforcement corrodes in the presence of neutral pH values and of chloride ions. The corrosion process induces several processes that are summarized in Figure 1 [1]: reduction of bar cross section, of steel ductility, of steel-concrete bond and generates the cover cracking. All of them lead into a reduction of the load-carrying capacity of the concrete section.

C. Andrade and G. Mancini (eds.), *Modelling of Corroding Concrete Structures*,
RILEM Bookseries 5, DOI 10.1007/978-94-007-0677-4_12, © RILEM 2011

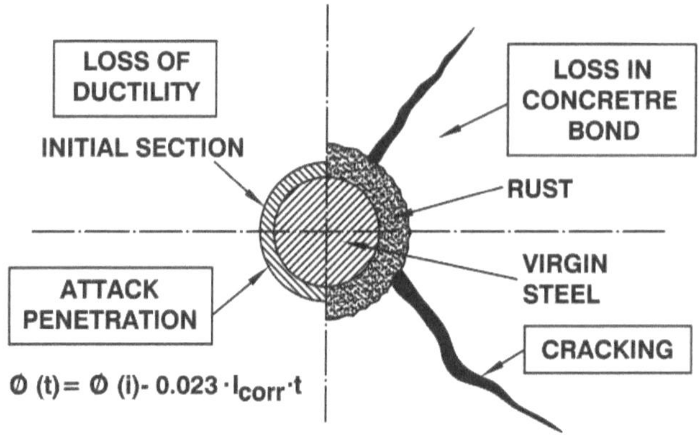

Figure 1. Consequences of corrosion [1]

From these effects, that of cover cracking is receiving important attention [1-12] in the last years due to the calculation of the evolution with time of crack widths in the concrete surface is an indication of the progressive loss in bond properties of the steel/concrete composite action and of the structural load-bearing capacity. From these works the main conclusion was that the beginning of the cracking depends principally on the relation concrete cover / diameter of the bars, the quality of the concrete and his tensile strength, being also very important the corrosion arte or the amount of current applied [13].

Regarding the nature of the expansive character of the iron oxides, after the numerical modelling by Molina et al [2] in the beginning of the 90's, the main question has been focussed on the value taken by the models on the expansion ratio between parent steel and oxide generated. In the work of Andrade and Molina [1, 2] either the experimental results or the numerical model, indicated that the first crack appears very early, when only few μm of corrosion have been produced, which indicated that the specific volume of oxides is about two times higher than the apparent steel volume and that the bulk modulus of the rust has to be lower than 2GPa. In the numerical model the properties of the rust where made to change linearly from those of the steel to those of the rust, which was considered to perform like liquid water because as was said [2] *"is one of the main components of the rust under these conditions"*.

In spite of the confirmation of the by the further models of the need to use low Young modulus of the rust freshly generated, there are numerous recent works stressing the importance to test the mechanical properties of the rust [14, 15] in order to find the most appropriate parameters to feed the models. Also it has been considered essential to establish the expansion ratio of the oxides trying to link it to the particular type of oxide produced.

In the present paper several tests are reported to show the morphology of the oxides generated, and their nature that is diverse and always mixed, as the iron oxides are several and their production in the corrosion process is progressive and very much dependent on the availability of oxygen and water in the media.

Experimental

Several tests were performed in order to illustrate the morphology of the oxides formed when the corrosion in presence of chlorides starts:

1. Tests in saturated calcium hydroxide solution with 0.5 of sodium chloride
    a. A bar is introduced in an alkaline solution (saturated calcium hydroxide) with 0.5M of sodium chloride and left to corrode in order to monitor the nucleation and growing of the oxides in the pits generated
2. Tests in mortar with chlorides
    a. A standarized mortar is made having 2% by weight of cement of sodium chloride. The mortar is cured and maintained under tap water. After one week of corrosion the bar is removed for its visual observation.
3. A hole is made in a cylindrical concrete specimen and a chloride solution is introduced in the hole. The arrangement is shown in Figure 2. The specimen having this bar inside in the hole, which will act as anode, is introduced in a 0.5 M NaCl solution. A stainless steel mesh is also introduced in the solution in order to act as cathode. The bar and the cathode are connected to a power supply and current is made to circulate in order to generate rust in large quantity. Several currents were applied, from very high (from hundreds to several few mA/cm^2). The oxides generated have served to study the colour and morphology and consistency of the oxides and also on how they dry, age and react when changed of solution.

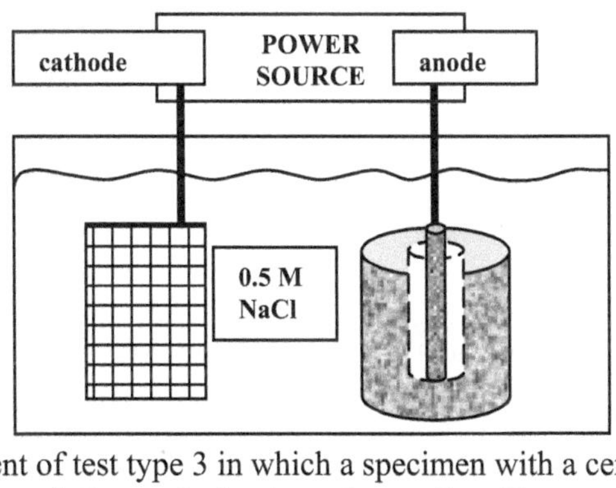

Figure 2. Arrangement of test type 3 in which a specimen with a central hole filled with sodium chloride solution and a bar is made anodic with respect to a mesh of stainless steel which is the cathode

a. A piece of aged oxide of black colour obtained in test type 3 was introduced in two solutions: 1) the solution around the concrete specimen of Figure 3 which had a pH value of 9.3, 2) distilled water and 3) saturated calcium hydroxide.

4. Tests in which the rust is removed from the bar of test type 3 and left to dry at the atmosphere in order to follow the morphological changes during drying.

The main measurement has been the simple visual observation of

- Colour of the oxides
- Morphological aspect
- Volumetric appearance

Results

In the first type of test pits are developed due the presence of chlorides in the alkaline solution as is shown as an example in Figures 3 left and right.

Figure 3. Left: a pit with the oxides of dark color growing upwards of a bar. Right: other stage of the process in which the oxides which have falled down can be appreciated in the bottom of the flask

Regarding the *color*, at the beginning the oxides are dark, green or black, although seldom a white product can also be appreciated. After, they turn to be brown and finally orange as can be appreciated of the rust falled down in the right part of the Figure 4. The *morphology* is like small "mountains" which grow until they collapse and fall to the bottom of the flask (Figure 3 right). The *aspect* is viscous and if they are touched with a glass stick, these protuberances detach from the bar and are partially adhered to the glass stick and other part fall down. They do not have any mechanical consistence. Finally, the volumetric appearance is expansive as was expected. They are as hollowed protuberances or as small pieces of lamellas, all maintained together. When detached from the bar, they seem as small agglomerates finely divided.

The appearance of the oxides after being corroding in very wet mortar (*test type 2*) can be appreciated in Figure 4. The oxides are of orange color and they show to be hydrophilic as they maintain attached the water. The rust spots look as water drops in which the oxides are small particles in suspension.

Figure 4. Aspect of the rust in pits formed in a mortar having admixed sodium chloride. The spots are like water drops having the oxide particles in suspension

In *test type 3* a considerable amount of oxides were produced. These oxides generated in the hole of the specimen were the bar was placed were of different colors, being at the beginning black or dark and evolving to brown and orange as the oxide entries in contact to the solution (Figure 5 left).These oxides did not remained attached to the bar but they migrate out of the hole and spray in all the solution contaminating it with these oxides. The oxides were as a gel or the aggregation of particles that are maintained agglomerated but they disperse (Figure 5 right) very quickly when the solution is stirred with a glass stick. Any mechanical consistency has been detected in the oxide in this phase.

However after two months of this accelerated corrosion, the oxide near the upper part of the hole and in suspension in the solution started to permanently agglomerate and form stable granules that progressively grow. These granules (Figure 6 left) were either fully dark or brownish not being easy to find a particular reason for the difference in color. Regarding they mechanical properties, some can be disaggregated by pressing with the fingers and other however remain cannot be pressed remaining with consistency.

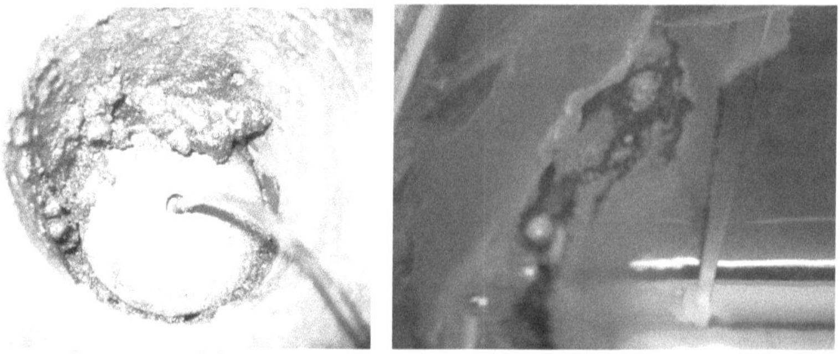

Figure 5. Left: aspect of the oxides when generated in the bar inserted hole of the concrete specimen of test type 3. Right: aspect of the oxide particles in suspension in the solution of that test type

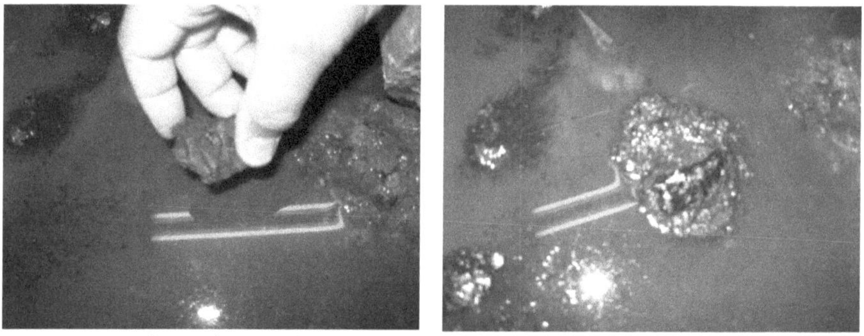

Figure 6. Left: A piece of dark oxide (likely magnetite) formed in test type 3 just before being introduced in the solution that is slightly alkaline. Right: the starting of the reaction of the dark oxide in the solution, it is observed bubbles around it as it a gas generates (likely hydrogen)

The piece of agglomerated oxide shown in Figure 6 left, that was of dark color, was placed in another flask containing an amount of the brownish chloride solution surrounding the concrete specimen whose pH was around 9.3. Immediately that this granulate entered in contact to this solution, a strong reaction was detected which dissolved the granulate and generated a certain “foam around it due the evolving of a gas (likely hydrogen), as can be appreciated in Figure 6 right and Figures 7 left and right, where consecutive phases of this reaction.

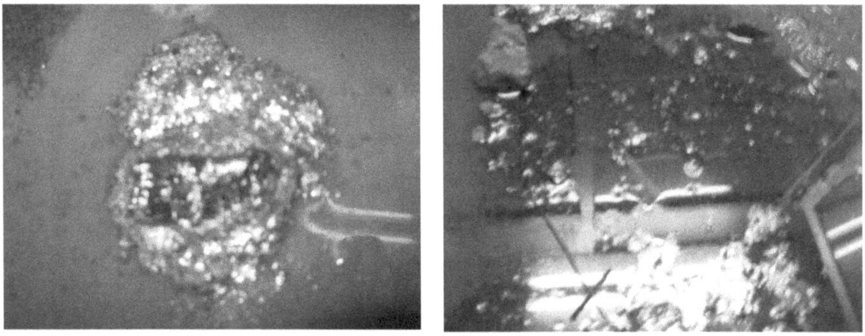

Figure 7. The progression of the reaction of figure 6 after few minutes. Left: the dissolution with generation of gas is evident and Right: the dissolution is complete leaving a spot of black color with a diameter much higher than the initial doiameter of the oxide piece

In order to explore whether this reaction may occur also when in contact to neutral water, another granulate was introduced in another flask with distilled water as shown in Figure 8. The abrupt reaction was not detected and the dark oxide was mixed with the water loosing its mechanical consistency and showing a clear dark green color instead of green. This color was maintained for more than one week in spite of the direct contact of the oxide with the atmosphere. After more than 20 says the upper part of the oxide-water mixture stared to exhibit a brown color in the upper part but maintained the dark-green color just below it because a "membrane – like layer was detected in the surface that seems to totally avoid the contact of the air with the oxide-water mixture still in liquid state underneath.

Thinking then, that it is the alkalinity which provoked the reaction, another piece of dark oxide was introduced in a saturated calcium hydroxide solution of pH around 12.6. The reaction was also produced but much less violent. Then it seems that the most rapid reaction of dissolution is produced at pH values alkaline but not so much.

A last test was made (*type 4*) in which part of the oxide "swimming" in the solution of test type 3 was placed in a flat surface to leave it to dray at the atmosphere in the open air. The aspect of the oxides is shown in Figures 9 left and right. In Figure 9 right is starting to dry and the following stages of the drying are depicted by Figure 10.

Figure 8. The dark oxide of figure 6 is introduced in a flask with distilled water. Left: no generation of gas is detected but the oxide collapses and produces a suspension of green color. Right: after around 20 days the top part has become brown (likely due to the formation of Fe(III) oxides and a superficial membrane is generated which avoids the contact of the external air with the green mixture which remains wet and green

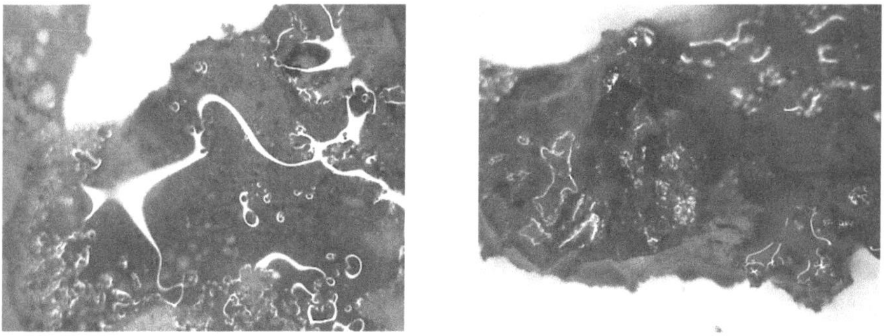

Figure 9 Left: a portion of oxide generated in test type 3 which exhibits the water it carries. Right: the spot starting to dry in contact with air in the laboratory

When completely dried (Figure 11 right) the oxide maintained the volumetric aspect (width and height) but showing the cracks formed due to the shrinkage of the water loosing. However, when touching this spot with the finger, it collapsed as can be observed in Figures 11 left and right. The volume disappeared and the oxide collapses in multitude on small particles and spray around.

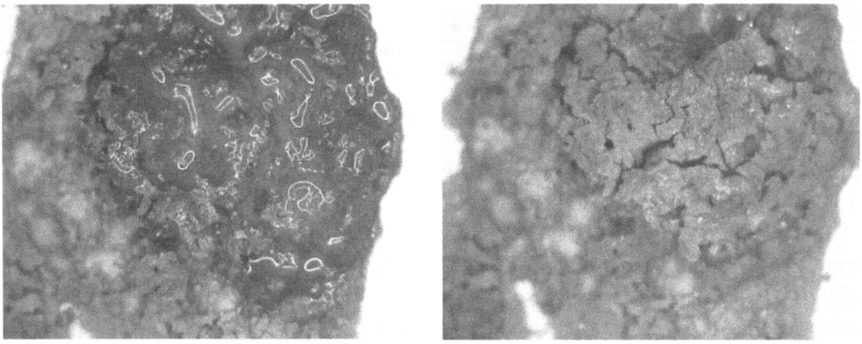

Figure 10. Left: further stage in drying of oxide of figure 10. Right: when it completely dried. The cracks formed are clear and it maintains the shape

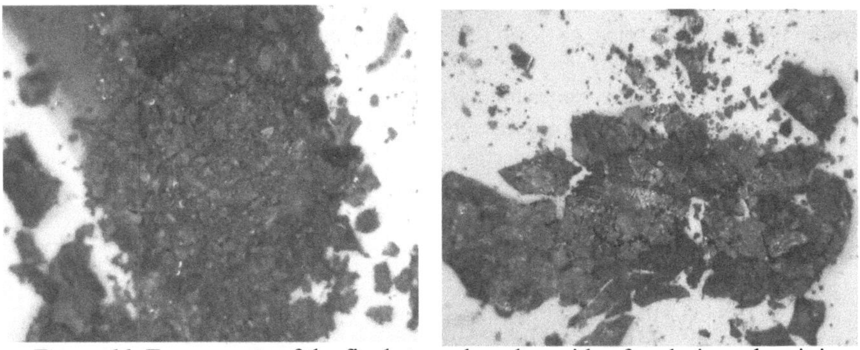

Figure 11. Two aspects of the final step when the oxide after drying when it is touched with a finger: it collapses (left) and sprays in multiple isolated particles (right)

When new water is put in contact to these oxide particles, they do not swell or retain the water in the same manner than originally which means that the effect is irreversible. As soon as the rust dries its properties and expansive character completely change.

Discussion

The iron can form not less than 16 varieties of oxides (Fe_xO_y), hydroxides ($Fe(OH)_x$) and oxy-hydroxides (FeOOH). They are some having only one of the two valences of iron, Fe(II) and Fe(III) or they can be the misture of the two valences as is typical in the magnetite. In order to realize the multiple paths that the composition of the rust can take, the list of the most common iron oxides is: 1) *wüstite* (FeO), 2) *magnetite* (Fe_3O_4), 3) *hematite* (α-Fe_2O_3), 4) *beta phase*, (β-Fe_2O_3), 5) *maghemite* (γ-Fe_2O_3), 6) *epsilon phase*, (ε-Fe_2O_3). The *hydroxides* are iron(II) ($Fe(OH)_2$ and iron(III) ($Fe(OH)_3$) and the Oxide/hydroxides are: 1) *goethite* (α-FeOOH), 2) *akaganéite* (β-FeOOH), 3) *lepidocrocite* (γ-FeOOH), 4) *feroxyhyte* (δ-FeOOH), 5) *ferrihydrite* ($Fe_5HO_8 \cdot 4H_2O$). These oxides fare the base of the metallurgy of the iron and as such ind many applications in the industry [16].

The iron minerals (Figure 12) can be easily found in the nature in many places as it is very common. These minerals can be a single species of a mixture of the different oxides that iron can form. It has to be realized that these minerals have been formed in geological ages and have suffered many transformations and aging. Although they can be the same mineral species than the oxides formed during the iron corrosion process, it is important to realize that they are very different in morphology and mechanical properties than the corrosion rust, as has been shown in present work.

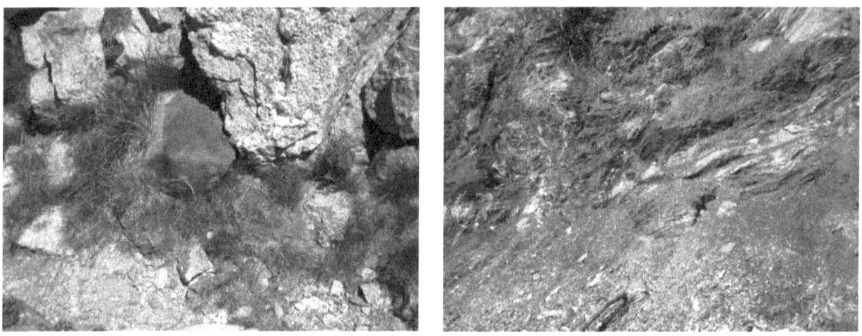

Figure 12. Two views of natural iron minerals in the Pyrenees. Left: an isolated stone, Right: the mineral appearing in veins in the rock.

When the oxides are obtained from a solution or formed in a corrosion process in presence of water it depends on the pH, the ionic strength and of the composition of the solution whether one or several of the possible oxides is obtained. In the corrosion process usually a mixture of several of them is always found, as has been reported in previous work [3, 13, 17, 18].

It is not the intention of present work to entry in the mechanistic aspects of the corrosion of the iron, but some elemental aspects will be commented in order to illustrate the results found in this work. The iron metal is oxidized to Fe(II) as a first elemental step (1):

$$Fe \rightarrow Fe^{+2} + 2e^{-} \quad (1)$$

Subsequent reactions are (2)

$$Fe^{+2} + H_2O \rightarrow Fe(OH)^{+} + H^{+} \rightarrow Fe(OH)_2 + H^{+} \quad (2)$$

$$Fe^{+2} \rightarrow Fe^{+3} + e^{-} \quad (3)$$

When the ions Fe^{+2}, Fe^{+3} go to the solution, they cannot remain as such as several processes develop immediately: hydration, protonation/deprotonation and solvation 13.

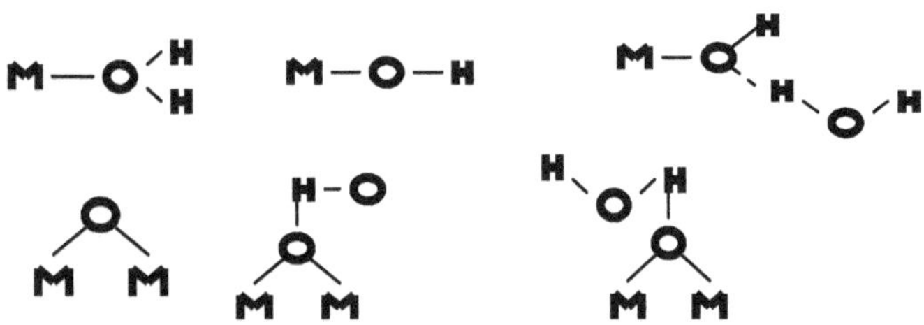

Figure 13. Different possibilities of bonds of a metal with the water. Protonation/deprotonation occurs through hydrogen bridges

The ions remain in solution for pH values very acid lower than around 2.5 but for higher pH values it precipitates by forming elemental monomers, usually of octahedral shape. These individual octahedrons may agglomerate to form dimmers, trimmers, etc. (Figure 14) ending in the formation of the several oxide/hydroxide varieties depending on pH, ionic strength and composition of the solution.

These agglomerates can remain as individual entities or flocculate as gel particles, electrically charged or not, by aggregation, maintaining water molecules in the first shell of solvation. For being a solid entity they need to loose the water and condensate and age to reach a crystallographic continuous structure.

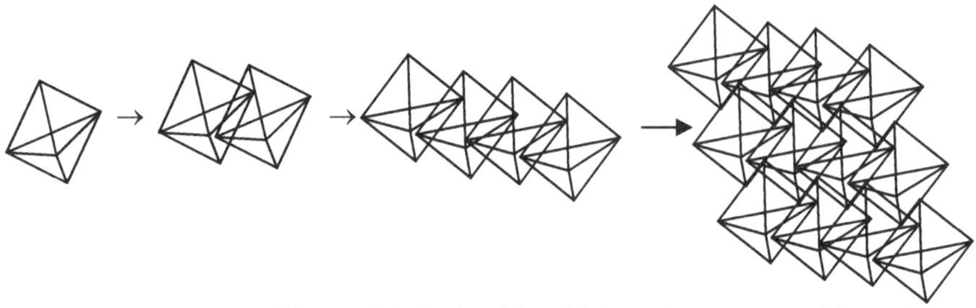

Figure 14. Monomer of iron oxide/hydroxide which agglomerates and flocculates during aging to form larger particles which passing through being a gel may end in the crystal structure

All these theoretical well known aspects of the physico-chemistry of iron oxides have been observed in present experimentation. When starting to corrode in solution, the iron gives rise of ferrous hydroxide (green and white colors) and magnetite. In many occasions amorphous gel-like "green rust" [17, 18] has been found. The rust evolve later in presence of oxygen to ferric oxides/hydroxides although the presence of the reduced species has been always detected in present results.

The rust freshly formed is like a gel (see Figures 3 to 10) without mechanical consistency and resistance, apart from that of the water, as the oxide particles remain in suspension in it. Their morphology is not unique as continuously evolves and if the water evaporates by drying, the oxide can collapse by simple low pressing. In Figure 15 has been tried to reproduce what we observed in the formed pits. The oxide particles flocculated and aggregated are maintained in its apparent volume by the water retained in between them. They seem like the petals of a flower which may collapse if the water evaporates. The expansive character of the rust formed during corrosion, it seems to come then from its ability to retain water molecules, which in turn depends on the nature of the mixed oxides types formed.

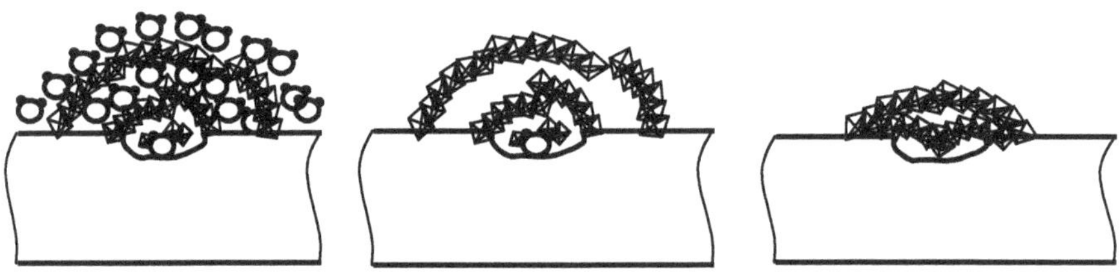

Figure 15. Interpretation of the sequence of oxide drying of figures 10-12. In the left, the oxide particles forming a like a "sheet" (or petals of a flower) with the inter-particle water molecules. In the center, the structure having dried. In the right being collapsed after being touched with a finger

In the case of the reinforcement corrosion, it can be deduced then that the expansion produced by the generation of oxides will depend on the moisture around the bar and its continuous growing will depend on the water availability. As soon as the concrete would dry, the oxide looses the water and shrinks in an irreversible manner. Regarding the accelerated tests applying a current, the concrete has enough moisture for the oxides to induce expansion, however this expansion would vary depending on the current applied as if the current is very high, there would be more species of Fe(II) than of Fe(III) as the oxidation is slower than the generation of oxides. If the current is low more Fe(III) species can be produced during the corrosion. Then the best is to use very moderate currents for the reproduction of realistic conditions.

The assumption in the work published in 1993 [1, 2] seems then validated in present work, that is the best approximation for the adscription of a young modulus value to the rust is to use the properties of the water, as the oxides particles isolated and flocculating in the pore solution in the first stages of the corrosion process and in the accelerated testing applying a current. It seems that it is not reproducing the process the comparison with the properties of the natural iron minerals.

Conclusions

This work had not the intention to thorough study the composition and nature of the formation of oxides generated in the reinforcement corrosion but to document the shape appearance and mechanical consistency of the freshly formed oxides with regard to the modeling of expansion due the corrosion. The observations made can be summarized as follows:

1. The oxide spots had the appearance of viscous protuberances of several shapes that, at the magnifying glass, can appear as a flower having small petals containing water in between them. It has not mechanical consistency as can be easily removed and falls down by its own weight. The small agglomerates detached move in the solution and act as a gel.
2. When the oxides are dried at open air maintain the shape but it collapses if touched. They have not mechanical consistency. When re-wetted they do not recover the original shape and do no swell.
3. When aging they flocculate and agglomerate and can form granulates which not too much mechanical consistency.
4. The granulates of dark or black color are very reactive in solutions slightly alkaline and less reactive in saturated calcium hydroxide. They mixed with distilled water but do not react or dissolve.
5. The expansive character is confirmed to be attributed to the water adhered and physically bonded to the particles in the first phase of the process, as they oxides seem hydrophilic. However after drying the swelling is not

reversible and they simply are dispersed in the solution or react and dissolve by liberating a gas (likely hydrogen).

Acknowledgements

The authors would like to acknowledge the financing of the Ministry of Science and Innovation for the INGENIO 2010-CONSOLIDER Project on "Safety and Durability of Structures: SEDUREC".

References

[1] Andrade, C., Alonso, C. and Molina, F. J.: Cover cracking as a function of rebar corrosion: Part I – Experimental test. Materials and Structures, 26, pp. 453-464, 1993.
[2] Molina, F. J.Andrade, C., Alonso, C.: Cover cracking as a function of rebar corrosion: Part II – Numerical model. Materials and Structures, 26, pp. 532-548, 1993.
[3] Tuutti, K.: Corrosion of steel in concrete. Swedish Cement and Concrete Research Institute, Stockholm, Sweden, 1982.
[4] Bazant, Z. P.: Physical model for steel corrosion in concrete sea structures – application. Journal Structural Division, ASCE, 105, ST6, pp., 1155-1166, 1979.
[5] Allan, M. L. and Cherry, B. W.: Mechanical simulation of corrosion induced cracking in reinforced concrete, Corrosion/89, Conference paper No. 377, NACE, Houston, Texas, 1989.
[6] Rasheeduzzafar, S. S., Al-Saadoun and Al-Gahtani, A.S.: Corrosion cracking in relation to bar diameter, cover and concrete quality. Journal of Material in Civil Engineering, Vol. 4 (4), Nov. 1992.
[7] Bedu, P.: Volumetric changes of cement paste under exposure to the simulated corrosion products of steel and their influence on cracking susceptibility. M. Sc. Eng. Thesis, Florida Atlantic University, Boca Raton, Florida, 1993.
[8] Sagüés, A. A. and Torres Acosta, A. A.: Concrete cover cracking and corrosion expansion of embedded reinforcing steel. Proceedings of the third NACE Latin American region corrosion congress on rehabilitation of corrosion damaged infrastructures, Castro, P., Troconis, O. and Andrade, C. eds., pp. 215-229, 1998.
[9] Martín-Perez, B.: Service life modeling of RC highway structures exposed to chlorides. Ph.D. dissertation, Dept. of Civil Engineering, University of Toronto, 1998.
[10] Torres Acosta, A.A.: Cracking induced by localized corrosion of reinforcement in chloride contaminated concrete. Ph.D. Thesis, University of South Florida, Florida, USA, 1999.

[11] Vidal, T., Castel, A. and Francois, R.: Analyzing crack width to predict corrosion in reinforced concrete, Cement and Concrete Research 34, pp. 165-174, 2004.
[12] Leung, K.Y.: Modeling of concrete cracking induced by steel expansion. Journal of Materials in Civil Engineering, May-June, 2001.
[13] Andrade, C., Alonso, C., Rodriguez, J. and Garcia, M.: Cover cracking and amount of rebar corrosion: importance of the current applied accelerated tests. In concrete repair, Rehabilitation and protection, R. K. Dhir and M. R. Jones eds., E&FN Spon, London, UK, pp. 263-273, 1996.
[14] Korea.
[15] Valerie.
[16] Deer, Howie and Zussman *An Introduction to the Rock Forming Minerals*, Longman, pp. 424-433, 1966.
[17] Sagoe-Crentsil, K. K. and Glasser, F. O.: Steel in concrete: Part II. Electron microscopy analysis. Magazine of Concrete Research, 41, 149, pp. 213-220, 1989.
[18] Sagoe-Crentsil, K. K. and Glasser, F. O.: Constitution of green rust and its significance to the corrosion of steel in Portland cement. Corrosion, 49, 6, pp. 457-463, 1993.

Severely Corroded Reinforced Concrete with Cover Cracking: Part 1. Crack Initiation and Propagation

Dario Coronelli[1], Kamyab Zandi Hanjari[2], Karin Lundgren[2] and Enrico Rossi[1]

[1] Dipartimento di Ingegneria Strutturale, Politecnico di Milano, Italy
[2] Department of Civil and Environmental Engineering, Chalmers University of Technology, Sweden

Abstract. In many corroding RC structures, it is not uncommon that cover spalling and delamination have occurred. Previous research has been mainly concerned with lower corrosion levels leading to cover cracking. Moreover, the main focus of the available knowledge concerns the corrosion of the main reinforcement; while the corrosion of the stirrups is often overlooked. In an experimental investigation, corrosion attack causing crack initiation, propagation and cover delamination are measured. The specimens have the shape of a beam-end and are corroded with an accelerated method. The location of the bar, middle and corner placement; the amount of transverse reinforcement; the corrosion level of longitudinal reinforcement and of transverse reinforcement are studied. The specimens after corrosion are also used in pull-out bond tests; the results are presented in a companion paper subtitled "Part 2. Anchorage Capacity". The first test results of the ongoing experimental campaign are presented. The crack patterns are analysed, showing differences between specimens with or without stirrups and when stirrups are corroding or not. Finally, the effect of corrosion was simulated as the expansion of corrosion products in a finite element model and the results, mainly the crack pattern, were compared with the test results. The conclusions address the importance of taking into consideration both high corrosion levels and the corrosion of the stirrups for the assessment of deteriorated structures.

Introduction

Corrosion of steel reinforcement is one of the most common causes of deterioration of reinforced concrete. Some existing concrete structures like parking garages, harbours and bridges show significant corrosion; it is not uncommon that cover spalling has occurred.

C. Andrade and G. Mancini (eds.), *Modelling of Corroding Concrete Structures*, RILEM Bookseries 5, DOI 10.1007/978-94-007-0677-4_13, © RILEM 2011

Previous research [1] has been mainly concerned with corrosion levels leading to cover cracking; while relatively little attention has been devoted to higher levels of corrosion causing cover delamination. This is a practically important problem, both for assessment of the residual load-carrying capacity of corrosion-damaged concrete structures, but also for lifetime design of new structures.

Moreover, the main focus of the available knowledge concerns with the corrosion of the main reinforcement; while the corrosion of the stirrups is often overlooked. Field investigations have showed that cover delamination is more probable in areas of tightly spaced stirrups. A rather common approach in modelling the effect of the corroded stirrups is to consider the loss of the cross-sectional area. However, if the effect of corroded stirrups on crack initiation, crack propagation and cover delamination is not considered, this will lead to overestimation of the load-carrying capacity of the corroded structure.

Accelerated corrosion tests are widely used in the laboratory to study the mechanical properties of deteriorated RC specimens. The chemo-physical effects are different from those of natural corrosion [2]. From the mechanical point of view, some concerns exist regarding spurious bond deterioration obtained with high current densities. This study assumes that a sufficiently slow accelerated corrosion test can produce cross-section loss, concrete cracking and bond deterioration approximating natural corrosion, as far as the mechanical effects on RC structures are concerned.

This first of a two part paper presents the accelerated corrosion tests (part 1) on beam-end specimens used for bond tests (part 2). The specimens and the corrosion setup are described. Specimens with or without stirrups were cast; for the former, a part of the specimens had the stirrups protected from the corrosion process, whereas the others had both main steel and stirrup corrosion. Results shown in this paper regard the concrete cracking, crack patterns and reinforcement corrosion.

Experimental Setup

The specimens have the shape of a beam-end after inclined shear cracking (Figure 1). The concrete specimens were reinforced with the main longitudinal reinforcement of 20 mm diameter and the transverse reinforcement of 8 mm. Since delamination of concrete cover due to corrosion was desired, a relatively small concrete cover, 1.5 times the main bar diameter, was used. The main bars are in contact with the concrete over a 210mm length. The concrete was an ordinary type with average cubic strength 37.5 MPa without chlorides and 34.3 MPa in the mix with 3% chlorides of the specimens with corrosion.

The influences of the location of the anchored bar, middle and corner placement; the presence or not of transverse reinforcement; the corrosion level of longitudinal reinforcement and corrosion of transverse reinforcement were studied.

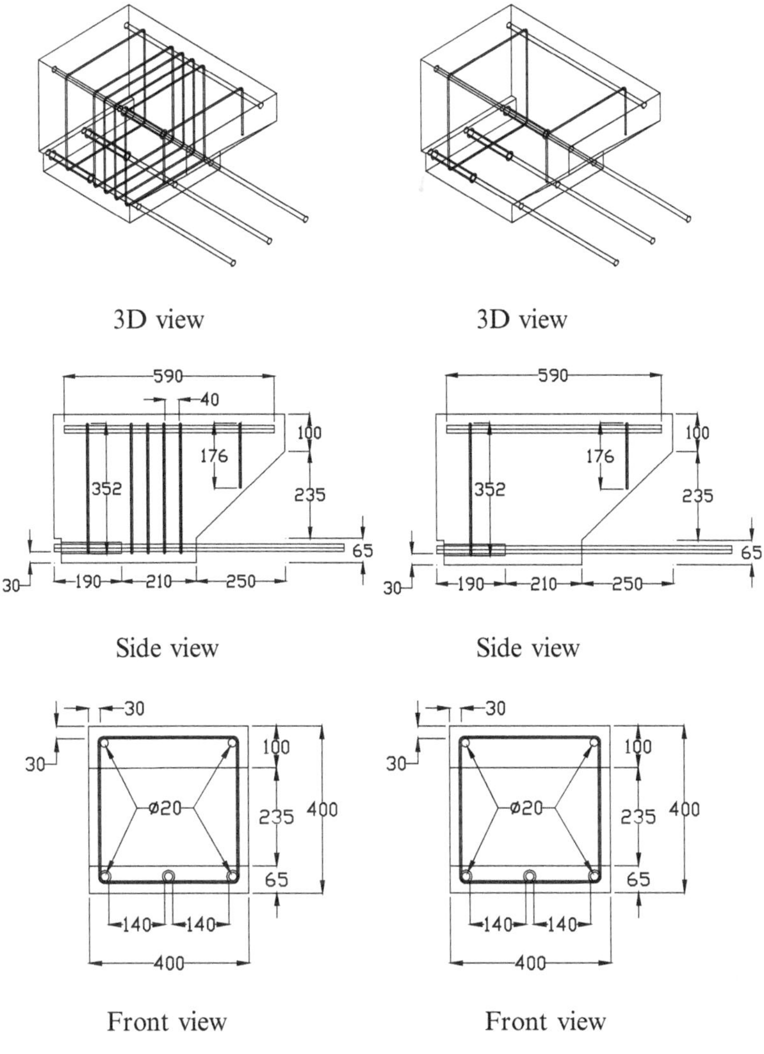

Figure 1. Specimen geometry and reinforcement with transverse reinforcement in the bonded zone of the main reinforcement (left) and without (right)

The specimens were corroded with an electrochemical method, using impressed current (Figure 2). The current flowed through the main bars across the concrete cover to a cathode placed at the bottom of the beam, inside a tank containing a solution of 3% chlorides. The stirrups in specimens Type A were isolated using electrical tape to avoid corroding. The current density was low (average value 1.43 A/m^2). Test results shown here were corroded for 7 months, obtaining approximately 1% weight loss for each month. Amongst artificial corrosion tests in the literature, this can be considered a low value. Other researches used faster rates, by even one order of magnitude, but spurious mechanical concrete-steel bond deterioration has been measured for high current density values [2].

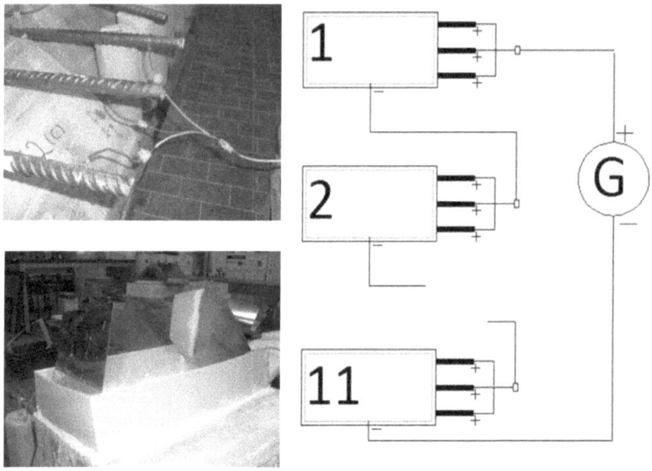

Figure 2. Accelerated corrosion setup

Three different corrosion levels were investigated, related to the crack propagation:

- Level 1 corresponded to cracks propagating along the main reinforcement; the corrosion level is around 1-2%;
- Level 2 corresponded to the reaching of approximately 10% corrosion;
- Level 3 aiming to reach the delamination of the covers (tests still on their way).

The specimens are of three different types, in relation to the reinforcement arrangement and corrosion:

- type A: main bars corroding while the stirrups are protected by insulating tape;
- type B: specimens without stirrups, only main bar corrosion;
- type C: main bars and stirrups corroding.

Results

Corrosion attack values were determined theoretically using Faraday's law and *a posteriori* by weight loss measurement. The differences between the two on average were approximately 10%.

Crack widths on the bottom (in the corrosion setup the top) and side cover were measured during the corrosion process using a microscope with a resolution of 0.04mm up to corrosion Level 1. Cracks that were not accessible by the former method, were measured by post-processing digital photographic images of the specimens before the load testing.

Corrosion level 1

Between corrosion initiation and level 1 all specimens showed longitudinal cracks along the main bars. The first corrosion cracks opened around 30-50 microns attack. The cover around each bar is cracked radially; either the side or bottom or both covers were cracked. The first cracks to open were on the bottom cover (top cover during corrosion); for these measurements at intervals were made (Figure 3).

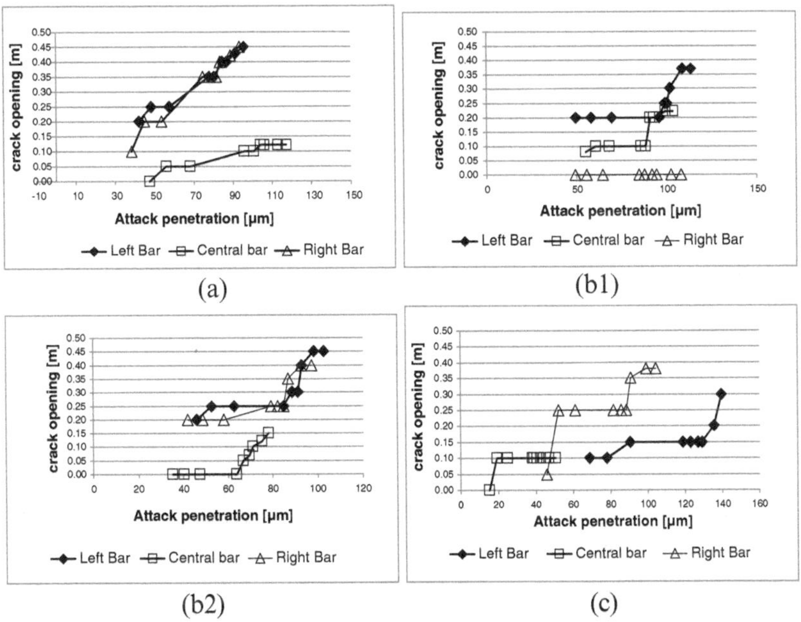

Figure 3. Bottom cover crack opening measurements: (a) with stirrups (not corroding) (b1-b2) without stirrups; (c) with stirrups (corroding)

Corrosion level 2

Moving from Level 1 to Level 2, in some specimens part of the radial cracks joined to form a delamination surface connecting two or three bars (Figure 3). The measurements of the bottom cover cracks were suspended because the accumulation of corrosion products made these observations not accurate.

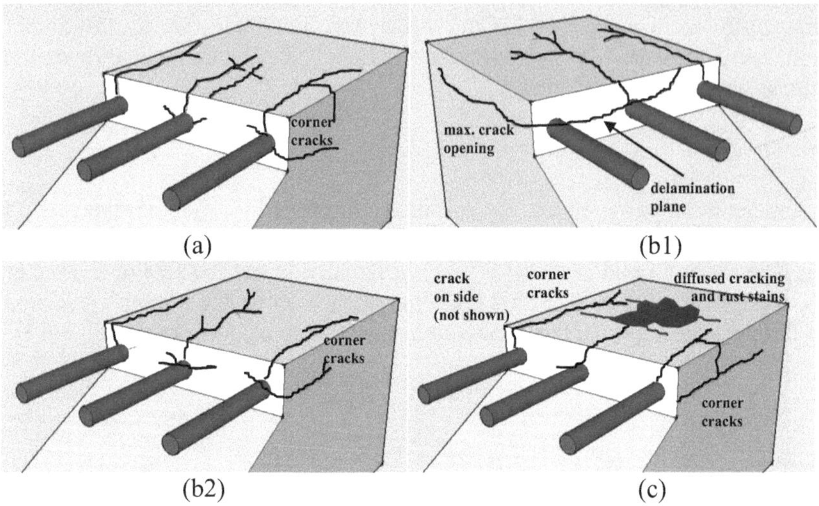

Figure 4. Crack patterns, Level 2: (a) with stirrups (not corroding); (b1-2) without stirrups; (c) with stirrups (corroding; diffused cracks on the top of the specimen not sketched, for clarity of the drawing)

Specimens with non corroded stirrups (Type A)
Longitudinal cracks showed in bottom and side covers (Figure 4a). The crack pattern on the front cover is:
- cracks radiating from the bar to the closest point of the outer surface;
- cracks lying in a horizontal plane;
- cracks in inclined planes forming a "V-shaped" pattern.

On the side covers, longitudinal and transverse cracks form.

Specimens without stirrups (Type B)
The cracks run mainly along the longitudinal reinforcement (Figure 4b). These specimens showed fewer and wider cracks than those with stirrups. In one of the specimens a delamination plane formed connecting two bars, with a maximum crack width of 1.4mm (Figure 4-b1).

Specimens with corroded stirrups (Type C)
Specimens C showed two different crack patterns (Figure 4c):

- bottom cover cracking, with corrosion products showing in big stains on the outer surface; many small cracks open, both longitudinally and in other directions (not shown in the drawings);
- initiation of delamination cracks, forming a plane across the bars, and front and side cover cracking.

On the whole the presence of stirrups causes a more complex crack pattern than for specimens without stirrups, both when these are non corroding or corroding. The corrosion level for the C specimens is nominal, because the amount of current involved in the main bars corrosion and that for the stirrups are still unknown, until gravimetric measurements are made on the bars after the tests.

Corrosion level 3

The ongoing corrosion process is moving towards this level, around 20% corrosion. The formation of delamination planes is initiated in most specimens. The opening of the longitudinal cracks on the bottom cover slows down. The pressure of the corrosion products is acting on a much wider plane, rather than inside the bar hole as it was initially.

Numerical Modelling

The beam-end specimens were modelled in detailed 3D finite element (FE) analysis. Due to symmetry, half of the specimen was model with 10 mm element size in the finite element program DIANA. Four-node, three-side isoparametric solid pyramid elements were used for concrete, transverse and longitudinal reinforcement. A constitutive model based on non-linear fracture mechanics using a smeared rotating crack model based on total strain is applied for concrete; see [1]. The crack band width was assumed to be equal to the element size; this was later verified in the analyses. For the concrete in tension and compression the models by [2] and [3] were adopted, respectively. The reinforcing steel was modelled based on an isotropic plastic model with Von Mises criterion. The material propertied used in the analyses are given in Table I.

Table I. Material properties used in the analyses

Mix	Concrete				Reinforcement		
	f_{cc}* [MPa]	f_{ct} [MPa]	G_F [N/m]	E_c [GPa]	f_y [MPa]	E_c [GPa]	f_u [MPa]
Reference specimens	29.7	2.33	64.3	29.42	510	200	610
Corroded specimens	27.7	2.19	61.2	28.74	510	200	610

* The values given are cylinder strength calculated from cubic strength.

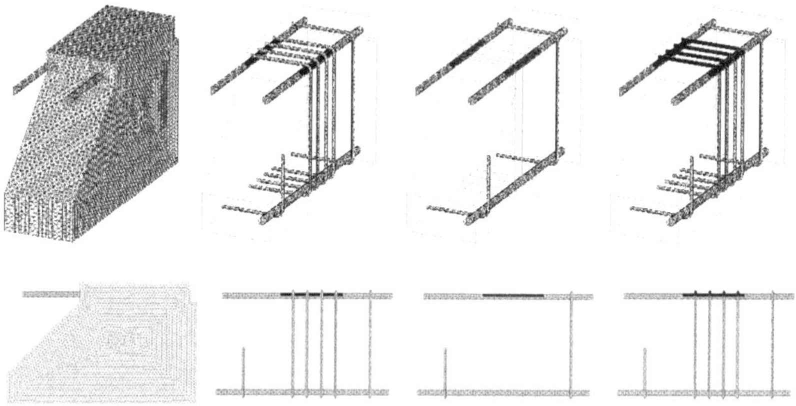

Figure 5. FE model of the beam-end specimen; the shaded bars were subjected to corrosion

The bond and corrosion models used in the analyses have been earlier developed by Lundgren [4, 5]. The modelling method specially suits for detailed 3D FE analyses, where both concrete and reinforcement are modelled with solid elements. The bond model is a frictional model, using elasto-plastic theory to describe the relations between the stresses and the deformations. The corrosion model takes into account the effect of corrosion as the expansion of the corrosion product. The bond and corrosion models were implemented into interface elements which were used to model the interaction between the concrete and reinforcement.

The ratios of volumetric expansion of different typical oxides with respect to the virgin material, given in the literature [6], varies between 1.7 for FeO and 6.15 for $Fe(OH)_3 3H_2O$. While the value of 2.0, suggested by Molina *et al.* [7], is frequently used in numerical analysis of corroded concrete [8, 9, 10, 11], Bhargava *et al.* [12] proposed a value of 3.39357 based on the available published experimental data. The composition of the rust produced in the tests presented here is not identified yet. Since the corrosion model used in these analyses has been calibrated with a value of 2 for volumetric expansion of rust; therefore, the same value was chosen for all analyses presented here.

Similar to the experiments, the longitudinal bars were subjected to corrosion attack from bottom cover, i.e. half of the cross-section is affected by corrosion; see Figure 5. The top leg of the stirrups was subjected to corrosion all around the cross-section. The vertical leg of the stirrups was corroded up to the half of the longitudinal bar section. Unlike the experiment, the same corrosion penetration was imposed on all bars; this level of corrosion corresponds to the corrosion penetration of the tested bar in the pull-out test.

An incremental static analysis was performed using a Newton-Raphson iterative scheme to solve the non-linear equilibrium equations. In a phased analysis, first the corrosion attack was applied. Then the external load was gradually applied on the tested bar as prescribed displacement. The outcome of the latter phase is presented in the companion paper subtitled "Paper 2. Anchorage Capacity".

Results

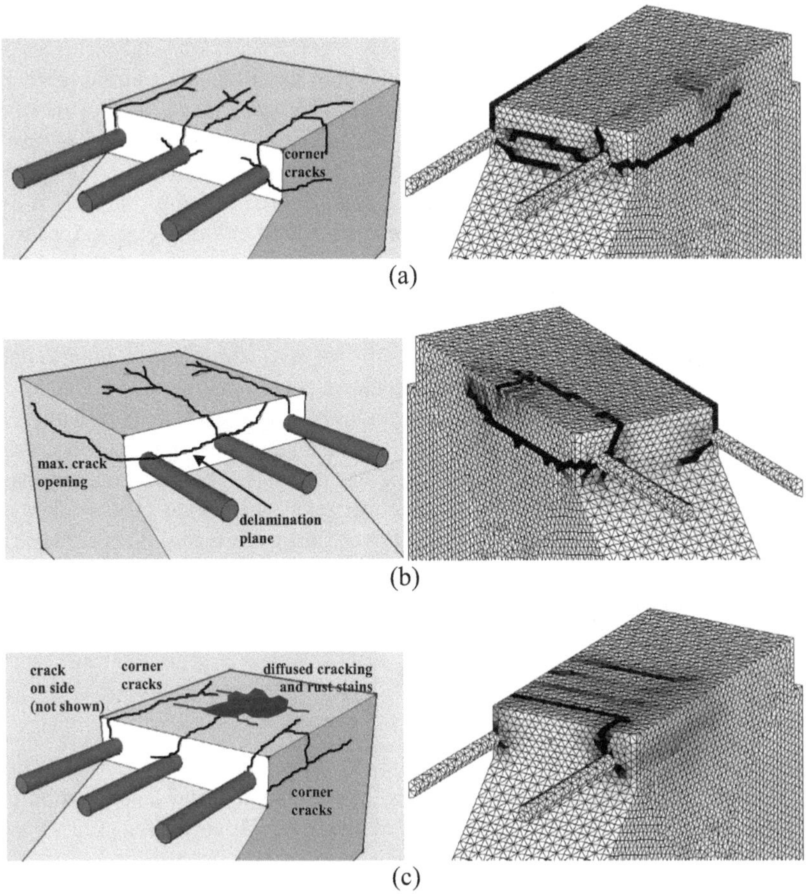

Figure 6. Comparison of experimental and numerical (half of the specimen) crack patterns: (a) Type A; (b) Type B; (c) Type C.

The longitudinal cracking and the delamination planes forming are simulated in Figures 6a,b with and without stirrups. The corrosion of stirrups (Figure 6c) fractures the cover transversely, with a damage pattern frequently observed in real structures. In the tests this part of the cover is densely cracked (not shown in left figure).

Three main cover spalling patterns can be observed in the numerical analysis:

- The corrosion-induced cracks, initiated from corner and middle bars, propagate across the cross-section and form a delamination plane. This type of pattern is dominant in the specimens of Type A (non-corroding stirrups).
- In the absence of stirrups, the corrosion-induced cracks around the corner bars propagate in a direction with the least resistance. Therefore, a corner cover spalling takes place; this is clearly seen in the specimens of Type B (without stirrups).
- A different pattern is observed when both longitudinal and transverse bars are corroding, specimens of Type C. This rises to a situation in which wide cracks appear transversely due to corroding stirrups before any of the former spalling patterns occurs. Here a more local cover spalling pattern, mainly spalling of the concrete between stirrups, shapes. As the concrete cover is smaller over the stirrups, the splitting cracks appear for relatively low corrosion penetration.

Generally, the crack patterns achieved in the numerical analyses agree well with the observation from experiments. In the present paper, quantitative comparison of the cracks, i.e. crack width, is not intended. But it is important to note that the amount of corrosion penetration needed to cause cover spalling was rather low in the numerical analyses compared with what was observed in the experiments. This is believed to be due to rust flowing into the cracks and slowing down the rate of the splitting pressure built up around the corroding bar. This phenomenon has a more significant effect for larger corrosion attacks when wide cracks allow for more rust to escape. The present version of the corrosion model does not include the flow of rust into the cracks but this will be one of the main concerns in further development of the model.

Conclusions

An experimental program on bond of corroded bars in reinforced concrete has been started, investigating high corrosion levels, cover cracking and delamination and the corrosion of the stirrups. The results of the artificial corrosion process have been presented in this first of two parts of the paper.

A numerical 3D finite element model has been setup to simulate the tests. The results correspond to the experimental observations, opening the way to the modelling of the pull-out bond tests performed on the same specimens (Part 2 of this paper).

The aim is to assess both experimentally and numerically the bond of the reinforcement in heavily corroded concrete specimens. In particular the study aims at understanding whether neglecting the effect of corroded stirrups on crack initiation, crack propagation and cover delamination may lead to overestimation of the load-carrying capacity of the corroded structure.

References

[1] Diana, DIANA Finite Element Analysis, User's Manual, release 9.1, TNO Building and Construction Research, Delft, Netherlands, 2006.

[2] D. A. Hordijk, Local Approach to Fatigue of Concrete, Doctoral thesis, Delft University of Technology, Delft, Netherlands, 1991.

[3] E. Thorenfeldt, A. Tomaszewicz, J. J. Jensen, Mechanical properties of high-strength concrete and applications in design, Conference on Utilization of High-Strength Concrete, Stavanger, Norway, 1987.

[4] K. Lundgren, Bond between ribbed bars and concrete. Part 1: Modified model, Magazine of Concrete Research, 57 (7) (2005) 371-382.

[5] K. Lundgren, Bond between ribbed bars and concrete. Part 2: The effect of corrosion, Magazine of Concrete Research, 57 (7) (2005) 383-395.

[6] Y. Liu, R. E. Weyers, Modeling the time-to-corrosion cracking in chloride contaminated reinforced concrete structures, ACI Materials Journal, 95 (6) (1998) 675-681.

[7] F. J. Molina, C. Alonso, C. Andrade, Cover Cracking as a Function of Rebar Corrosion. 2. Numerical-Model, Materials and Structures, 26 (163) (1993) 532-548.

[8] D. Coronelli, P. Gambarova, Structural assessment of corroded reinforced concrete beams: Modeling guidelines, Journal of Structural Engineering, 130 (8) (2004) 1214.

[9] K. Lundgren, Modelling the splitting effects of corrosion in reinforced concrete, Computational Modelling of Concrete Structures, Euro-C Conference, St. Johann, Austria, 2003, Balkema, 491-500.

[10] K. Zandi Hanjari, Load-Carrying Capacity of Damaged Concrete Structures, Lic. Thesis, Department of Civil and Environmental Engineering, Chalmers University of Technology, Gothenburg, 2008.

[11] K. Lundgren, A. S. S. Roman, H. Schlune, K. Z. Hanjari, P. Kettil, Effects on bond of reinforcement corrosion, International RILEM Workshop on Integral Service Life Modeling of Concrete Structures, 5-6 November 2007, Guimaraes, Portugal, 2007, RILEM Publications S.A.R.L, 231-238.

[12] K. Bhargava, A. K. Ghosh, Y. Mori, S. Ramanujam, Model for cover cracking due to rebar corrosion in RC structures, Engineering Structures, 28 (8) (2006) 1093-1109.

Severely Corroded Reinforced Concrete with Cover Cracking: Part 2. Anchorage Capacity

Kamyab Zandi Hanjari[1], Dario Coronelli[2] and Karin Lundgren[1]

[1] Department of Civil and Environmental Engineering, Chalmers University of Technology, Sweden
[2] Dipartimento di Ingegneria Strutturale, Politecnico di Milano, Italy

Abstract. There is a growing need for reliable methods of assessing the load-carrying capacity and remaining service life of corroded structures. In an ongoing research by the authors, issues that have not been investigated in the methods and models available today to calculate the remaining load-carrying capacity of the corroded structures are identified. Two main issues; i.e. high amount of corrosion leading to cover spalling and the effect of corroding stirrups, were investigated in an experimental program. Pull-out tests were carried out on beam-end specimens with long embedment length to study the anchorage capacity of a corroded bar. The specimens were subjected to electrochemical corrosion process leading to different corrosion penetrations prior to mechanical loading. Details concerning electrochemical corrosion setup, corrosion-induced cracking and numerical modelling of a corroding bar are presented in a companion paper subtitled "Part 1. Crack initiation, crack propagation and cover delimitation". Three types of specimens, with stirrups, without stirrups and with corroding stirrups, were subjected to pull-out test. The test results showed a significant influence of stirrups not only on corrosion-induced cracking but also on anchorage capacity and failure mode in the pull-out test. Finally, the corrosion and mechanical testing phases were simulated in a finite element model using the corrosion and bond models earlier developed by Lundgren [1,2]. The outcomes of the numerical modelling help to further understand the effect of high corrosion penetrations and presence of stirrups on failure modes observed in the experiments.

Introduction

When studying the anchorage of a ribbed bar in structural concrete members, the anchorage capacity is strongly influenced by the actual confinement conditions. In

C. Andrade and G. Mancini (eds.), *Modelling of Corroding Concrete Structures*,
RILEM Bookseries 5, DOI 10.1007/978-94-007-0677-4_14, © RILEM 2011

general, confinement is a result of the surrounding concrete, stirrups and transverse pressure [3]. The corrosion of reinforcement leads to volume expansion of the steel, which generates splitting stresses in the concrete influencing the bond between concrete and reinforcement. For a larger corrosion penetration, the splitting stresses may lead to cover cracking and, finally, spalling of the concrete cover. In the extreme case, when cover spalling occurs, the resisting mechanism in the cross section is altered and stirrups are the main factor providing confinement to the main reinforcement and resulting in residual anchorage capacity.

In earlier works within this field, several models of the corrosion and bond have been developed and implemented in the finite element. In the model by Lundgren [1, 2, 4], the splitting stresses are introduced and the bond stress depends not only on the slip but also on the normal stress between the reinforcement and the surrounding concrete. In addition the model is capable of taking into account the effects of the transverse pressure. However, the model is only calibrated for the effect of corrosion on the main reinforcement. To the authors' knowledge, there exists no model taking into account the effect of corroded stirrups on cover cracking and cover spalling and to study the anchorage capacity after cover delamination. This is the main aim of the research carried out.

The detailed 3D bond and corrosion model, developed by Lundgren, was used to analyze test specimens with severe corrosion. The type of the specimen was similar to the ones tested by Magnusson [5], in which the specimens had a shape of a beam-end after inclined shear cracking. The details concerning the specimens, electrochemical corrosion and test set-up are presented in a companion paper subtitled “Part 1. Crack Initiation, Crack Propagation and Cover Delamination”.

The bond model is a frictional model and the corrosion model takes into account the effect of corrosion of main reinforcement and stirrups as the expansion of the corrosion products. The concrete was modelled using 3D solid elements with a constitutive model based on non-linear fracture mechanics using a rotating crack model based on total strain. Three-dimension solid elements with a constitutive model based on Von Mises yield criterion with associated flow and isotropic hardening were used to model the main reinforcement and stirrups. The results computed by the model were compared with the experiments.

Experimental Setup

Pull-out tests were carried out on beam-end specimens with long embedment length to investigate the global bond behaviour of an anchored bar when the concrete cover has cracked and spalled off due to corrosion. In total, twenty two beam-end specimens were cast using a concrete grade of C30/37. The concrete was mixed in two batches, with and without 3% sodium chloride, and cast into beam-end moulds of which eleven specimens were made without sodium chloride and eleven were cast with 3% sodium chloride concrete. Six cylinders and five cubes were also prepared from each concrete batch for compression and splitting tension tests. The specimens were kept in a laboratory environment until 28 days after which they were demoulded and kept in a curing room at 20°C and 50% RH. More details are provided in the companion Part 1.

The specimens had a shape of a beam-end after inclined shear cracking, see Figure 1. For the beam-end specimens, since delamination of concrete cover due to corrosion was desired, a small concrete cover, 1.5Ø, was used. The influences of the location of the anchored bar, middle and corner placement; the amount of transverse reinforcement; the corrosion level of longitudinal reinforcement, cracked cover and delaminated cover; and corrosion of transverse reinforcement were studied, see Table I. The beam-end specimens were cast with the main longitudinal reinforcement of 20 mm in the horizontal position at the bottom of the moulds, and with the transverse reinforcement of 8 mm. More details of the specimen geometry and reinforcements are given in the companion Part 1.

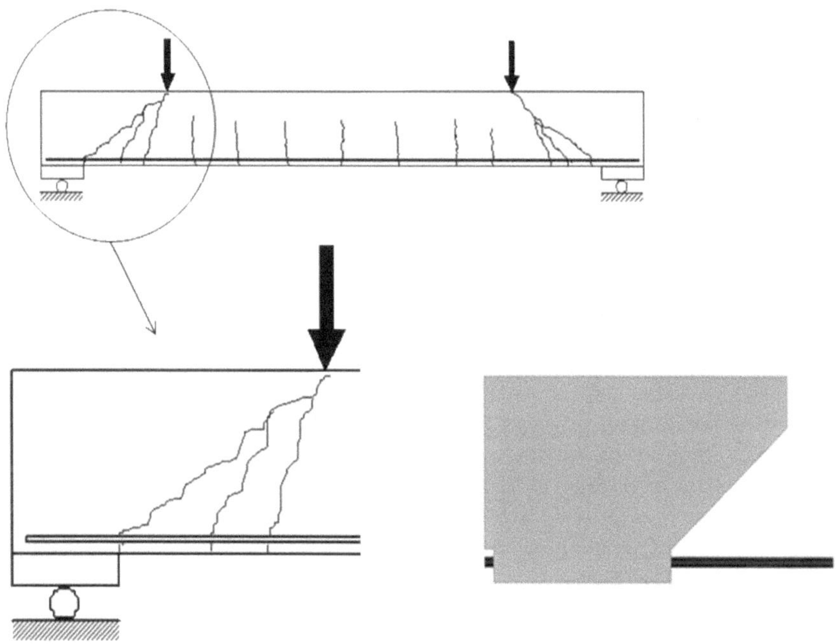

Figure 1. Schematic illustration of the beam-end specimen

The test program is summarised in Table I. Pull-out test was carried out on reference specimens and corroded specimens to different levels:

Table I. Test program

Specimens	Position of tested bar(s)	N° of specimens		
		With non-corroded stirrups⁽¹⁾ (Type A)	Without stirrups⁽¹⁾ (Type B)	With corroded stirrups⁽¹⁾ (Type C)
Reference; 0% corrosion	Middle bar	1,1,1	1,1,1*	×
	Corner bars	1,1,1*	1,1	×
Cracks along the main bar; 1-2% corrosion	Middle bar	1	×	×
	Corner bars	1	×	×
Propagation of cracks; Approx. 10% corrosion	Middle bar	×	1	×
	Corner bars	1	1	1
Corrosion leading to cover delamination	Middle bar	1*,1*	×	1*
	Corner bars	1*	×	1*

(1) Along the embedment length
* Not subjected to pull-out test yet

- Level 1 corresponded to cracks propagating along the main reinforcement; the corrosion level is around 1-2%;
- Level 2 corresponded to a corrosion level of approximately 10%;
- Level 3 aiming to reach the delamination of the covers (tests still on their way).

The specimens were tested in a test rig specifically designed for these tests. The test-setup is sketched in Figure 2. Deformation control was adopted to permit measurements in the post-peak behaviour. The loading was controlled in the displacement of the active end of the main bar. Initially, the deformation rate was about 0.10 mm/minute; after the maximum load capacity reached the deformation rate was increased in steps. In each test either the middle bar or the two corner bars of the specimen was subjected to displacement.
The tensile force in the bar was measured using one load cell for each main bar tested; i.e. two load cells in a corner bar test and one in a middle bar test. Instrumentation was provided to measure the relative displacement of the main bars at both the active and passive ends relative to the stable faces of the specimen; see Figure 2.

Numerical Modelling

The beam-end specimens were modelled in detailed 3D finite element (FE) analysis using FE program DIANA; see Figure 3. The analyses were carried out in two phases. In the first phase, the corrosion attack was applied in time steps as the expansion of the corrosion product. In the latter phase, the external load was gradually imposed on the tested bar as prescribed displacement. The principal aspects of the FE model, concrete model, interaction of the concrete and reinforcement (bond model), corrosion model and the material properties are provided in the companion paper subtitled "Part 1. Crack Initiation, Crack Propagation and Cover Delamination". Here, a short description of the second phase of the analysis simulating the pull-out testing condition is given below.

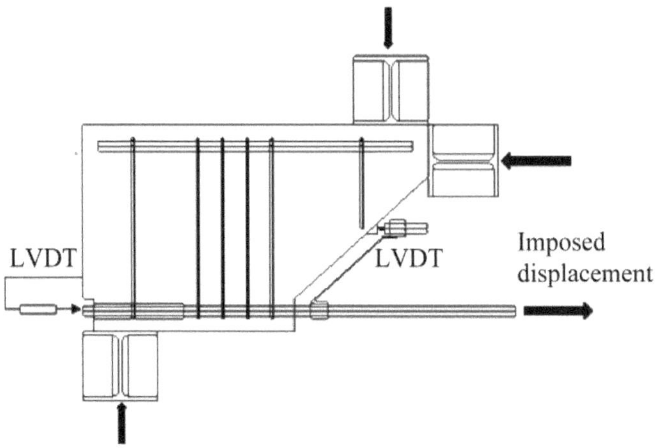

Figure 2. Test setup and instrumentation.

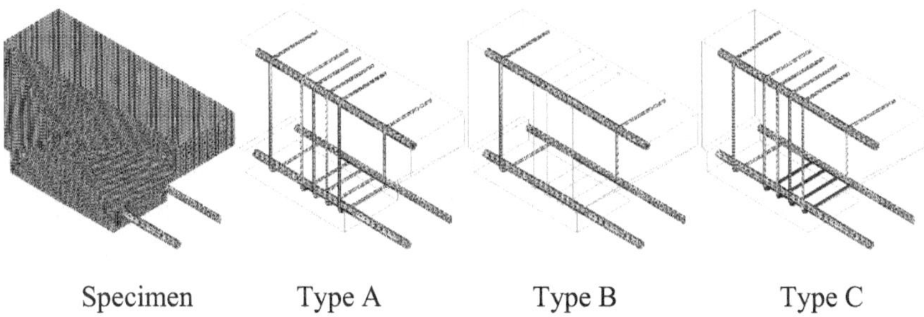

Figure 3. The FE model for the beam-end specimens.

Due to symmetry, half of the specimen was modelled. The boundary conditions at the supports and the symmetry plane are shown in Figure 4. All the nodes at the symmetry plane were supported for displacement in y-direction. Similar to the experimental setup, the load was imposed on either middle bar or the two corner bars as prescribed displacement.

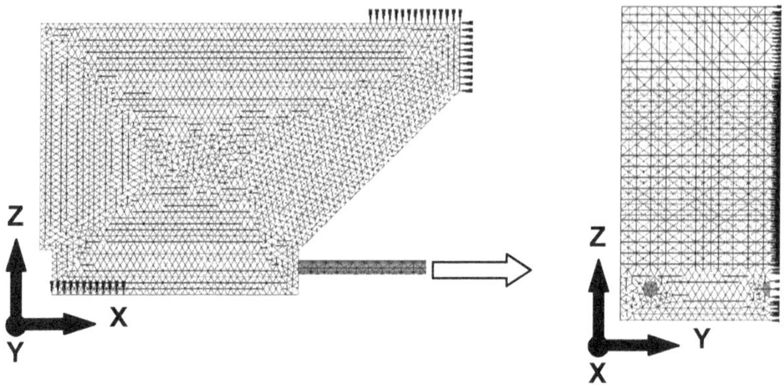

Figure 4. Boundary conditions at supports and symmetry plane.

Results

The test results together with the results from numerical modelling are presented and compared here. The results are limited only to specimen types A and B up to the second level of corrosion (about 10% cross-section loss); specimens with cover spalling and specimens of type C are still subjected to electrochemical corrosion. An overview of the test results in terms of anchorage capacity normalized with respect to reference specimens versus corrosion attack is presented in Figure 5. The anchorage capacity achieved in the pull-out test of reference specimens without stirrups is roughly 65% of the capacity observed in specimens with stirrups. For specimens with stirrups, small bond deterioration was observed in spite of rather large corrosion penetration, about 0.4 to 0.5 mm. However, such corrosion caused relatively large bond deterioration in specimens without stirrups, around 50% reduction in anchorage capacity. This highlights the importance of stirrups as the main source of confinement after corrosion-induced cover cracking. From the results available in this stage of the program, slightly smaller effect of stirrups is seen on anchorage capacity of the middle bar. This is due to the fact that stirrups provide more confinement, at the bend, to the corner bar compared to middle bar. Generally, the bond deterioration observed in the specimens is within the expected range when compared with the test results available in the literature.

In the test carried out on anchorage in corner regions, the two bars were loaded simultaneously. However, as displacement was controlled using two LVDTs, one on each bar, it was possible to register the individual behaviour of the bars. A comparison of the anchorage capacity and the slip at the maximum anchored force is given for test result and numerical modelling in Table II. The tests showed a significant importance of the bar position; i.e. higher anchorage capacity for middle bars. Depending on the level of corrosion and presence of stirrups, different types of crack patterns at failure were observed (see Figure 6):

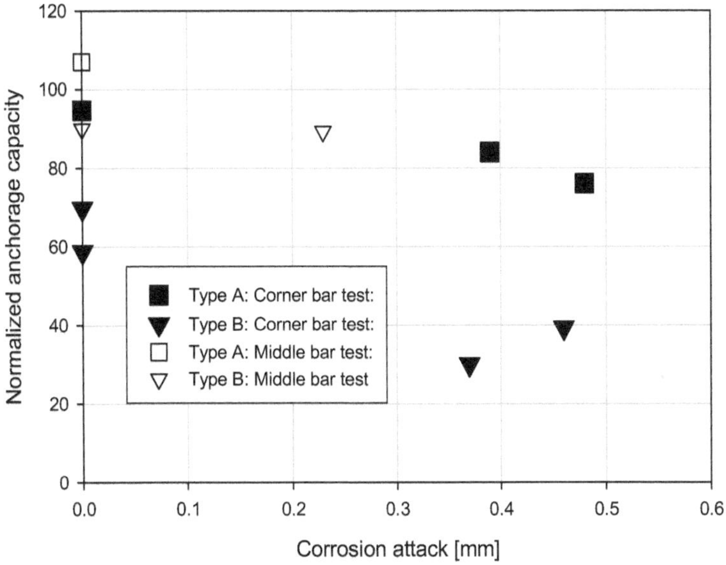

Figure 5. Overview of the test results in terms of anchorage capacity normalized with respect to reference specimens versus corrosion attack

(a) (b) (c)

Figure 6. Different crack patters at failure: (a) shear, (b) bond shear, and (c) bond

(a) Shear: inclined cracking starting from support, developing particularly in specimens with stirrups with low corrosion attack.
(b) Bond shear: cracks running along the bar and turning parallel to inclined side of the "nose", mostly developing in non-corroded specimens.
(c) Bond: splitting cracks parallel to the bar, mainly in corroded specimens.

It should be noted that neither in the tests nor in the numerical modelling the longitudinal reinforcement yielded. As far as numerical results are concerned, no yielding of the stirrups took place; 350 MPa was the maximum stress in stirrup observed in the analyses.

Table II. Experimental and numerical results

Specimen	Tested bar	Corrosion condition	Experiment		FE modelling		Crack pattern at failure*
			τ_{max} [MPa]	Slip at τ_{max} [mm]	τ_{max} [MPa]	Slip at τ_{max} [mm]	
Type A	Corner	Reference	5.05	0.36	4.80	0.12	BS
		Corroded	5.52	0.13	4.23	0.06	BS or S
	Middle	Reference	8.11	0.17	7.60	0.32	B
		Corroded	6.44	0.09	6.93	0.26	B
Type B	Corner	Reference	4.34	0.08	4.42	0.12	BS
		Corroded	2.64	0.02	3.18	0.03	B
	Middle	Reference	7.05	0.13	7.47	0.59	BS
		Corroded	6.08	0.06	6.22	0.05	B

* S: shear, BS: bond shear, and B: bond

Specimens with non corroded stirrups (Type A)

The crack propagation in non-corroded corner bar test was different from that observed in non-corroded middle bar test. In the corner bar test, the cracking started with the development of a transverse crack at the end of the bonded zone. At a higher load, this crack further propagated and formed a shear crack; this corresponds to the first peak in the bond-slip diagram shown in Figure 7(a). Thereafter, several transverse cracks initiated and inclined toward the loaded end forming a bond shear crack pattern; corresponding to the second peak. Meanwhile, longitudinal cracks initiated from the loaded end stopped when reached the first stirrup. The two peaks in the load were also observed in the numerical modelling. The corroded specimen tested for corner bar showed higher capacity, see Figure 7(b), and different crack pattern at failure in comparison with the reference specimen. This might be due to relatively low amount of corrosion leading to slightly higher bond resistance. In the middle bar test of reference and corroded specimens, the failure occurred when a dominant longitudinal crack and two inclined cracks running to the corner bars formed.

Specimens without stirrups (Type B)

The crack propagation in non-corroded corner bar test was similar to that observed in non-corroded middle bar test; however it was different from what was seen in specimens with stirrups (Type A). In these specimens, the cracking started with the development of dominant longitudinal crack appeared on the bottom cover all

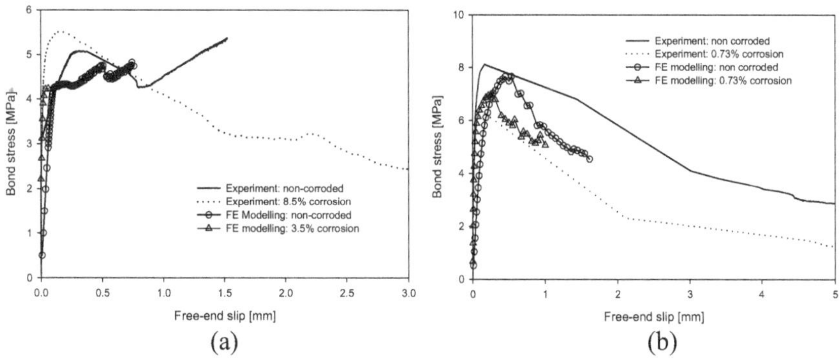

Figure 7. Bond stress versus free-end slip for (a) corner bar and (b) middle bar test

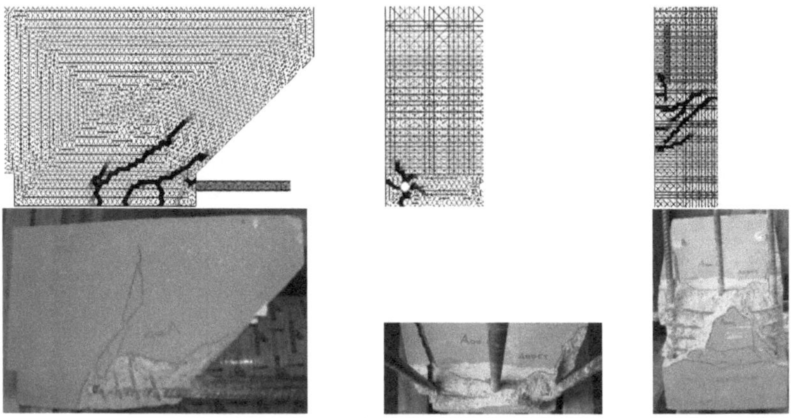

Figure 8. Crack pattern: reference specimen Type A, corner bar test

along the bonded length. This was followed by extensive inclined cracking forming a bond shear crack pattern. The corrosion in specimens without stirrups, prior to mechanical loading, led to a wide longitudinal crack along the bonded length. This crack appeared on both bottom and side covers around a corroded bar in corner region. This resembles a corner cover spalling situation, although the corner cover was not completely fallen off as the amount of corrosion was relatively low. These corrosion-induced cracks were widened when pull-out forced was imposed on the bar. The final failure took place with a typical bond (splitting) crack pattern. The bottom cover totally fallen off in a corner bar test; see Figure 10. This agrees very well with what was seen in the numerical analysis. The deterioration of the capacity is also well estimated by numerical analysis.

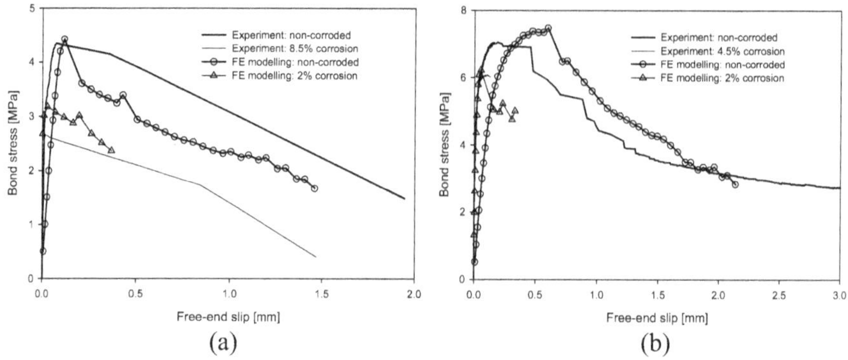

Figure 9. Bond stress versus free-end slip for (a) corner bar and (b) middle bar test

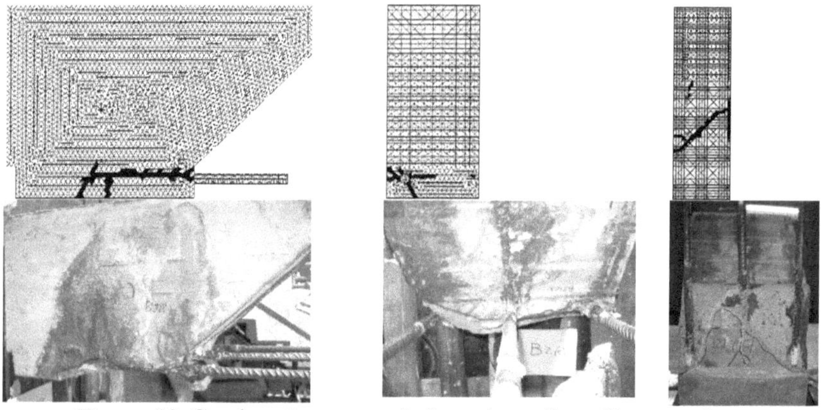

Figure 10. Crack pattern: corroded specimen Type B, corner bars test

Conclusion

In an ongoing research, the effect of high amount of corrosion and presence of transverse reinforcement on anchorage capacity of an anchored bar in a corner and middle position has been studies experimentally and numerically. Beam-end specimens were subjected to electrochemical corrosion process prior to anchorage test. While the results of the corrosion phase are presented in the companion Paper 1, the anchorage test results together with the numerical simulations are discussed in this paper.

The test results showed a significant effect of the stirrups, position of the tested bar and the effect of corrosion on the anchorage capacity and the failure mode. The result extracted from a 3D model of the specimens and the earlier developed bond and corrosion models, by Lundgren, show a good fracture correspondence both in corrosion phase and in pull-out phase. The outcomes of the numerical modelling are useful to further understand the effect of high corrosion penetration and presence of stirrups on failure modes observed in the experiments.

In the continuation of the presented research work, the effect of corroding stirrups on corrosion-induced cracking and anchorage capacity will be studied. Also higher corrosion levels close to 20% of the main bars will be analysed. This will be, later on, used for further verification and calibration of the 1D bond model earlier developed by the authors [6].

References

[1] K. Lundgren, Bond between ribbed bars and concrete. Part 1: Modified model, Magazine of Concrete Research, 57 (7) 371-382, 2005.

[2] K. Lundgren, Bond between ribbed bars and concrete. Part 2: The effect of corrosion, Magazine of Concrete Research, 57 (7) 383-395, 2005.

[3] Fib. Bond of reinforcement in concrete, State-of-art report, Fédération internationale du béton, prepared by Task Group Bond Models, Lausanne fib bulletin 10, 2000.

[4] K. Lundgren, Three-dimensional modelling of bond in reinforced concrete: theoretical model, experiments and applications, Ph.D. Thesis, Department of Structural Engineering / Concrete Structures, Chalmers University of Technology, Göteborg, Sweden, 1999.

[5] J. Magnusson, Bond and Anchorage of Ribbed Bars in High-Strength Concrete, Ph.D. Thesis, Division of Concrete Structures, Chalmers University of Technology, Göteborg, 2000.

[6] K. Lundgren, A. S. S. Roman, H. Schlune, K. Z. Hanjari, P. Kettil, Effects on bond of reinforcement corrosion, International RILEM workshop on Integral Service Life Modeling of Concrete Structures, 5-6 November 2007, Guimaraes, Portugal, 2007, RILEM Publications S.A.R.L, 231-238.

Modelling the Stiffness Reduction of Corroded Reinforced Concrete Beams after Cracking

Arnaud Castel[1], Dario Coronelli[2], Raoul François[1] and David Cleland[3]

[1] UPS, INSA, LMDC (Laboratoire Matériaux et Durabilité des Constructions), Université de Toulouse, 135, Avenue de Rangueil, F-31077 Toulouse cedex 04, France
[2] Dipartimento di Ingegneria Strutturale, Politecnico di Milano, Piazza Leonardo da Vinci 32, 20133 Italy
[3] School of Planning, Architecture and Civil Engineering, David Keir Building, Queen's University Belfast, Belfast BT9 5AG, United Kingdom

Abstract. The serviceability of reinforced concrete structures (i.e., bending stiffness) is mostly affected by the corrosion of the tension reinforcement (steel cross-section and the steel-concrete bond reduction). This paper presents two models aiming to predict the stiffness degradation of corroded reinforced concrete beams particularly after cracking. The two modelling approaches are: beam Macro-Finite-Elements (MFE) using cross-section analysis developed at the LMDC Toulouse and 2D nonlinear finite element analysis developed at Politecnico di Milano. Both, the steel cross-section and bond reduction versus corrosion rate are taken into account. The models are validated by simulating tests on a 23 year old corroded beam from long term experiments still in progress at LMDC. Results show a good agreement between both simulations and the experimental responses and highlight the predominant influence of the steel-concrete bond loss.

Introduction

The main cause of ageing damage in reinforced concrete structures is reinforcement corrosion. Corrosion damage can be detected visually as coincident cracks along the reinforcement, and leads to a significant reduction of the bar cross-section and loss of bond between reinforcement and concrete [1, 2].
In normal conditions, reinforced concrete elements subjected to bending are always supposed cracked in their tensile zones because the tensile strength of concrete is low. Consequently, the usual mechanical models for reinforced concrete design do not take into account the tensile concrete, since it does not significantly influence the load-bearing capacity of a structure [3]. However the tensile concrete located

C. Andrade and G. Mancini (eds.), *Modelling of Corroding Concrete Structures*,
RILEM Bookseries 5, DOI 10.1007/978-94-007-0677-4_15, © RILEM 2011

between two flexural cracks contributes to the flexural stiffness of the structural member. Indeed, the bond between the reinforcing bars and the concrete is still active in these areas and leads to a mechanical interaction between the steel bars and the concrete. This is called the "tension-stiffening" effect, and is well known to structural engineers. The purpose of the paper is to quantify this "tension stiffening" effect and then the effect of the steel-concrete bond loss due to corrosion on the global stiffness. Indeed, the bond reduction leads to the loss of tension-stiffening effect.

A first model, labelled Macro-Finite-Element MFE, is presented, based on a reduced-inertia approach, which takes tension stiffening and corrosion effects on bond into account. This first modelling approach is developed at the LMDC Toulouse, France. The second model is a 2D nonlinear finite element analysis developed at Politecnico di Milano, Italy. The models are validated by simulating tests on a 23 year old corroded beam from long term experiments still in progress at LMDC Toulouse.

Experimental Program

This paper is based on a large experimental program initiated in 1984 at the LMDC in Toulouse (France), dedicated to the study of reinforcement corrosion [4]. 36 RC-beams were cast and stored in a chloride environment under sustained loading. At different periods, tests were carried out to collect experimental data such as cracking maps, chloride content, and mechanical behaviour under service loading. The main interests of the program are: the beam dimensions which are representative of real structures, the sustained loading state applied (beams are cracked), the natural corrosion process (no accelerated corrosion using electrical field), and the availability of numerous control members.
This paper focuses on one 23 years old corroded beam denoted B2CL1 and three 23 years old control beams denoted BT, BT2 and BT3. All beams were loaded in 3-point flexure at the design service load (M_{ser} = 20 kN·m). The flexural load was maintained using a special loading device [4] throughout the experiment. The service load is the maximum load applied to both beams during all the experiments. The reinforcement lay-out for all beams is shown in Figure 1.

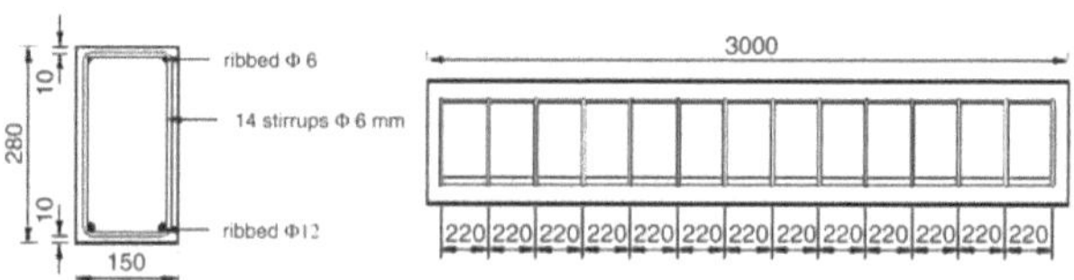

Figure 1. Layout of the reinforcement (all dimensions in mm) for type B beams

The reinforcement was provided by ribbed bars with a 500 MPa yield strength. The mechanical characteristics of the aged concrete were: compressive strength = 63

MPa, tensile strength (through splitting tests) = 6.8 MPa, and elastic modulus = 35 GPa. Water porosity was 15.2%.

Macro Finite-Element Model

The model adopted after cracking is presented in Figure 2 in the case of four-point-flexure. The MFEs are beam elements, characterized by their homogenised moment of inertia I_a. The MFE length L_{elem} is equal to the distance between two consecutive cracks due to flexure.

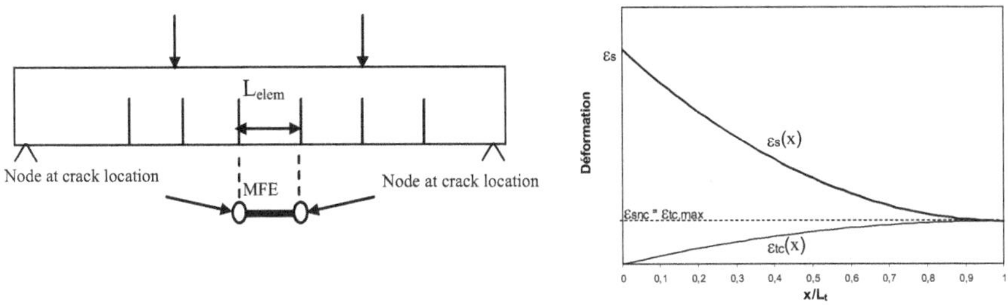

Figure 2. Macro Finite Element adopted for cracked beams: typical non linear distribution of the strain of the reinforcing bars and the concrete between two consecutive cracks

Steel strain distribution between two cracks

To take into consideration the bond action between the cracks for the ordinary bars, the steel strain is maximal at the cracks locations and equal to ε_s, and then decreases along the MFE, as shown in Figure 2, to reach a minimum value of ε_{snc}. The length necessary for the ordinary reinforcing bars to redevelop the full tension force in the concrete, lost by the occurrence of the crack, is the transfer length L_t [5]. The strain variation of the ordinary reinforcing steel is assumed non linear along the transfer length in accordance with experimental observations [6]. The function distribution of the steel strain $\varepsilon_s(x)$ is given by Eq. (1). Figure 2 also shows the non linear strain distribution in the concrete $\varepsilon_{tc}(x)$ where $\varepsilon_{tc,max}$ is the maximum tension concrete strain. Another assumption is that the variation of the depth of the neutral axis is linear over the transfer length between the value in the cracked cross-section y_{0c} and that calculated before cracking y_{0nc}. The function distribution of the neutral axis $y_0(x)$ is given by Eq. (2).

$$\varepsilon_s(x) = \varepsilon_s - (\varepsilon_s - \varepsilon_{snc}) [2(x/L_t) - (x/L_t)^2] \quad (1)$$

$$y_0(x) = (y_{0nc} - y_{0c})(x/L_t) + y_{0c} \quad (2)$$

Calculation of the steel strain σ_{snc}

Two beam cross-sections are considered (Figure 3): a first cross-section is cracked then the steel works under stress σ_s, and a second section, located between two flexural cracks in the middle (uncracked section), where the steel works under stress σ_{snf}. It is assumed that both cross-sections are loaded with the same bending moment M_{ser}. For this calculation, the distance between two consecutive cracks is considered equal to $2L_t$, then, at half distance between the cracks, the steel strain is equal to ε_{snc}.

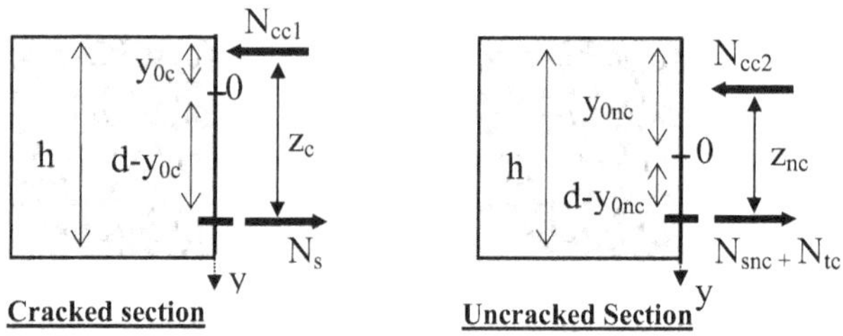

Figure 3. Internal loads in cracked and non cracked cross-sections

In Figure 3, $N_s = \sigma_s A_s$ is the tensile load in the steel in the cracked section, $N_{snc} = \sigma_{snc} A_s$ and N_{tc} are the tensile loads in the steel and in the concrete respectively in the uncracked section. The leverage of the internal forces in cracked and non cracked sections are respectively noted z_c and z_{nc} which are calculated using classical reinforced concrete models (neglecting tensile concrete for z_c and before cracking for z_{nc}). d and h are the effective and the total depth of the cross-section respectively. The resultant of the tensile stresses in concrete N_{tc} is fixed at the level of the steels in accordance with CEB-FIP model code [5]. N_{tc} is equal to $\sigma_{tcmax}.A_{tc,ef}$ where $A_{ct.ef}$ is the effective tensile active concrete section. The effective tensile active concrete section is given by Eq. (3). This equation has been developed thanks to an accurate concrete stress distribution modeling between bending cracks. All details are available in [7].

$$A_{ct,ef} = b\left(\frac{(d - y_{0nc} - a/2)}{2} + a + (h - d - a/2)\left[1 - \frac{(h - d - a/2)}{2(d - y_{0nc} - a/2)}\right]\right) - A_s \quad (3)$$

where "a" represents the height of tensile steels (in case of the presence of several re-bar layers), A_s is the steel cross-section and b is the cross-section breadth. As both sections are assumed loaded with the same bending moment and as the internal load resultants are equal to zero, Eq. (4) is deduced.

$$z_{nc}/z_c(A_s\,\sigma_{snc} + A_{tc.ef}\,\sigma_{tc,max}) = A_s\sigma_s \quad (4)$$

Bond condition between steel and concrete is given by Eq. (5) where n is the instantaneous coefficient of equivalence matches the ratio between the elastic modulus of the steel and the instantaneous elastic modulus of the concrete. To take into account a bond strength degradation due to corrosion, a scalar bond damage parameter D_c is introduced which varies from 0 (no corrosion) to 1 (total debonding due to corrosion as the concrete stress is equal to zero: total loss of tension stiffening effect).

$$(1-D_c)\,\sigma_{snc}=n\sigma_{tc,max} \tag{5}$$

From Eqs. (4) and (5), Eq. (6) is deduced allowing to calculate the steel stress σ_{snc}.

$$\sigma_{snc}=\frac{\sigma_s}{\frac{Z_{nc}}{Z_c}\left[1+\frac{(1-D_c)A_{tc,ef}}{nA_s}\right]} \tag{6}$$

Calculation of the average inertia I_a of the MEF

Curvatures in cracked and uncracked cross-sections and the average curvature are respectively: $\chi_c=\varepsilon_s/(d-y_{0c})$, $\chi_{nc}=\varepsilon_{snc}/(d-y_{0nc})$ and $\chi_a=\varepsilon_{sa}/(d-y_{0a})$. The average steel strain ε_{sa} and the average neutral axis y_{0a} are calculated along the half cracks spacing $L=L_{elem}$ using Eqs. (7) and (8) respectively.

$$(L/2)\varepsilon_{sa}=\int_0^{L/2}\varepsilon_s(x)dx \tag{7}$$

$$(L/2)y_{0a}=\int_0^{L/2}y_0(x)dx=((y_{0c}-y_{0nc})(L/2))/(2L_t)+y_{0c} \tag{8}$$

As $M_{ser}=E_cI_a\chi_a=E_cI_c\chi_c=E_cI_{nc}\chi_{nc}$, where I_c and I_{nc} are the inertia of the cracked and uncracked cross-sections, the average inertia of the MEF is given by Eq. (9).

$$I_a=\frac{(d-y_{0a})I_cI_{nc}}{(d-y_{0c})(1-C_H)I_{nc}+(d-y_{0nc})C_HI_c} \tag{9}$$

where

$$I_{nc}=\frac{Z_{nc}}{Z_c}\left[1+\frac{(1-D_c)A_{tc,ef}}{nA_s}\right]\frac{(d-y_{0nc})}{(d-y_{0c})}I_c \tag{10}$$

C_H is the homogenization coefficient. As recommended by the CEP-FIP model code, when the stabilized cracking situation is reached, the ratio $L_{elem}/2$ to L_t can be fixed to 0.75. If that condition is assumed, $C_H=0.5625$ and the calculation of the average inertia I_a (Eq. (9)) does not depend on any parameter regarding the cracks spacing.

Corrosion damage

Interesting Numerical investigations describing the bond behaviour between concrete and corroded reinforcement are already available in the literature [8]. But, in order to develop a model which can be useful to conventional engineers, an empirical approach is preferred and used in the analytical MEF model. The deterioration of the bond due to corrosion of the reinforcement is linked to the section loss of the bars. The appearance of cracks (parallels to steel bars) and their width are directly correlated to the amount of corrosion products present due to the oxidation process. The loss of bond is correlated with the cracking process because of the loss of confinement of the bars in the surrounding concrete. The bond damage variable is then expressed as a function of the steel cross-section loss ΔA_{sm} (Eq. (11)). ΔA_{sm} is the average steel cross-section loss of the reinforcing bars between two consecutive cracks then along a MFE length L_{elem} (about 200 mm). This average cross-section loss ΔA_{sm} is assumed more representative of the amount of corrosion products leading to the concrete cracking than the maximum steel cross-section loss which is very localised (pitting corrosion: about 1 mm length)).

$$D_c = 1 - \left(\frac{A_s - \Delta A_{sm}}{A_s - \Delta A_{s0}}\right)^n \qquad \text{for } \Delta A_{sm} > \Delta A_{s0}, \text{ elsewhere } Dc = 0 \qquad (11)$$

where ΔA_{s0} is the section loss threshold that initiates the first crack [9], ΔA_{sm} is the average section loss of the reinforcing bar along L_{elem} [10], n = 5 is an empirical parameter describing the quantitative variation of the progressive debonding versus the section loss of the reinforcement [6]. This damage variable is also included in the calculation of the neutral axis evolution in uncracked cross-section versus corrosion propagation: $y_{0nc}(D_c) = y_{0c} + (1-D_c)(y_{0nc} - y_{0c})$. Figure 4 shows a typical average inertia I_a reduction due to the bond loss resulting from corrosion propagation.

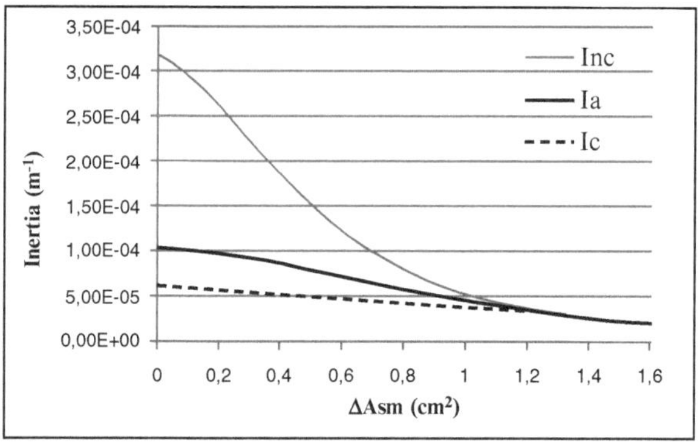

Figure 4: Average inertia reduction due to the bond loss resulting from corrosion propagation for beam B2CL1

Before first concrete cracking, the inertia is not affected. After concrete cracking, the inertia reduction is quite fast. For high corrosion values, the average inertia is equal to the inertia in cracked cross-section I_c as the tension stiffening effect is completely lost. The steel cross-section reduction is also taken into account in the calculation of both I_{nc} and I_c. Consequently, both the steel cross-section loss and the steel-concrete bond reduction are taken into account in the reduced average inertia calculation. Again, the average steel cross-section loss ΔA_{sm} is used and assumed more appropriate than the maximum local cross-section loss. Indeed, local maximum pitting cross-section loss is of course the main factor affecting the ultimate capacity. But local pits do not affect significantly the overall stiffness of beams under service loading [11].

Beams overall stiffness calculation

The calculation of the overall stiffness of reinforced concrete members is performed by assembling the MFE defined between each consecutives cracks (Figure 1). Each MFE is a finite element with 4 unknowns (two node displacements at each end, transversal displacement and rotation) and the mechanical formulation is "BEAM element". For each MFE, a specific value of steel mass loss due to corrosion is used to calculate the reduced average inertia according to the location of the MFE along the beam. After MFE assembling, the beam obtained is with a variable inertia. This method allows taking into account the non uniform distribution of corrosion along the reinforcing bars. For beam B2CL1, the corrosion distribution taken into account in the calculation of the average inertias is shown in Figure 5.

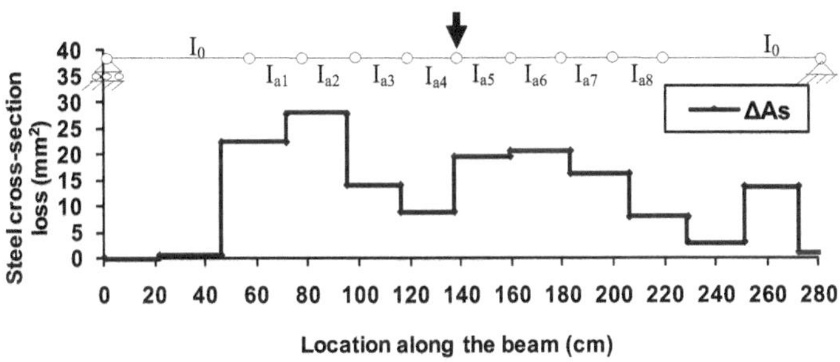

Figure 5: Corrosion and MFE distribution along beam B2CL1

Figure 5 also shows the MFE distribution adopted. The cracks spacing is about 200 mm and correspond to the MFE length L_{elem}. In uncracked region (near the supports), the inertia is equal to I_0 calculated before cracking. Corrosion effect on the overall stiffness is neglected in these areas. More details about beam B2CL1 corrosion and cracking are available in [10].

Finite Element Analysis

Finite element model

The model here described performs a non-linear finite element (FE) analysis, and has been formulated to model corroded members [12] and specifically developed to analyse shear critical elements [13]. The main issues in structural modelling of corroded members are the evolution of geometry and material properties and the changing in bond-slip properties. An analysis at cross-section level should be replaced by a whole member analysis in case of deteriorated or corroded structures, to study the effects of bond deterioration and increasing slip values at the steel-concrete interface.

The 2D plane stress analysis uses separate elements and non-linear models for steel, concrete and bond (Figure 6). The steel bars, wires and stirrups are modelled by two-nodes truss elements while concrete is modelled by iso-parametric eight-nodes elements; steel is connected to concrete through mono-dimensional four-nodes contact elements. The concrete model is based on the smeared crack approach. A rotating crack model is adopted with equal principal stress and strain directions. Relations between stress and equivalent uniaxial strains are used along each principal direction [5], with compression and tension strength parameters obtained from a biaxial failure envelope shown in Figure 6e [13]. The tension response of concrete takes into consideration the fracture energy of the material, following the crack-band model for mesh-objectivity; the size of the mesh is fixed dividing the beam depth with an element size close to 60mm. The fracture energy of the concrete was calculated in relation to the concrete grade, following Model Code 90. The cyclic behaviour of the material is described on the basis of the model by Stevens et al [14].The bond stress-slip model proposed by MC90 is used (Figure 6b), with the parameters for ordinary bars in concrete. The cyclic behaviour is based on the model by Eligehausen et al [15]. For the steel the elasto-plastic model by Menegotto and Pinto [16] has been implemented.

Coronelli and Gambarova [12] provided guidelines in structural modelling of corroded members; some aspects of which are here recalled. Steel cross-section reduction in longitudinal bars and stirrups, and the changes in the ductility of steel bars owing to pitting corrosion should be considered. In the finite element model the uniform corrosion over a given length of the reinforcement coincides with an average reduction of the cross-section of several elements the steel bars and stirrups. If a localized attack occurs the residual cross section at the pit is input and the ductility reduction is reached by enforcing lower ultimate strains of the corroded bars compared with the sound ones.

The corrosion spalling is modelled by reducing the thickness of concrete elements symmetrically with respect to the main vertical chord of the section. At the same time the cracking of the compressed chord due to the corrosion of the top bars is described by reducing the concrete strength of the concrete cover elements.

To define the very low bond strength for the highly corroded beams at study, reference is made to the results obtained in [17] on bars with delaminated covers. Concerning the slip, to determine the values s_1, s_2, s_3, corresponding to the different points defining the model in Figure 6b, the formulation is based on the CEB-FIB Model Code 90.

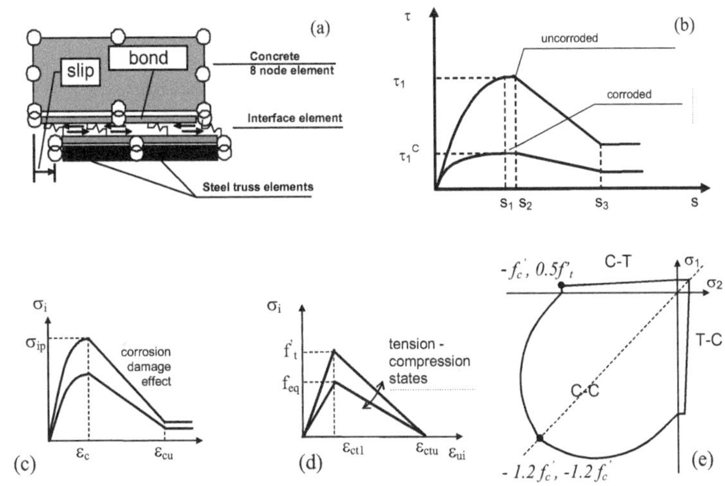

Figure 6: Finite element approach: (a) assembly of different elements; (b) constitutive relation for interface elements; (c) concrete in compression; (d) concrete in tension; (e) biaxial failure envelope

Models of beams of LMDC B series

The mesh was drawn including the whole beam length, in order to model the unsymmetrical distribution of damage along the reinforcement. The dimension of the 8 node concrete elements was approximately 40mm. The pitting of the reinforcement was modelled with the maximum attack along a 40mm length, calculated as the sum of the attack of the two main tensile bars. The aged concrete properties at the time of the test were considered. The bond properties were determined as explained in the preceding section. The test load-displacement response is obtained as the result of a first loading, followed by unloading and reloading. Only the last reloading branch will be shown in the results section.

Results and Discussion

The test results are compared to the global response of both models in Figure 7 for beams B2CL1 and all control beams. In Figure 7a, experimental results show scatters in the control beams bending stiffness. In spite that the three beams have the same reinforcement layout and were cast using the same concrete at the same time, the deflections measured at 20 kN ranges between 1.55 to 1.80 mm. Consequently, the decrease in B2CL1 overall stiffness ranges between 53 to 75% in comparison to the control beams.

The numerical results of both models are in close agreement to test result for the corroded beam B2CL1 (Figures 7a and 7b). The overall stiffness of the corroded beam is very well approximated, particularly for EF response. For the control beams, according to the scatters observed, MFE response is quite close to experiments (Figure 7a). FE response gives a quite important underestimation of the bending stiffness for the control beams (Figure 7b).

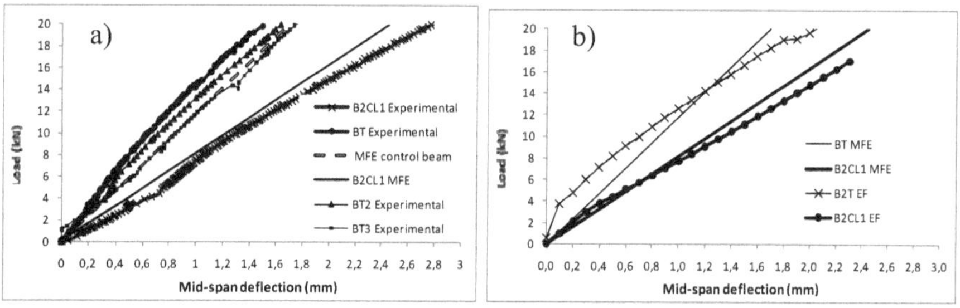

Figure 7: Comparison between model predictions and experimental results

It is important to clarify the respective influence of the steel cross-section reduction and the steel-concrete bond loss on the overall stiffness reduction.

A basic reinforced concrete calculation is performed considering only the effect of the steel cross-section reduction. The maximum cross-section reduction due to corrosion is used (27.95 mm^2 according to Figure 5) and considered constant all along the beam (very conservative option). The reduction in the inertia in cracked section I_c is then equal to about 10%. The corroded beam stiffness reduction observed experimentally ranges between 53 to 75% and the stiffness reduction predicted by the MFE and FE models was 45 and 35% respectively. Consequently, the steel cross-section reduction seems to have a relatively low effect on the stiffness. The steel-concrete bond loss (tension stiffening effect reduction) is the main factor affecting the overall bending stiffness under service loading.

Conclusions

Two models aiming to predict the overall stiffness of corroded beams including the effect of both the steel cross-section loss and the steel-concrete bond reduction have been presented. The results of both models have been successfully compared to the experimental response one 23 years old corroded beam.
Simulations show that the beams overall stiffness reduction due to chloride induced steel corrosion is mainly due to the steel-concrete bond loss.

References

[1] Castel, A. François, R. and Arliguie, G., (2000), *Mater. Struct.*, Vol. 33, pp. 539-544.
[2] Castel, A. François, R. and Arliguie, G., (2000), *Mater. Struct.*, Vol. 33, pp. 545-551.
[3] Rodriguez, J. Ortega, L. M. Casal, J., Diez, J. M., (1996), In: *Congress on concrete in the service mankind*, Proceedings of the 4th International Conference, Dundee UK.
[4] François, R. Arliguie, G., (1999), *Mag. Concr. Res.*, Vol. 51, n. 2, pp. 143-150.
[5] CEB-FIP model code, (1999), Structural concrete: basis of design volume 2, Updated knowledge of the CEB-FIP model code 1990.
[6] François, R. Castel, A. and Vidal, T., (2006), *Mater. Struct.*, Vol. 39, pp. 571-584.
[7] Castel, A. Vidal, T., François, R., (2006), *Mater. Struct.*, Vol. 39, n. 1, pp. 103-113.
[8] Fischer, C., Ožbolt, J., Gehlen, C., (2010), *Proceedings of FraMCos-7 - 7th International Conference on Fracture Mechanics of Concrete and Concrete Structures*, Jeju, Korea.
[9] Vidal, T. Castel, A., François, R., (2004), *Cem. Concr. Res.*, Vol. 34. pp. 165-174.
[10] Zhang, R. Castel, A., François, R., (2010), *Proc. ICE Const. Mater.*, Vol. 163, n. 2, pp. 97-108.
[11] Zhang, R. Castel, A., François, R., (2009), *Mater. Struct.*, Vol. 42, n. 10, p. 1407-1416.
[12] Coronelli D., Gambarova P., (2004), *ASCE Journal of Structural Engineering*, August, pp. 1214-1224.
[13] Coronelli D., Mulas M.G., (2006), *ACI Struct Journal*, Vol. 103, No. 3, May–June, pp. 372-382.
[14] Stevens, N., Uzumeri, S., Collins, M., (1987), SP No. 87, University of Toronto, Toronto, Ontario, Canada, 201 pp.
[15] Menegotto, M., Pinto, P., (1973), *Proceedings of IABSE Symposium on the Resistance and Ultimate Deformability of Structure Acted by Well Defined Repeated Load*, Lisbon, pp. 15-22.

[16] Eligehausen, R.; Popov, E. P.; and Bertero, V. V., (1983), Report No. UCB/EERC 83-23, Earthquake Engineering Research Center, University of California, Berkeley, Berkeley, Calif., 169 pp.
[17] Regan, P.E., Kennedy Reid, I.L., (2009), *Studies and Researches* – Graduate School in Concrete Structures – Fratelli Pesenti, Politecnico di Milano, Italy, Vol. 29, pp. 245-275.

Bond Response in Structural Concrete with Corroded Steel Bars. Experimental Results

M. Prieto, P. Tanner and C. Andrade

Instituto Eduardo Torroja de Ciencias de la Construcción (IETcc-CSIC), Serrano Galvache, 4, 28033 Madrid, Spain.

Abstract The growing interest in upgrading existing reinforced concrete structures or extending their service life, and in ensuring greater durability in new designs, has led to a need for resistance models that take deterioration processes into account to verify structural safety. Bond activation between reinforcing steel and concrete is of cardinal importance in this context. A number of experimental studies have been conducted in recent years on bond failure, which normally leads to brittle behaviour. The findings have diverged rather widely, however, due primarily to differing test conditions. The present paper presents an experimental programme for eccentric pull-out tests in which specimens were subjected to both accelerated and natural corrosion in an attempt to surmount these inconsistencies. It also introduces an embedded fibre-optic sensing system with corrosion-resistant fibre Bragg grating sensors and discusses some of the findings.

Introduction

Resistance

Strength models that take deterioration processes into consideration must be developed to verify the structural safety of new and existing construction. The strength of corroded reinforced concrete structures depends on the cross-section of the corroded steel, its stress-strain diagram, the bond between it and the concrete and corrosion-induced concrete cracking. Since the failure of force transfer between the reinforcement and the surrounding concrete normally leads to brittle behaviour, bond is one of the most important aspects to consider and it is the basis of the structural performance of reinforced concrete. A number of experimental and numerical studies on corroded steel bar bonding have been conducted in recent

C. Andrade and G. Mancini (eds.), *Modelling of Corroding Concrete Structures*, RILEM Bookseries 5, DOI 10.1007/978-94-007-0677-4_16, © RILEM 2011

years (e.g., [1, 2, 3, 4, 11, 12]), although the differences in test conditions have led to widely dispersed results. In an attempt to overcome these inconsistencies, in the present study eccentric pull-out tests were conducted in which specimens were subjected to accelerated and natural corrosion, to study bonding in structural concrete with corroded steel bars. The experimental procedure is described, including the data acquisition by means of an embedded fibre-optic sensing system with corrosion-resistant fibre Bragg grating sensor. Some of the results forthcoming to date are also discussed.

Force Transfer between Corrosion-Free Reinforcement and Concrete

Interface properties

Bonding materializes in structural concrete due to the contact between the steel reinforcement and the concrete. Therefore, bond response may vary with the properties of this interface. Depending on these properties, three different mechanisms may be come into play: adhesion (where slipping is relatively minor), friction (prevalent when confinement pressure is present or in the unloading phase) and rib action (present only where ribbed bars are used). The former two influence bond strength in plain bars, while the third determines the maximum bond strength that can be attained for ribbed bars. Depending on rib geometry, moreover, different failure modes can be found [6]: pull-out, splitting-induced pull-out accompanied by crushing and/or shearing-off and splitting accompanied by slip on the rib faces.

Anchorage of reinforcing bars

Where ribbed bars are used, once the stress exceeds the bond strength afforded by adhesion and friction, force is transferred across the ribs. For a simplified description of the internal forces at failure, the actual stress field can be regarded to be an axisymmetric stress field [7] in which the diagonal deviation of the compression field is absorbed by a concrete tension ring (Figure 1a) or stirrups (Figure 1b). Anchorage fails either because the concrete tension ring fails or the stirrups yield.

Bond strength

In the absence of stirrups, for ribbed bars the maximum bond strength is achieved due to the ribs and depends on anchorage length, concrete tensile strength and the thickness of the tension ring. A theoretical assessment of the stress field in Figure 1a [7] found a linear relationship between bond strength and ring thickness.

The present study attempts to verify whether this relationship is suitable for the test set-up used in this work and to determine bonding of corroded steel bars in the absence of stirrup for different degrees of corrosion.

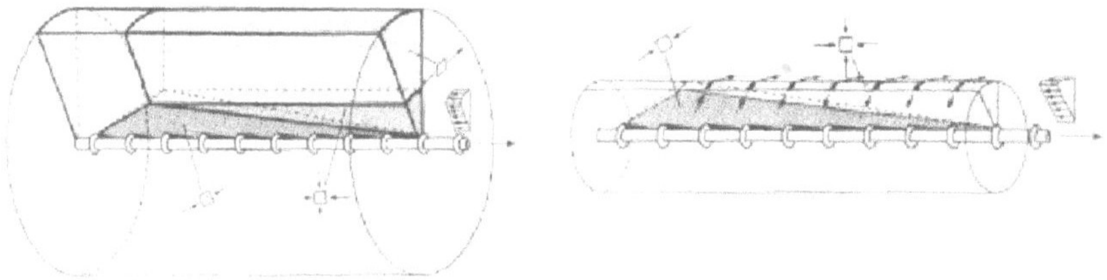

Figure 1. Stress field describing the stress state in the anchorage zone: a) diagonal deviation absorbed by concrete tension ring and b) diagonal deviation absorbed by confining reinforcement [7].

Previous Studies on Bond with Corroded Steel Bars

Overview

Corrosion can affect bond development and with it the transfer of longitudinal stresses. Factors that affect bond strength include the weakening of the steel bar confinement due to cracking on the concrete cover and/or stirrup corrosion, development of corrosion products at the interface, and, in the case of ribbed bars, reduction of the relative rib area due to cross-section loss in the steel. The effect of corrosion on the steel-to-concrete bond has been studied both experimentally and numerically. Numerical studies are beyond the scope of this paper at this time. A number of experimental studies involving tests on ribbed bars without stirrups are briefly reviewed below.

Experimental studies

A wide variety of test specimens with intact and corroded steel have been used to study bond behaviour. While the number is far too large to be analyzed here, the importance of choosing the right size and shape of specimen and suitable corrosion acceleration techniques in bond testing cannot be overstated.

Due to the fact that natural corrosion needs a long time to develop, current is usually applied to accelerate the process. In a previous study [10] was suggested to use a maximum current density of 100-200 $\mu A/cm^2$. Furthermore, a comparison of findings using current densities ranging from 3 to 100 $\mu A/cm^2$ [10] showed that the higher the current density, the higher was the penetration rate needed to achieve a given crack width.

The use by some authors [2, 4, 12, 13, 14] of current densities higher than the maxima found under real conditions may have yielded bond behaviour values that may not be representative of field conditions.

Specifically, some of these studies [4, 12, 14] used short specimens with a length-to-bar diameter ratio of $L/\phi \leq 5$ for the bond strength trials. These specimens are useful for studying local bond-slip behaviour, and slip and bond stresses can be regarded to be constant along the length of the bar. Long specimens ($L/\phi > 5$) have also been used [1, 2, 3, 13] for bond tests. These specimens provide helpful information on anchorage behaviour and bond stress distribution along the embedded length of the bar, while serving as a reference for FE modelling. The main focus of the present study is bond stress behaviour along this embedded length, using long specimens.

Test Planning

Pull-out tests

Traditional bond tests measure bond strength at pull-out-induced failure. In pull-out tests, the bar is centrally embedded, and the c/ϕ ratios are so high that failure normally takes place because of pull-out. In beam tests, pull-out failure is usually the result of high amount of transverse reinforcement.

A bond test has been developed [15] and used [1, 3] that reproduces the situation of a reinforcing bar in a beam with a variable bending moment and a constant shear force. The advantages of this test are that, in addition to simulating the actual condition of the bars, it can be used to reproduce splitting failures, and it accommodates testing of four bars in each specimen and comparison of the results by bar location (top and bottom casting position). For all these reasons, this was the test set-up chosen for the present research. Compared to previous studies, in this case, an additional PVC pipe was placed on the load side of the specimen to prevent undesired boundary effects. The bond test set-up is shown in the figure below.

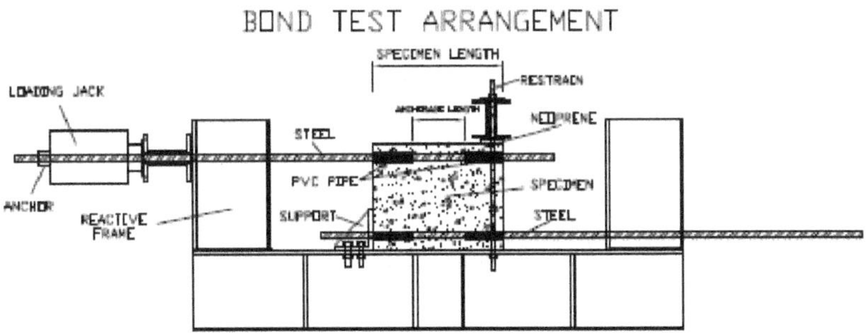

Figure 2. Test set-up

Specimens

Specimens with a 400x400-mm cross-section and 500 or 600 mm long, depending on bar diameter, were used (Figure 3). Long specimens were used and the same length-to-diameter ratio was maintained throughout, namely, anchorage length was 11ϕ in all specimens thus preventing the steel from yielding. To prevent pressure induced by the test set-up and load-side boundary effects, the bars were protected with an embedded PVC pipe on both sides of the specimen. The ratios used for the concrete tension ring ranged from 2.1 to 4. The concrete specimens were reinforced at the corners with four bars measuring 10, 12 or 25 mm in diameter. Two bars were placed at the top and two at the bottom (casting position). The following figures show the test specimens with and without stirrups.

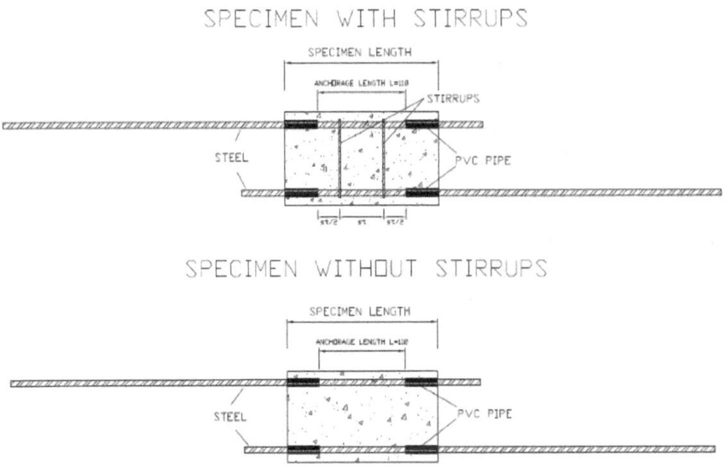

Figure 3. Set-up for specimens with and without stirrups

Data acquisition

The mechanical properties of the materials, i.e., concrete splitting and compressive strength and steel yield and ultimate strength, were found at the same age as when the eccentric pull-out tests were conducted. The following data were acquired during these tests:

- Relative bar slip measurement. Three linear variable displacement transducers (LVDT) with a maximum range of 6 mm. were placed on each side of the specimen.
- Steel strain measurement along the anchorage length. With the fibre optic system used, three Fibre Bragg grating (FBG) strain gauges were positioned along the anchorage length. A Micron Optics sm130 optical sensing interrogator with a wavelength range of 1 510 to 1 590 nm was used for measurements.
- Crack width measurement. Three omega-shaped displacement transducers with a range of ±2 mm were placed on the concrete surface in the same position as the FBG to monitor crack width.
- Load measurement. A 310-kN loading jack was used. The load was applied at a constant 3 kN/min in several steps.

In this investigation, an embedded fibre-optic sensing system with fibre Bragg grating sensors (FBGs) was used to measure strain along the length of the embedded bar, in order to monitor the load transfer from the bar to the concrete. This type of sensor was chosen over electrical sensors because its small dimensions allow fibre embedment without affecting the mechanical behaviour to be measured. Moreover, FBGs have excellent resolution and range and can be multiplexed, i.e., several sensors can be used in the same fibre. The fibres themselves also act as both sensing elements and a signal propagation conduit. Finally, one of the most important reasons for using FBGs is that they are water- and corrosion-resistant and impervious to harsh weather conditions, electromagnetic interference and ground loops.

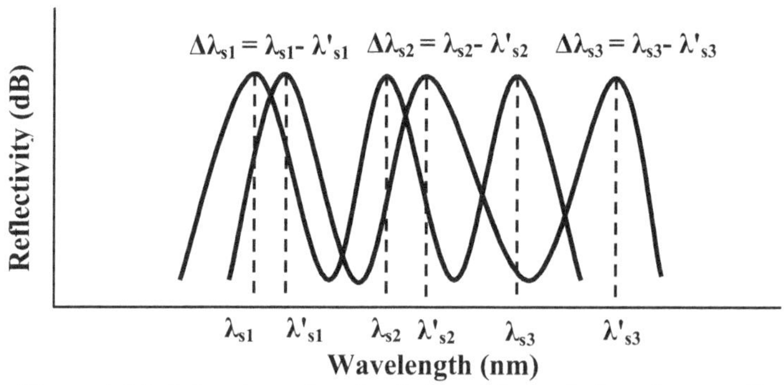

Figure 4. Variation in peak reflectivity with applied strain according to multiplexed FBG system used in this study

When strain is induced in an FBG sensor, either due to mechanical stress, thermal expansion or a combination of the two, its grating pitch changes in such a way that its reflected wavelength varies in proportion to the strain. Strain can therefore be found by measuring the variation in the reflected wavelength.

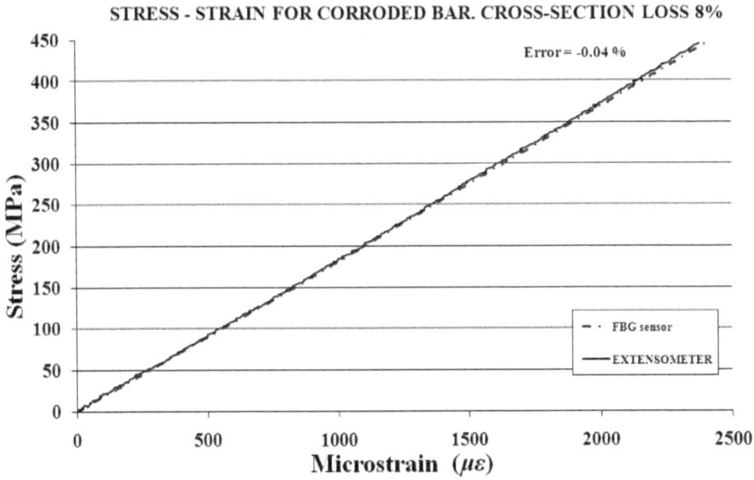

Figure 5. FBG sensor versus conventional strain gauge measurements during a tensile test on a corroded bar

In this investigation, preliminary tensile tests were conducted on non-corroded and corroded bars to perfect a fibre optic system for measuring steel strain in corroded bars during the eccentric pull-out test. The optical fibre was secured with epoxy resin in a 1x2-mm slot in the longitudinal rib of the bar along the anchorage length. The measurements taken with the fibre are compared to the results found with an ordinary strain gauge in the following figure.

Test programme

Eight specimens were prepared for this study, three measuring 400x400x500 and five 400x400x600. Two of these were control specimens, not exposed to corrosion; five were prepared with corrosion levels ranging from 2 to 12%, and one was exposed to natural corrosion, likewise as a control. The characteristics of the experimental programme are shown in the following table:

Parameter	**Value**
Type of test	Eccentric pull-out
Type of bars	Ribbed bars
Nominal steel yield strength	500 MPa
Concrete compressive strength	25 to 40 MPa

Type of splicing	No splicing
Bar position	Top and bottom
Anchorage length	11ϕ Straight
Thickness of concrete tension ring (a/ϕ)	2,1, 3 and 4
Longitudinal bar diameter	10, 12 and 25 mm
Stirrup diameter	10 mm
Confinement	None, stirrups and external pressure
Type of corrosion	Accelerated and natural
Corrosion-induced cross-section loss	0%, 2%, 8%, 12%

Testing

Three concrete mix compositions were used to obtain compressive strengths ranging from 25 to 40 MPa. To induce corrosion, 3% NaCl by weight of cement was added to the concrete mix for the specimens used for natural and accelerated corrosion. After curing in damp burlap for seven days, the specimens were stored under no special temperature or humidity conditions to an age of 28 days to ensure that the material had reached a reasonable compressive and splitting strength before proceeding to accelerated corrosion. The control specimens not exposed to corrosion were stored under the same conditions until they were tested. For eighteen months, while this study is ongoing, the natural corrosion specimen will be subjected to wet-dry cycles for comparison to the accelerated corrosion procedure. In this specimen, corrosion rate is periodically measured by means of a corrosion rate meter, Gecor 8, to monitor the steel cross section loss along time.

Further to a procedure described elsewhere [1, 3, 9], corrosion was accelerated in the other five specimens by applying a constant current density with a galvanostat through counter-electrodes placed on the concrete surface of the specimen. A wet sponge was used to produce electrical contact between the counter-electrodes and the specimen. A current density of 100 $\mu A/cm^2$ was used in two of the specimens, and of 200 $\mu A/cm^2$ in the other three. Two current densities were used to compare the findings to the results for the control specimen exposed to natural corrosion. With only two exceptions [1, 3] that used the same value, the current densities applied in this study were much lower than cited for bond tests in the literature.

Results and Discusion

Only the maximum bond strength values found to date are shown in this paper. A linear regression analysis was performed to confirm, for the test set up used in this work, the relationship between normalized bond strength (maximum bond strength divided by concrete tensile strength) and the ratio between the radius of the concrete tension ring, *a*, and the bar diameter, ϕ, as shown in the graph below. The data used to plot this graph were taken from the literature (beam tests [7], compiled test findings [8], pull-out tests on non-corroded specimens [1, 2, 3, 4]) and the results of the pull-out tests on non-corroded specimens conducted for the present study.

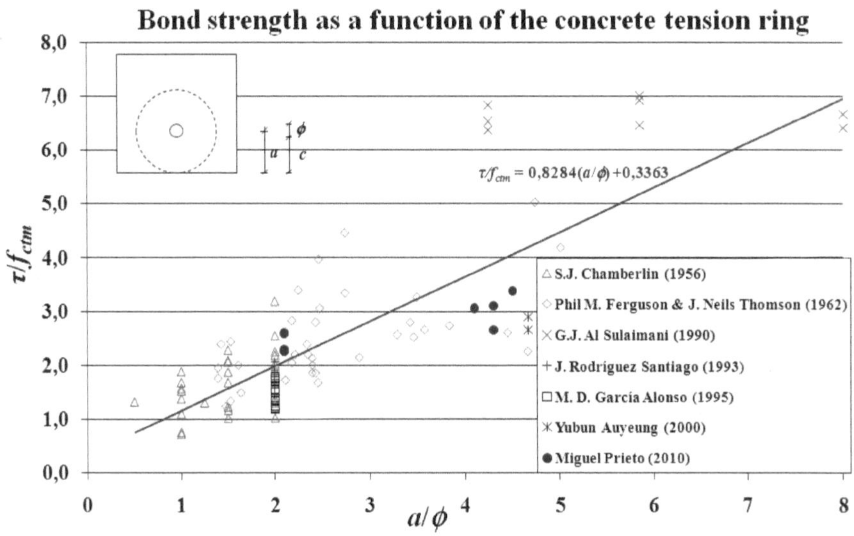

Figure 6. Normalized bond strength versus concrete tension ring thickness

The corroded steel bar bond test results (without stirrups) reported in the literature [1, 2, 3, 4, 12] and the values observed in the present study to date were evaluated also to determine the existence or otherwise of a relationship such as shown in Figure 6.

Normalized bond strength for non corroded bars, for corroded bars with cross section loss up to 5% and cross section loss between 5 and 10%, is plotted against tension ring thickness in Figure 7. Linear regression analysis was performed for the results available from literature for each of these categories. Furthermore, distribution fitting taking into account the number of test results was performed, showing the log-normal distribution the best fit. The results of the present investigation to date are within the scatter observed for each category of results wich, on the other hand, is rather wide with coefficients of variation (CoV) of 0.23 for non corroded bars, 0.78 for corroded bars with cross section loss up to 5% and 0.82 for corroded bars with cross section loss between 5 and 10%. This dispersion is typical of bond tests due to the large number of influencing variables.

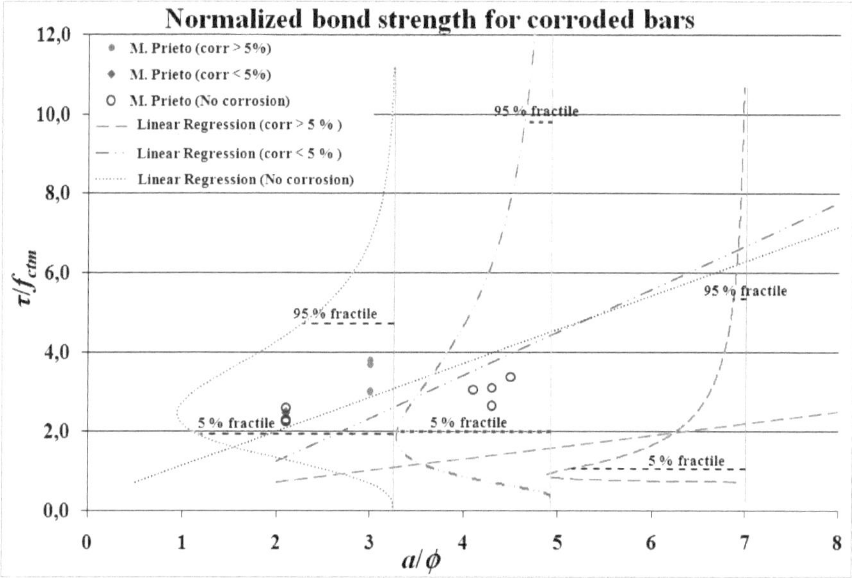

Figure 7. Normalized bond strength for corroded bars versus concrete tension ring thickness and degree of corrosion

The comparison of the results from the present study to the findings reported by other authors [1, 2], in which the same accelerated corrosion procedure, long specimens and nearly the same test set-up were used, show for nearly the same concrete tension ring ratio a clearly higher normalized bond strength than the results reported in the two papers cited. These differences may be due to the variations in the test set-up: in the present study, an additional PVC pipe was used to prevent undesired boundary effects on the load side.

Final Remarks

This paper discusses an experimental programme of eccentric pull-out tests with specimens exposed to accelerated and natural corrosion. An embedded fibre-optic sensing system with corrosion-resistant fibre Bragg grating sensors, developed to measure strain along the anchorage length of corroded bars, is also introduced. Some of the experimental results found to date are given and compared to the findings reported by other authors in terms of normalized bond strength plotted against concrete tension ring thickness and degree of corrosion. Although the results are widely scattered, typical of bond tests due to the large number of influencing variables, a relationship between normalized bond strength for corroded bars without stirrups and concrete tension ring thickness for each category of corrosion has emerged. The detailed analysis of the fibre optic measurements of load transfer should contribute to a better understanding of bond mechanics in case of corrosion. Furthermore, the results for the specimens not yet tested (three exposed to accelerated and one to natural corrosion), together with the findings presently available and a broader parametric study with a numerical model based on these tests, is expected to lead to a fuller explanation of bond behaviour.

Acknowledgment

The present study was funded under the INGENIO 2010-CONSOLIDER Project, "Safety and Durability of Structures: SEDUREC".

References

[1] García Alonso, M.D. (1995). *Aportaciones al comportamiento resistente de estructuras de hormigón armado afectadas por la corrosión de sus armaduras*, Thesis of Polytechnical University of Madrid (UPM), E.T.S.A., Madrid.
[2] Auyeung, Y. and Balaguru, P. and Chung, L. (2000), *ACI Structural Journal*, vol. 97, n. 2, p. 214.
[3] Rodriguez, J. et al. (1995). *The residual service life of reinforced concrete structures. Task 3.2. Relation between corrosion and bond deterioration.* Brite Euram Project BREU-CT-0591.
[4] Al-Sulaimani, G.J. and Kaleemullah, M. and Basunbul, I.A. and Rasheeduzzafar (1990), *ACI Structural Journal*, vol. 87, n. 2, p 220.
[5] Fernandez, M. and Hars, E. and Muttoni, A. (2005). *Bond mechanics in structural concrete. Theoretical model and experimental results*, Draft, IS-BETON, Ecole Polytechnique Fédérale de Lausanne.
[6] FIB (2000). *Fib bulletin 10: Bond of reinforcement in concrete*, Lausanne.
[7] Muttoni, A. and Schwartz, J. and Thürlimann, B. (1997). *Design of Concrete Structures with Stress Fields*, Birkhäuser Verlag, Berlin.
[8] Chamberlin, S. J. (1956). *Proceedings of Journal of the American Concrete Institute*, vol. 53, n. 6, p. 113.
[9] Andrade, C. and Alonso, C. and Molina F. J. (1993). *Materials and Structures*, vol. 26, p. 453.
[10] Andrade, C. and Alonso, C. (1996). *Construction and Building Materials*, vol. 10, n. 5, p. 315.
[11] Alonso, C. and Andrade, C. and Rodriguez, J. and Diez, M.J. (1998). *Materials and Structures*, vol. 31, n. 7, p. 435.
[12] Fang, C. and Lundgren, K. and Chen, L. and Zhu, C. (2004). *Cement and Concrete Research*, vol. 34, n. 11, p. 2159.
[13] Almusallam, A.A. and Al-Ghatani, A.S. and Aziz, A.R. and Rasheeduzzafar (1996), *Construction and Building Materials*, vol. 10, n. 2, p. 123.
[14] Cabrera, J.G. (1996), *Cement and Concrete Composites*, vol. 18, n. 1, p. 47.
[15] Chana, P.S. (1990), *Magazine of Concrete Research*, vol. 42, n. 151, p. 83.

Mechanical Behavior of Long-Term Corroded Reinforced Concrete Beam

Inamullah Khan, Raoul François and Arnaud Castel

UPS, INSA, LMDC (Laboratoire Matériaux et Durabilité des Constructions), Université de Toulouse, 135, Avenue de Rangueil, F-31077 Toulouse, France

Abstract. This paper presents both experimental results and modelling of a highly corroded beam subjected to chloride environment to assess the performance of long-term corrosion damaged beams. A 26 year old corroded reinforced concrete beam exposed to chloride environment was tested along with a control beam of same age till failure. Cracking maps and corrosion maps were drawn for corroded beam. Also force displacement curves for both beams and stress strain curves for steel bars were drawn. From the results it appeared that corrosion has a significant impact on load carrying capacity and deflection of beams. The modelling approach using Macro Finite Element (MFE) shows a good agreement with experimental results in terms of load capacity, stiffness of the beam and ultimate deflection.

Introduction

Corrosion is one the major causes of deterioration of concrete structures. Corrosion not only effects the mechanical properties of steel but the corrosion products induce the stresses in the concrete which easily exceed the limited tensile strength of concrete and cause cracking of concrete and thereby reducing the ultimate strength of concrete elements in structures[1] , decreasing the bond strength [2, 3] and hence reducing the service life of the structures.
When corrosion occurs it causes decrease in deflection, reduction in load carrying capacity and reduction in ductility which can lead to affect the safety and serviceability of concrete structures [4, 5]. Vidal et al [6] have investigated a corroded beam under the couple effect of load and chloride environment to study the corrosion evolution for a period of 17 years. He has observed that corrosion rate goes on increasing in the tension zone, but it remains constant in compression zone. Deflection measurement shows a significant decrease of stiffness over a period of 14 and 17 years due to corrosion increase. Yingang Du et al [7] reported that in addition to decreasing beam flexural strength, corrosion can change the

RILEM Bookseries 5, DOI 10.1007/978-94-007-0677-4_17, © RILEM 2011

failure mode of concrete beams and corrosion can alter a beam's load-deflection curve and flattens the response of over-reinforced beams, but makes those of under-reinforced beams become more brittle.

In the present paper, a 26 years old concrete beam exposed to chloride environment under a sustained loading was studied. The beam was tested till rupture along with a control beam of same age. After performing the tests, results of both the beams are compared.

Two mains important points must be outlined:

- The corrosion of the beam results of a natural process: chloride induced corrosion due to chloride ingress through both flexural cracks and concrete cover porosity without any accelerated process such as impressed current. The resulting corrosion is then very heterogeneous at all levels: from one rebar to another one, from one section to another one, on a given cross section. Very few studies deal with structural performance of such long term naturally corroded members. Rance beams [8] were mechanically tested at the age 40 years but the level of corrosion was too small to really influence the mechanical behavior [9]. Comparing to experimental work using accelerated corrosion process with impressed current [5, 10], the advantage of natural process is to test corroded members representative to on-site corroded structure. The inconveniency of this process is the limited number of elements tested at a given time due to the facts that a long time required to get the natural corrosion and the need to find a place in the laboratory for a longer duration (26 years in this case).

- The beam were corroded during 19 years on sustained loading which is closer to on-site structure than that of unloaded members corroded. Only 4 experiments program deal with corrosion under sustained load [11, 12, 13, 14] and all previous case used accelerated corrosion which lead to generalized corrosion, very different to on-site corrosion.

Experimental Context

The long-term experimental program was started in 1984 at Laboratoire Matériaux et Durabilité des Constructions (L.M.D.C.) in INSA-Toulouse (France). A set of 36 Reinforced concrete beams of similar dimensions (300×28×15 cm) as supplied by industry, cast with two different section types A and B were stored in a chloride environment under sustained loading. At the same time, another set of 36 Reinforced concrete beams of same composition serving as control beams were cast but stored under normal laboratory conditions (non-aggressive environment) in order to have a comparison. At different stages, experiments were carried out to collect the data such as of cracking map, chloride content, and mechanical behavior under service load. Some of them have been tested until failure to evaluate its ultimate capacity and inspect the distribution of rebar corrosion.

Reinforced concrete specimens

This paper only deals with the beam type A. The maximum cover provided in type A beams were 40mm and ordinary reinforcing steel (yield strength=500MPa) was used. The typical steel arrangement for type A beams is shown in figure 1. The compositions of concrete and is given in Table I. Water content was adjusted to obtain a slump of 7 cm. The average compressive strength and elastic modulus obtained on cylinder specimens (110×220 mm) were 45 MPa and 32 GPa at 28 days.

Table I. Concrete composition (kg/m3)

Mix component		
Rolled gravel (silica + limestone)	5/15 mm	1220
Sand	0/5 mm	820
Portland Cement: OPC HP (high perform)		400
Water		200

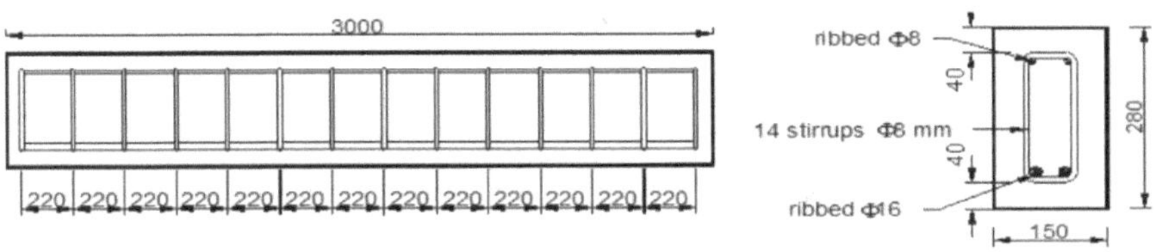

Figure 1. The layout of reinforced concrete beam type A

Beam exposure environment

The beams were kept in aggressive chloride environment. The aggressive environment is a salt fog (35g/l of NaCl corresponding to the salt concentration of sea water) generated through the use of four sprays located in each upper corner of a confined room (Figure 2). After 6 years of storage, the beams were subjected to wetting–drying cycles in order to accelerate the corrosion process:

0 to 6 years: continuous spraying under laboratory conditions (T°≈20°C),

6 to 9 years: cycles spraying under laboratory conditions (T°≈20°C), one week of spraying and one week of drying.

9 to 19 years: cycles spraying, one week of spraying and one week of drying, however the confined room was transferred outside, so the beams were exposed to the temperature of the south-west of France climate, ranging from −5°C to 35°C.

19 to 26 years: cycles have been stopped, unloaded, the beams submitted to the temperature of the southwest of France and had corroded naturally.

The control beams have the same concrete composition and lay-out of the reinforcement, but were stored in a 50% of R.H. and 20°C laboratory room.

Loading of beams

The beams were loaded in a three-point flexure by coupling a type A beam with a type B beam (see Figure 2). According to French standards, the level 1 loading (M_{ser1}=13.5kN m) for type A beams corresponds to working load determined at Service load limit state SLS. The beams stored under sustained level 1 loading are called A1. The level 2 loading (M_{ser2}=21.2kN m) for type A beams corresponds to the working load determined at ultimate load limit state ULS. The beams stored under sustained level 2 loading are called A2. For all of these beams, the upper surface corresponded to the bleeding surface.

During the first period of 6 years, the loading level was checked by an original device [15] using strain gauges and springs to allow a constant load in spite of creep of concrete. After 6 years, creep effects were smaller and then the load was not re-adjusted with time.

In this paper one corroded beam A2CL3 and one control beam A2TI are tested under three point flexion until failure in order to have a comparison between the mechanical properties of corroded and non corroded beam.

Experimental Program

The beam specimens were retrieved form the exposure chamber were disassembled and a concrete cracking map was performed at all surfaces of the beam. The beams were then tested in flexure until failure and afterwards an autopsy was performed to the fractured beam to obtain the corrosion map of the reinforcement (corroding areas and corrosion depth and finally part of the reinforcing steel were cut to perform a direct tensile stress. The experimental program is explained as following.

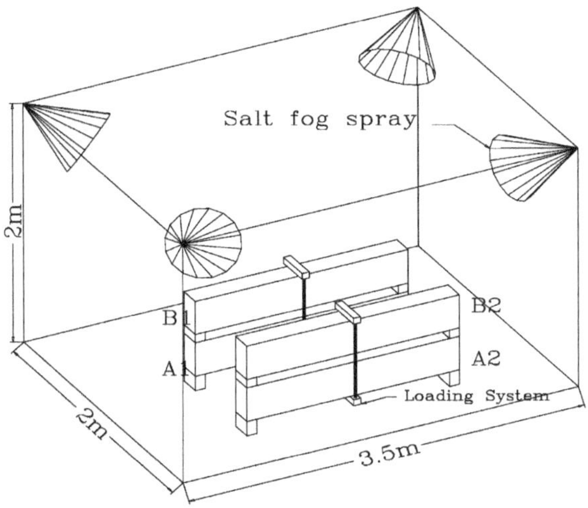

Figure 2. Loading system and conservation of beams

Cracking maps

Cracking maps were drawn with the locations of flexural transverse cracks and longitudinal corrosion cracking at four ages, 28 days, 6, 14, 17 and 26 years. The crack widths were also measured using a binocular lens with an accuracy of 0.02 mm.

Corrosion maps

Corrosion maps of corroded beams were drawn after the steel was recuperated from concrete. Corrosion maps show the extent of corrosion on longitudinal main bars FS (front surface bar) and BS (back surface bar).

Steel behavior

After performing flexure test till failure on both the beams, the steel bars were recuperated from concrete and tensile strength test was performed on the both corroded and non corroded steel. Stress strain curve are drawn for both of the steel bars.

Flexure load response

The beams A2CL3 and A2TI were tested under three point loading system until failure in order to have comparison in the moment carrying capacity of the two beams. Load deflection curves for both the beams are plotted and will be presented. Displacement was measured by a numerical sensor with an accuracy of 0.01 mm.

Experiments Results and Modeling

Cracking maps

The cracking maps of corroded beam A2CL3 are presented in Figure 3 after 26 years of exposure. Flexural transversal cracks, which resulted from the three-points-flexure, appeared in the tensile central part of the beam where the stresses in the concrete exceeded its tensile strength. Along with the flexural cracks, many corrosion cracks were also observed. On front face maximum width of corrosion cracks in central zone was recorded as 1.6 mm where as at the ends of the beam maximum crack width was 0.8 mm. On the other hand on back face the maximum crack width in the central zone was recorded as 2.7 mm and at the ends of beam it was recorded as 2.5 mm.

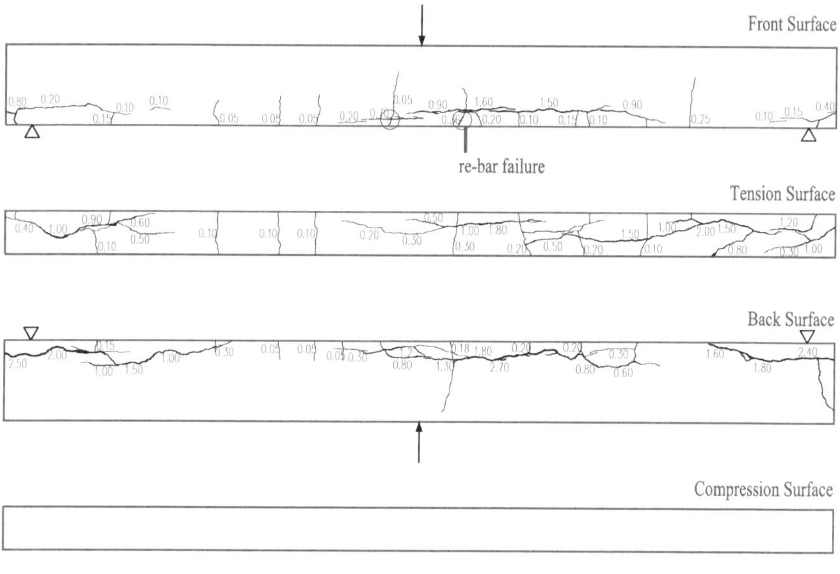

Figure 3. Cracking map of beam A2CL3 after 26 years

Corrosion maps

Corrosion maps of corroded beam A2CL3 is presented in figure 4. The maps are plotted in two directions, one in downward direction which is the side of steel directly exposed to cracks and the other is upward direction in which steel face has much larger cover. It can be seen that corrosion attack is non-uniform. Many corrosion pits were observed on the tensile bars and the maximum corrosion pit depth was measured as 5.1 mm on front tensile bar.

The downward direction was also the casting direction because “A” beams were casted with tensile bars as top bars due to the design of the loading system (Figure 2). So the defects due to settlement and bleeding are located above the tensile bars (top bar effect) and visible from upward direction. An apparent contradiction with previous results [16, 17] is the fact that corrosion is more developed at the top of tensile re-bars (visible from downward direction on Figure 4) at the opposite of the defects. Nevertheless this result could be explained by the two steps of corrosion process during the propagation period with environmental changes due to corrosion cracks development. As a result corrosion cracks are only developed in the concrete cover and then lead to corrosion increase on the bar surface located in front of the surface cover (Figure 5).

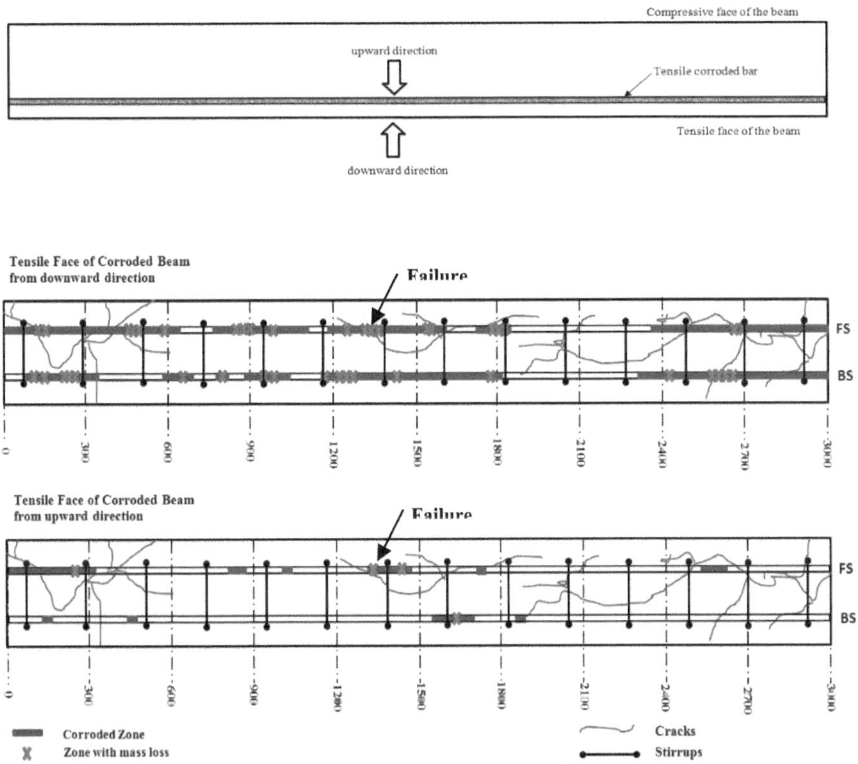

Figure 4. Distribution of corroded areas on the length of beam for tensile reinforcement

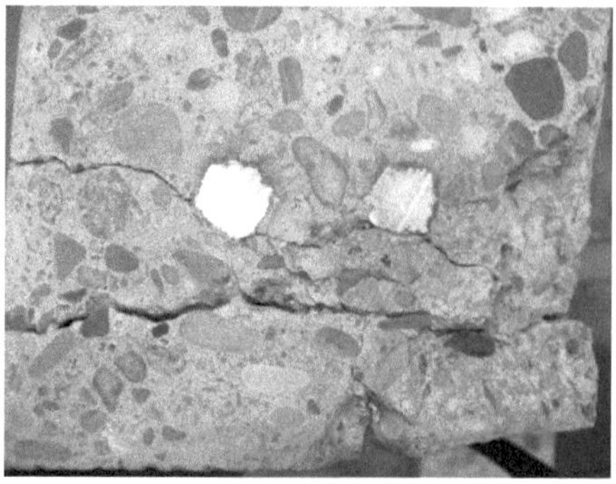

Figure 5. Corrosion cracks located between tensile bars and concrete cover surface

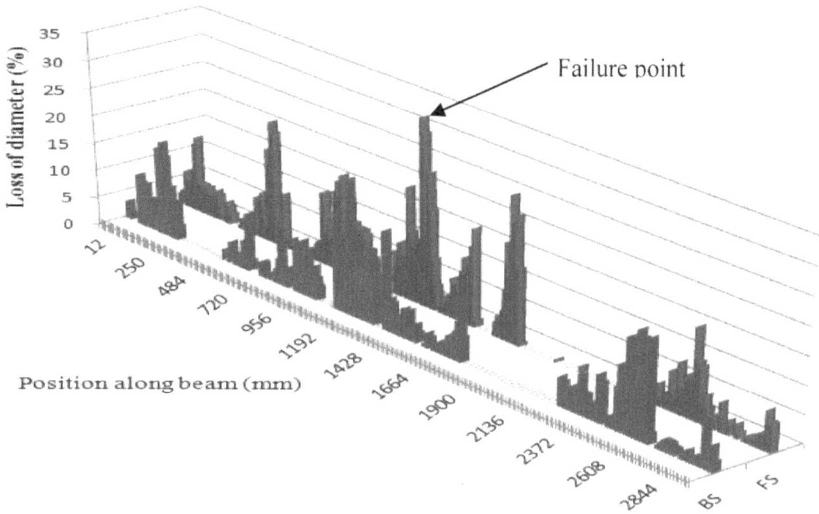

Figure 6. Loss of diameter of reinforcement along the length of beam A2CL3

Figure 6 shows the corrosion of tensile rebars of corroded beam A2CL3. The diameter loss was measured with the help of a vernier caliper after the complete removal of the corrosion products (using Clark's solution ANSI/ASTM G1-72). For the tensile rebars, the maximum diameter reduction of 33.3% occurred on the Front tensile rebar FS around the mid-span which was the failure point during flexure testing of beam. An average diameter reduction of 21.6% on both front and back side bars was measured at failure location. Most of the steel diameter loss was observed in central zone. In order to have a clear comparison of the diameter loss of steel between FS and BS, Figure 7 shows the loss of diameter near the failure point of beam.

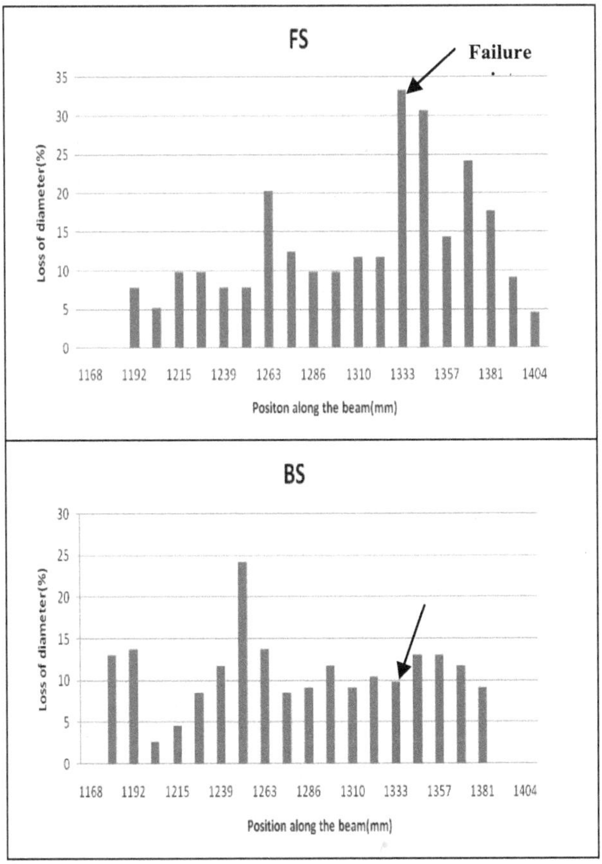

Figure 7. Loss of diameter of both Front side rebar FS and Back side rebar BS of the beam A2CL3 near failure point.

Steel behavior

The steel bars from both control and corroded beams were collected and were tested to measure ultimate tensile strength of corroded as well as control steel bar.

Only two typical curves are plotted on Figure 8. In the case of the corroded bar, the stress is calculated using the residual cross section calculated using the residual diameter measured by a vernier caliper. Figure 8 shows that corrosion has much more impact on ductility than on strength which in results shows a shorter yield plateau, much reduced ultimate tensile strength and easier occurrence of unanticipated brittle breakage, no indication of necking at time of breakage. The reduction of the ultimate failure strain of corroded steel is due to stress concentration at corrosion pits which reduced the ductility reserve [18, 19, 20].

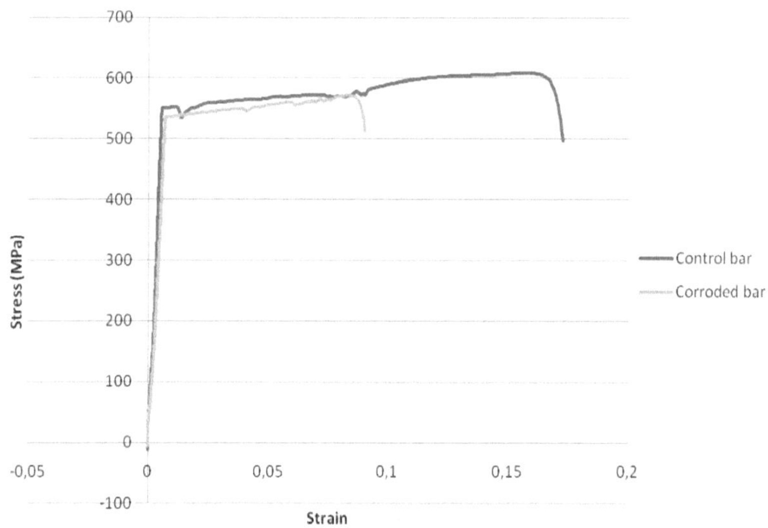

Figure 8. Stress strain curve of corroded and control steel bar

Another way to exhibit the loss of ductility is the recording of the striction area after testing. In Figure 9, the striction area is defined as the difference between initial cross-section and final cross-section divided by initial cross-section and then multiplied by 100. Also corroded bar showed a striction area lesser than that of control bar during the tensile strength testing of rebars as shown in Figure 9 while no such striction was observed on corroded bar during the flexure testing of corroded beam as shown in Figure 10.

From experimental results, the data used for the modeling are the following:
- Yielding stress = 550MPa for both sound and corroded steel
- Post-yielding hardening stiffness = 8.4 e9 Pa for both sound and corroded concrete

The ductility loss is modeled according to the analytical behavior proposed by Castel et al [18].

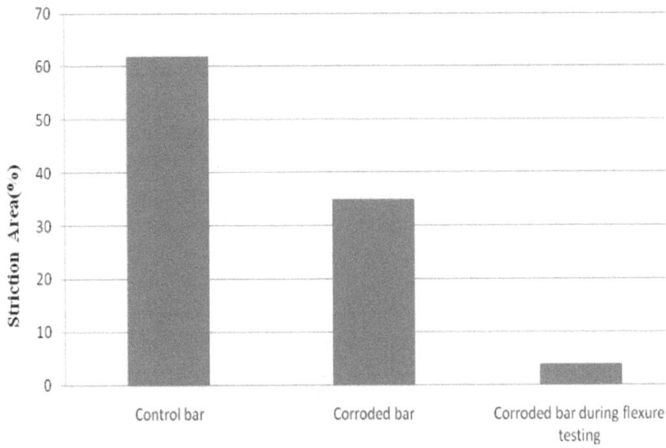

Figure 9. Striction area of control and corroded tensile re-bar

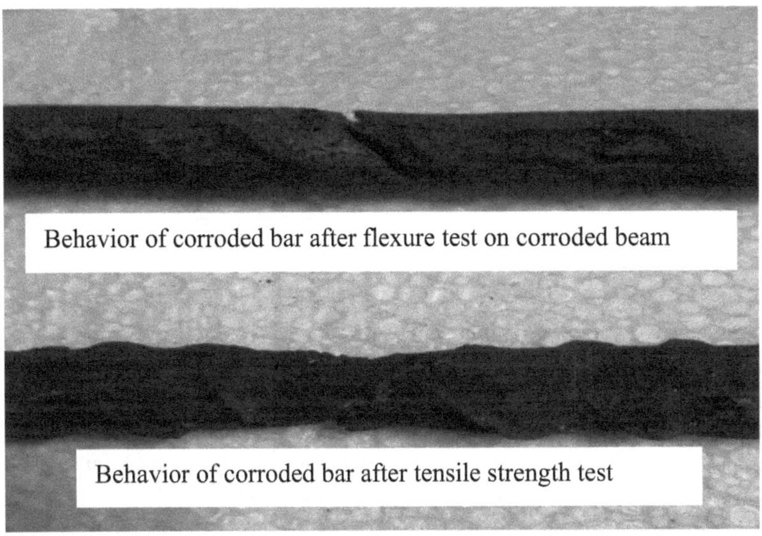

Figure 10. Failure of corroded bars: at top: brittle failure during the flexural test, at bottom: striction during a pure tensile test

Flexure Load Response

Effect of corrosion on ductility of beams. It is clear from the graph (Figure 11) that corrosion decreases the ultimate deflection of the beam. This implies that area under the curve of load deflection curves decreases with the corrosion. Since the area under the curve is an indication of the absorbed energy and ductility, the increase in the corrosion intensity decreases the absorbed energy and hence the ductility of the beams. This implies that the corrosion not only affects the strength of the beams but also induces brittleness in their behavior. Hence, the large deformations, which occur in under-reinforced flexural members prior to failure, will not occur in the case of severely corroded reinforcement, thereby eliminating the most desirable warning before failure of the structure.

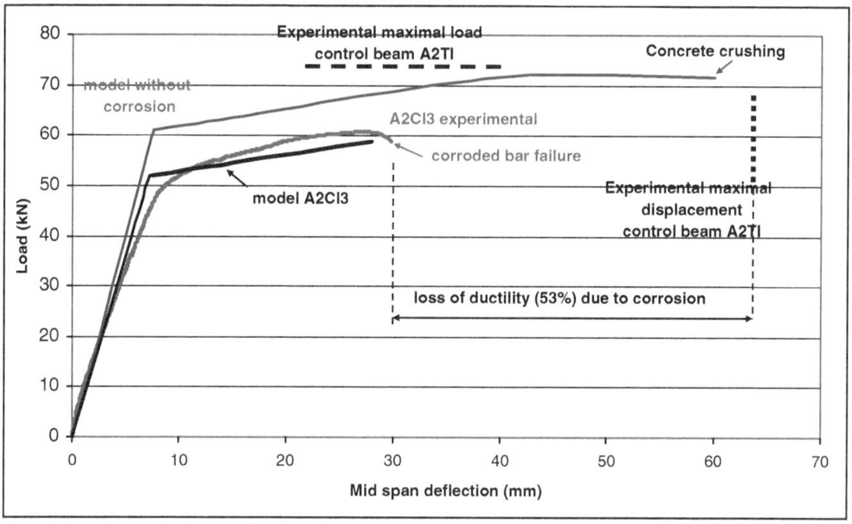

Figure 11. Load displacement curve of beam A2CL3 and maximal load and deflection for control beam A2TI

The maximum deflection in case of control beam A2TI was recorded as 64 mm where as in case of corroded beam its value was 30 mm. The total of 53% decrease in ultimate deflection of corroded beam was recorded as compared with the control beam which clearly indicates that corrosion directly affects the ductile behavior of reinforced concrete beam and hence corrosion can change the failure mode of concrete beams. The model use in this paper to take into account the loss of ductility is the one proposed by Castel et al [18]. Similar model was developed by Ouglova et al [19].

Effect of corrosion on load carrying capacity of beams. From Figure 11 it can be observed that load carrying capacity of corroded beam is also reduced as compared with control beam. The maximum ultimate load recorded in case of control beam A2TI was 73.6 kN with failure due to concrete crushing, whereas in case of corroded beam it was 60.8 kN with failure due to tensile corroded bar rupture. A reduction of 17 % in load carrying capacity of corroded beam was observed in reference with the control beam. The load bearing capacity is related to the maximal loss of steel cross-section which highlighted the need to get natural corrosion or at least to localized accelerated corrosion in the more stressed zone of tested members [13].

The decrease in load carrying capacity of corroded beams is mainly because of reduction of steel area due to pitting attack of corrosion on main tensile bars which lead to change the failure mode.

To calculate the yielding point in the MFE model, a reduction of 21.5% of the steel cross-section was assumed. As a result the reduction of the diameter (figure 6) is not constant around both tensile bars but the shape of the pits which are large in the failure zone allows this assumption. At the end of experiments, the loss of cross-section will be assessed by weight measurement to confirm this assumption.

4Effect of corrosion on stiffness before bending. Unfortunately the stiffness of the control beam A2TI could not be recorded and then could not be compared with the stiffness of the corroded beam A2CL3. Nevertheless, the loss of bond between the reinforcing bars and the concrete due to corrosion leads to a decrease of the "tension-stiffening" effect and then to a decrease the global stiffness. To model this effect, the approach developed at LMDC called MFE (Macro Finite Element approach) [21, 22] will be used. This approach could be used in global structural model and then used by engineers to predict the residual capacity. Another way to model the loss of bond due to corrosion is the used of interface finite elements taking into account the pressure generated by oxides formation and the local slip due to concrete damage [23, 24] or the plasticity theory [25].

The model consists to mesh the beams by assembling MFE defined between each consecutives flexural cracks. Details about the modeling process could be found in [21, 22]. For A2CL3 beam, the mesh is shown on Figure 12.
The decrease of bond due to corrosion leads to a decrease the inertia of the MFE. In [21, 22] the deterioration of bond was modeled thanks to a damage variable varying between 0 to 1 and related to the corrosion level.

$$D_c = 1 - \left(\frac{A_s - \Delta A_{sm}}{A_s - \Delta A_{s0}} \right)^5 \qquad \text{for } A_{sm} > A_{s0}, \text{ elsewhere } Dc = 0$$

where A_{s0} is the section loss threshold that initiates the first crack [21, 22], A_{sm} is the average section loss of the reinforcing bar along L_{elem}. The Dc value is calculated for each MFE and then lead to reduce the inertia of the MFE by increasing the transfer length. The assembling of MFE inertia and calculations are performed using COMSOL software.

According to Zhang et al [26], the propagation of corrosion in chloride environment follows two steps: localized corrosion (LC) leading to corrosion cracks then generalized corrosion (GC) when corrosion cracks are significantly developed all along the reinforcement. Then to model the mechanical behavior of corroded members two approaches could be considered: LC approach where only the loss of bond due to corrosion is taking into account and GC approach where both loss of bond due to corrosion and average loss of steel cross-section are taking into account.
For A2CL3 beam, despite large corrosion cracks, the propagation phase is still in the first steps of corrosion (LC) as checked by Figure 6 even if the central area could be considered in the beginning of GC phase.

Figure 11 shows both experimental result and modeling. The experimental stiffness of A2CL3 is slightly weaker than the one obtained from modeling, this could be due to the fact that the model takes into account only the loss of bond but not the loss of cross-section due to corrosion.

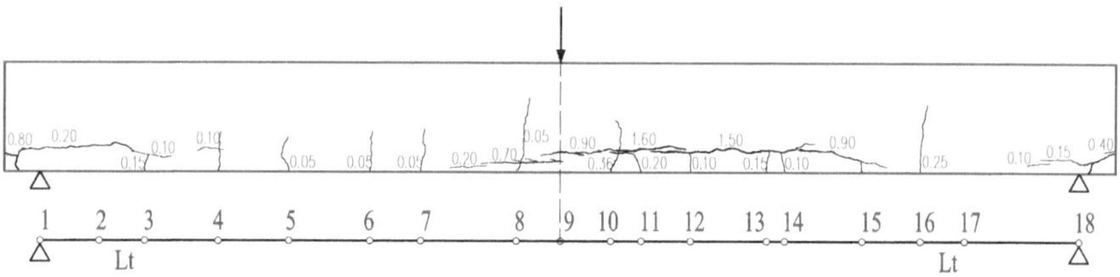

Figure 12. Finite Element mesh of the beam A2CL3: nodes correspond to flexural cracks

Conclusion

The mechanical behavior of a corroded beam exhibits three mains differences in relation to their behavior before corrosion: a decrease in load bearing capacity characterized by a change in the failure mode from concrete crushing to corroded re-bar failure, a reduction of ultimate deflection which is linked to the brittle behavior of corroded bar and a decrease of the global stiffness due to the corrosion of bars and the resulting loss of bond. The MFE approach developed at LMDC Toulouse allows to take into account these three aspects. Nevertheless, there are still some questions about the ductility reduction due to corrosion pits and calculation of the reduction of tension-stiffening effect due to corrosion.

Acknowledgment

The work is a part of the French research project called APPLET supported by ANR (Agence Nationale de la Recherche).

References

[1] Cabrera J.G. (1996) *Cement & Concrete Composites*, 18: 47-59.
[2] Fu X., Chung D.D.L. (1997) *Cement and Concrete Research*, 27(12): pp.1811-1815.
[3] Amleh L., Mirza S. (1999) *ACI Structural Journal*, 96(3): 415-423.
[4] Rodriguez J., Ortega L.M., Casal J. (1994) *International Conference on Concrete across borders, Odense, Denmark*, Vol. 2, pp. 315-326.
[5] Rodriguez J., Ortega L.M., Casal J. (1997) *Construction and Building Materials*, 11(4): 239-248.
[6] Vidal T., Castel A., Francois R. (2007) *Cement and Concrete Research*, 37: 1551-1561.
[7] Du Y., Clark L.A., Chan A.H.C. (2007) *ACI Structural Journal*, 104(3): 285-293.
[8] Poupard O., L'Hostis V., Catinaud S., Petre-Lazar I. (2006) *Cement and Concrete Research*, 36(3): 504-520.
[9] Vu N.A., Castel A., François R. (2010) *Engineering Structures* 32: 556-569.
[10] Mangat P.S., Elgarf M.S. (1999) *ACI Structural Journal*, 96(1): 149-158.
[11] Ballim Y., Reid J.C., Kemp A.R. (2001) *Magazine of Concrete Research*, 53(3), 171-181.
[12] Yoon S., Wang K., Weiss J., Shah S. (2000) *ACI Materials Journal*, 97(6): 637-644.
[13] Malumbela G., Alexander M.G., Moyo P. (2010) *Construction and Building Materials*, 24: 1051-1059.
[14] El Maaddawy T., Soudki K., Tooper T. (2005) *ACI Structural Journal*, 102 (5), 649-656.
[15] François, R., Ringot E. (1988) Capteur de force sur chevêtre de charge pour poutre en béton armé, GAMAC INFO, n°2-3, pp. 21-28 (in French).
[16] Castel A, Vidal T, Francois R, Arliguie G. (2003) *Mag. of Concrete Research*, 55(2): 151-159
[17] Soylev T.A., Francois R. (2003) *Cement and Concrete Research*, 33(9): 1407-1415.
[18] Castel A., François R., Arliguie G. (2000) *Materials and Strucuctures.*, 33: 545-551.
[19] Ouglova A., Berthaud Y, Foct F., François M., Ragueneau F., Petre-Lazar I. (2008) *Materials and Structures* 41: 969-980.
[20] Almusallam A., Al-Gahtani A., Aziz A., Rasheeduzzafart (2006) *Construction and Building Materials*, 10(2), 123-129.
[21] François R., Castel A.,Vidal T. (2006) *Materials and Structures*, 39: 571-584.
[22] Vidal T., Castel A., François R. (2004) *Cement and Concrete Research*, 34: 165-174.
[23] Coronelli D. (2002) *ACI Structural Journal*, 99(3): 267-276.
[24] Ragueneau F., Richard B., Cremona C. et al. (2010) *European Journal of Environmental and Civil Engineering*, 14 (6-7): 869-890.

[25] Lundgren K. (2002) *Magazine of Concrete Research*, 54(3): 165-173.
[26] Zhang R., Castel A., François R. (2010) *Proc. ICE Const. Mater.*, 163(2): 97-108.

Modelling and Nonlinear FE Analysis of Deteriorated Existing Concrete Structures Based on Inspection

Takumi Shimomura[1], Shigehiko Saito[2], Ryosuke Takahashi[2] and Akihiro Shiba[3]

[1]Nagaoka University of Technology, Niigata, Japan
[2]Yamanashi University, Kofu, Japan
[3]Sumitomo Mitsui Construction Co., Ltd., Tokyo, Japan

Abstract. In evaluating structural behaviour of deteriorated existing concrete structures by numerical analysis, it is important to properly detect and quantify material deterioration in the objective structure by means of conducted on-site inspection and to adequately consider them in terms of FE modelling. When the deterioration level of the structure is low, its structural behaviour can be estimated by a common numerical method with sound structure as much as the effect of deterioration is considered in constitutive models. In case of heavily deteriorated structure, however, the way of macroscopic modelling of deteriorated portion has great influence on analytical results. In addition, since an actual structure is generally composed of number of structural members having various levels of deterioration, it is essential to analyze whole the structure instead of part of the structure in order to know overall performance of the structure adequately. In this paper, a general procedure for nonlinear FE analysis of a deteriorated existing concrete structure based on on-site inspection is proposed and an integrated case study of analysis of a concrete bridge with reinforcement corrosion is demonstrated.

Introduction

This paper introduces a part of the significant results made by JSCE-331: a technical committee on structural performance of deteriorated concrete structures in Japan Society of Civil Engineers, which had worked from 2004 till 2009 [1, 2]. The grand objective of the project by JSCE-331 is to establish a technical scheme for evaluation of structural performance of deteriorated existing concrete structures namely due to reinforcement corrosion. In this paper, analytical phase of the project is focused.

C. Andrade and G. Mancini (eds.), *Modelling of Corroding Concrete Structures*, RILEM Bookseries 5, DOI 10.1007/978-94-007-0677-4_18, © RILEM 2011

In order to simulate structural behaviour of deteriorated concrete structures by nonlinear FE analysis, it is needless to say that appropriate constitutive models for deteriorated reinforcement and concrete are necessary. On this line, development of constitutive models for deteriorated reinforcement and concrete has been focused in this field so far, as well as laboratory test [1, 2]. Consequently, tensile properties of corroded reinforcement, for example, were formulated based on number of test data. However, the more difficult and really important point is how to properly detect and quantify material deterioration in the objective structure by means of conducted on-site inspection and how to adequately consider them in terms of FE modelling. When the deterioration level of the structure is low, its structural behaviour can be estimated by a common numerical method with sound structure as much as the effect of deterioration is considered in constitutive models, i.e., simply taking into account reduction of cross sectional area of reinforcement. In case of heavily deteriorated structure, however, the way of macroscopic modelling of deteriorated portion has great influence on analytical results. In addition, since an actual structure is generally composed of number of structural members having various levels of deterioration, it is important to analyze whole the structure instead of part of the structure in order to know overall performance of the structure adequately. In this paper, those problems which were found through the work by JSCE-331 are presented with an integrated case study of nonlinear FE analysis of a deteriorated existing concrete bridge.

Procedure of Evaluating Residual Structural Performance of Existing Concrete Structures by Nonlinear Analysis

General flow

Figure 1 shows the proposed flowchart for performance evaluation of existing concrete structures by means of nonlinear FE analysis. Preliminary diagnosis is carried out at first in order to judge the necessity of more detailed evaluation of the existing structure and to select an appropriate evaluation method. In the inspection in the preliminary diagnosis, therefore, information which is necessary to select evaluation method should be obtained. To meet this purpose, visual inspection, coupled with locally detailed inspection, is generally carried out. As result of the preliminary diagnosis, nonlinear FE analysis is selected as evaluation method for residual structural performance if it is considered necessary. However, it is also possible to employ simplified method or quasi-quantitative method, such as linear analysis, design equation and grading method [3], in case that those methods are enough. Though not discussed in this paper, cause of deterioration of the structure is investigated and deterioration process in the future is predicted in the preliminary diagnosis. When nonlinear FE analysis is selected, detailed inspection is carried out in the secondary diagnosis in order to obtain quantitative information necessary for FE analysis.

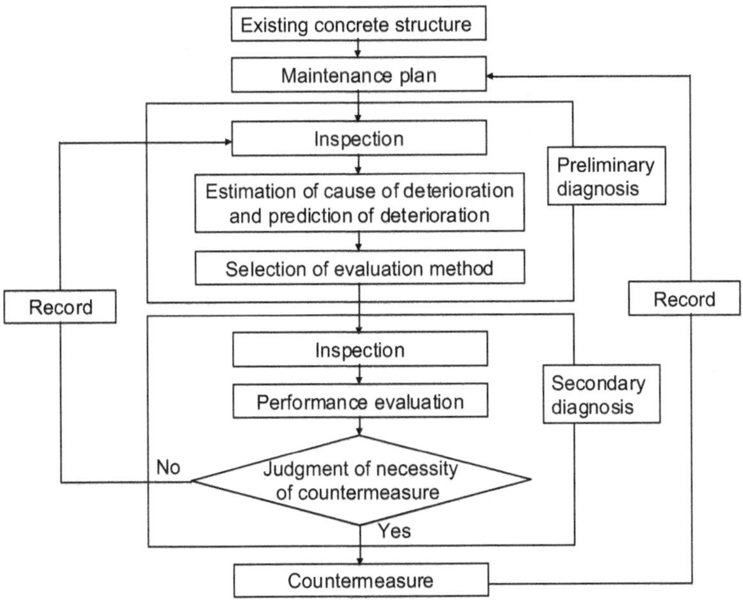

Figure 1. Proposed flowchart for performance evaluation of existing concrete structures by nonlinear FE analysis

Inspection of deteriorated structure

In order to calculate structural behaviour of deteriorated concrete structure by nonlinear FE analysis, constitutive models for concrete and reinforcement have to be enhanced so that they can appropriately take into account the influences of material deteriorations, namely reinforcement corrosion and crack due to ASR, corrosion and so on. In addition, it is emphasized that degree and location of material deterioration in the objective structure should be appropriately known and adequately represented in terms of FE model in the analysis.

In evaluating residual performance of concrete structure with reinforcement corrosion, state of reinforcement corrosion in the structure is the most important information. Nonlinear FE analysis has possibility to accurately evaluate residual structural performance when distribution of residual cross sectional area of reinforcement in the structure is known. However, since it is impossible to know distribution of residual cross sectional area of all reinforcement in the structure, we have to set portions to be inspected in the structure and select appropriate inspection method according to the degree of deterioration.

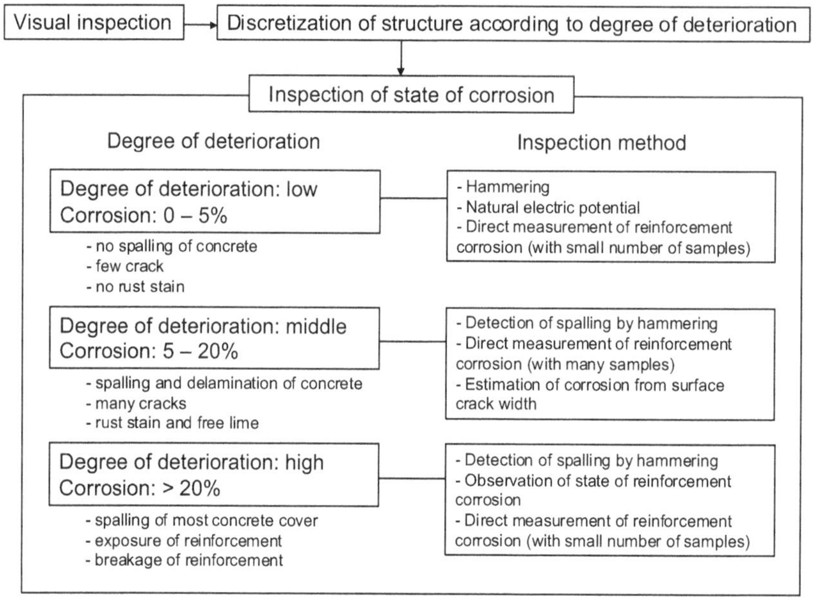

Figure 2. Inspection of state of corrosion according to degree of deterioration

Figure 2 shows the proposed inspection method of state of corrosion according to degree of deterioration. At first, the objective structure is discretized into some control volumes which can be regarded in same degree of deterioration based on visual inspection. Thereafter, state of corrosion in each control volume is investigated by inspection method indicated in Figure 2. The accuracy of inspection depends on method of inspection, setting of control volume and number of samples. In case that degree of deterioration is middle, in particular, number of samples might be of great importance because state of reinforcement corrosion embedded in concrete in this case is difficult to be confirmed. On the other hand, in case of heavily deteriorated structure, it is relatively easy to grasp state of corrosion since most concrete cover is already spalled off.

Table I shows proposed methods of modelling of members with reinforcement corrosion based on on-site inspection. For instance, mechanical properties of concrete in the structure can be measured with specimens drilled from the structure. Corrosion of reinforcements and their spatial variation in the structure have to be estimated from samples taken from the structure. Since information obtained from the structure is limited even by the detailed inspection, modelling of members should be done carefully so that safety side results shall be obtained by FE analysis. The way of consideration of corrosion in analysis depends on the type of model used. For example, the effect of bond deterioration due to corrosion is directly expressed in bond-slip relationship when discrete reinforcement model is used, whereas it is indirectly taken into account in terms of constitutive model of reinforcement and concrete in RC when smeared reinforcement model is used.

Table I. Modelling of members with reinforcement corrosion

modeling	item in modeling			method of modeling
configuration	dimension in analysis		spatial distribution of deterioration	3-D modeling, averaging of deterioration
	mesh for analysis	concrete	spalling, crack	delete of concrete element, reduction of bond
		reinforcement	bar arrangement, cover thickness	arrangement of reiforcement element
	boundary condition		support, connection, deflection	modeling of initial state
material	concrete		strength, elastic modulus	adoption of measured value, consideration of spatial variation
			stress-strain relationship in tension	consideration of bonding in terms of tension-stiffening model
			stress-strain relationship in compression	consideration of effect of corrosion crack
			shear stress transfer	consideration of effect of corrosion crack
			crack	reduction of strength and elastic modulus
	reinforcement		strength, elastic modulus	adoption of measured value or design value, consideration of
			corrosion, loss of cross sectional area	reducion of cross sectional area, consideration in stress-strain
	bond, anchorage		bonding strength, condition of anchorage	reduction of bonding stress as a function of corrosion

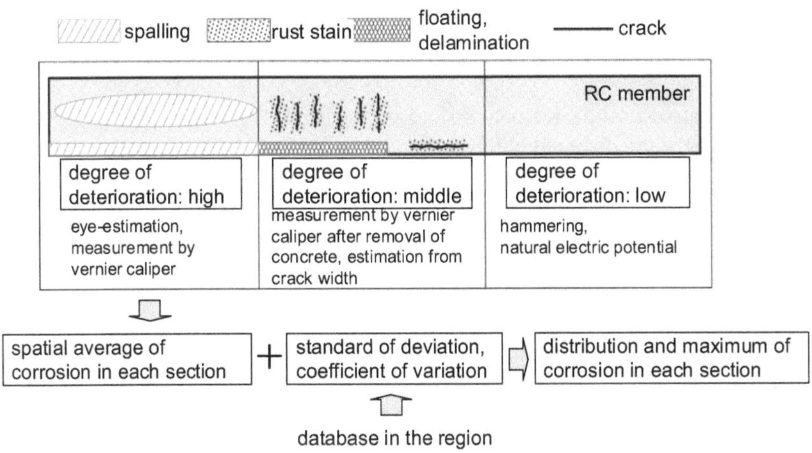

Figure 3. Example procedure of determination of distribution of corrosion

Figure 3 shows an example procedure of determination of distribution of corrosion in an existing concrete member with reinforcement corrosion. Material deterioration, such as corrosion and crack, in actual existing structure is generally not uniform but has spatial variation being partially induced as shown in Figure 3. In Figure 3, the objective member is divided into several sections depending on degree of deterioration. Corrosion of reinforcement in each section is measured by on-site measuring technique according to the degree of deterioration. For example, residual diameter of reinforcement is directly measured by venier caliper in some cases, while corrosion is estimated from corrosion crack width on the surface in some cases though the universal relationship between corrosion and surface crack width has not been established. Then, spatial average of corrosion in each section is obtained. Next, distribution and maximum corrosion in each section are calculated, which can be adopted as input data for FE analysis. Coefficient of variation in corrosion distribution, which is necessary to calculate maximum corrosion from average value, can be ideally estimated based on statistical database of measured corrosion in real structures under similar environment with the objective structure. Otherwise, it should be empirically assumed. It is significant that difference in modelling of the deterioration in the objective structure has sometimes great influences on the analytical result.

Even though only one or two dimensional analysis had been conducted in the initial design, three dimensional FE analysis is recommended in evaluating structural performance after deterioration. In particular, when concrete is partially spalled off or reinforcement corrosion has spatial variation, the member has to be analyzed by three dimensional FE analysis. However three dimensional effect of the material deterioration can be apparently neglected, two dimensional analysis is still acceptable.

Performance evaluation and judgment of countermeasure

In principle, structural performance shall be verified if the evaluated performance index can satisfy the designated limitation.

$$\gamma_i \cdot S / R \leq 1.0 \qquad (1)$$

where $_i$: safety factor, *S*: structural response evaluated, *R*: designated limitation. In order to verify structural performance rationally, performance index that can represent required performance directly as much as possible should be used. In conventional structural design, capacity of member cross section such as flexural capacity or shear capacity is adopted as a performance index for structural safety. In performance verification using nonlinear FE analysis, various performance indices can be adopted, such as crack width, deflection, member ductility and ultimate load carrying capacity. If it is clarified as result of the verification that the structure can not satisfy the required level of performance, some countermeasure should be operated, such as repairing or strengthening work.

Integrated Case Study on Evaluation of Residual Structural Performance of Existing Concrete Structure by Nonlinear FE Analysis

Objective structure

In order to verify the applicability of the proposed evaluation procedure, the working group in JSCE-331 conducted an integrated case study on evaluation of residual structural performance of deteriorated existing concrete structure. Figure 4 shows the objective structure here: a virtual RC simple girder bridge with three clear spans and five main girders near the coast. The bridge has been under service for 35 years under corrosive environment with airborne salt. It is, therefore, deteriorated with reinforcement corrosion due to salt attack. In this study, one of the clear spans with five main girders shown in Figure 5 was analyzed.

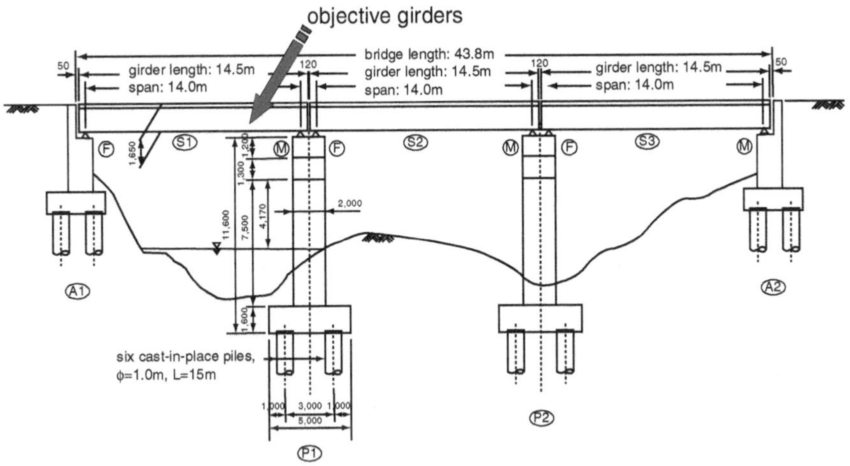

Figure 4. Overview of the objective RC simple girder bridge

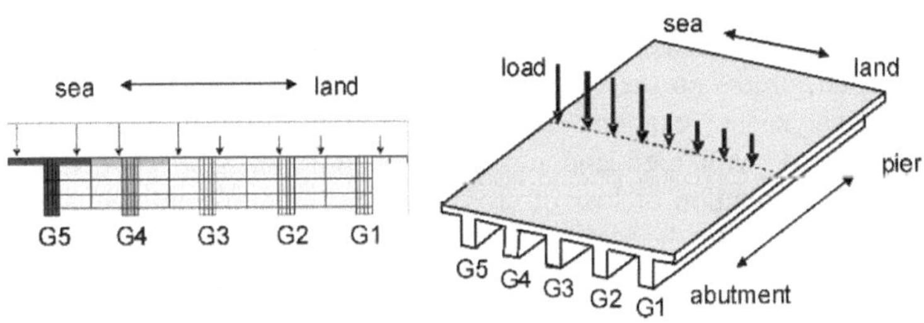

Figure 5. Objective girders

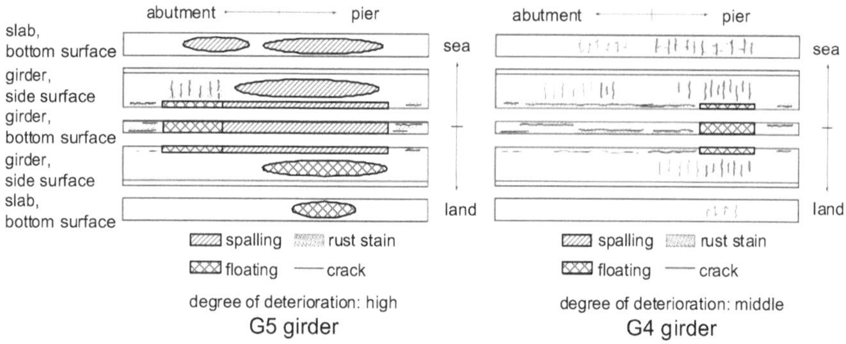

Figure 6. Panorama of the deteriorated girders

Two of the five girders are deteriorated as shown in Figure 6. The girder G5 which is located in the sea side is heavily deteriorated with spalling and floating of concrete. The girder G4 which is located inside G5 is moderately deteriorated.

Analytical cases

In the conducted analysis, structural safety of the superstructure of the bridge against failure of cross section is examined. Though shear reinforcements in the girders are significantly corroded, shear failure is hardly expected considering load condition. Development of longitudinal reinforcements is effective without serious deterioration. Hence, design equation or simplified structural analysis might be enough to estimate residual capacity. However, since failure mode of the structure can not be predicted by simple method because of heavy corrosion in G5, nonlinear FE analysis is employed to evaluate the residual structural performance. Figure 7 shows finite element mesh of the objective superstructure. Size and arrangement of mesh are determined taking into account deterioration portions and computation time. Live loads are applied in the centre of the span as shown in Figure 5. Dead load is applied in terms of body force.

Table II shows analytical cases. Case 1 is sound case as a reference. Case 2 is the objective case. Case 3, 4 and 5 are analyzed to study the influence of the position of deteriorated girders on the overall behaviours of the superstructure. Case 6, 7 and 8 are analyzed to examine the load-deflection curve of individual girders which are sound, with high and moderate deterioration respectively. Figure 8 shows spatial distribution of loss of cross sectional area of reinforcements in the deteriorated girders. In the deteriorated zone where residual bar diameter was directly measured, mean residual bar cross sectional area was determined by averaging maximum loss of cross sectional area of all reinforcements in the zone, while in other zone, reinforcement corrosion was estimated from the measured crack width. The evaluated maximum loss of cross sectional area of reinforcement was 56.5% in the highly deteriorated girder and 15.5% in the moderate one.

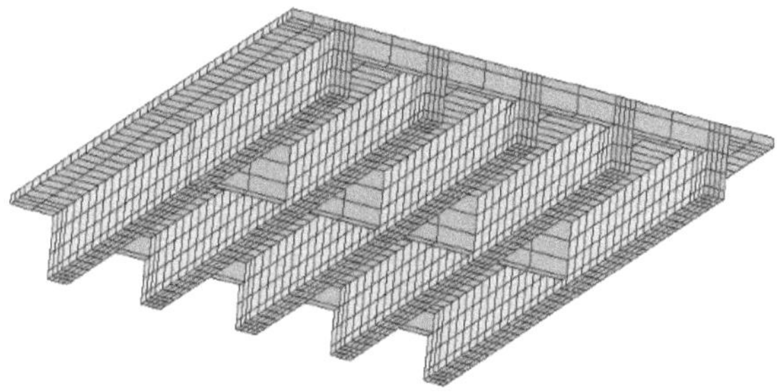

Figure 7. Finite element mesh

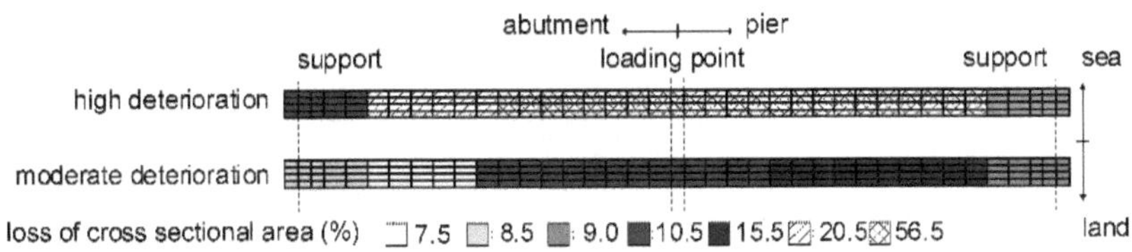

Figure 8. Loss of cross sectional area of reinforcements

Table II. Analytical cases

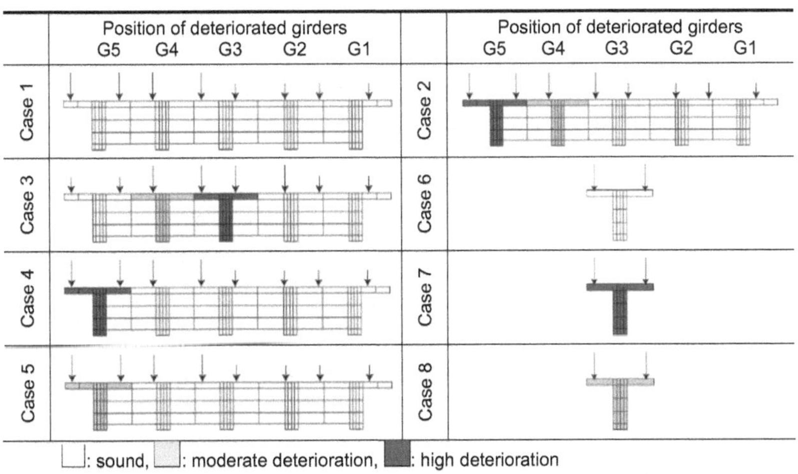

Numerical program and constitutive models

Three dimensional nonlinear FE program was used [4]. Constitutive models used in the analysis are shown in Table III. Since anchorage failure is not expected in the objective girders, smeared reinforcement model is adopted based on the assumption of deformation compatibility between reinforcement and concrete. Effect of corrosion is taken into account in terms of reduction of cross sectional area of reinforcement and reduction of tension stiffening effect of concrete. Tension stiffening of concrete in RC is expressed by following model [5].

$$\sigma_t = f_t \left(\frac{\varepsilon_{tu}}{\varepsilon_t} \right)^c \tag{2}$$

where $_t$: tensile stress of concrete in RC, f_t: tensile strength of concrete, $_{tu}$: cracking strain, $_t$: tensile strain of concrete in RC, *c*: the parameter representing bond effect, which is set as 0.4 for sound RC member. Reduction of tension stiffening effect due to bond deterioration by corrosion is expressed by changing the value of *c*. As shown in Figure 9, *c* is set as 1.2 when reduction of cross sectional area of reinforcement is over 30 % so that the softening curve shall correspond with -3 power model for plain concrete. When reduction of cross sectional area of reinforcement is less than 30 %, linear relation is assumed between *c* and corrosion ratio.

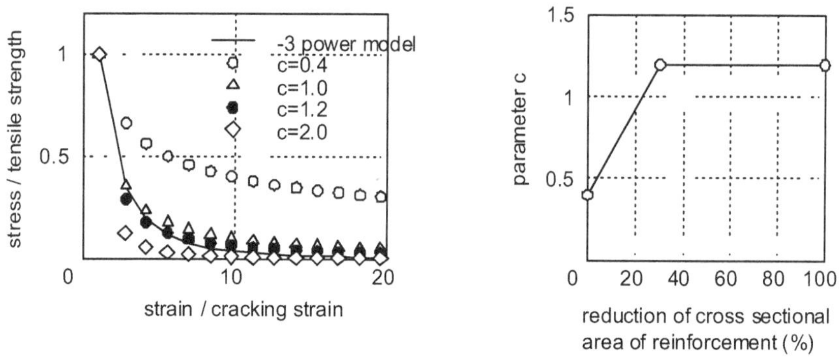

(a) stress – strain ralationship

(b) relationship between parameter *c* and reduction of cross sectional area of reinforcement

Figure 9. Tension stiffening model for corroded RC

Since there is no corrosion crack in compression zone in the objective structure and, in addition, flexural behaviour is dominant than shear under the given condition, corrosion crack is not expected to much affect its structural behaviour. Therefore, effect of corrosion crack is not considered in constitutive models in this case study.

Though the smeared modelling adopted here is very simple, it would be enough to evaluate load-bearing capacity of the superstructure. However, in case that deformation capacity or failure mode of heavily deteriorated structure is of importance, type of modelling and analytical method should be selected carefully.

Table III. Constitutive models

material	status	constitutive models
concrete, RC	before crack	three dimensional elasto-plastic fractural model [6]
	cracking criteria	compression-tension: Niwa model [7]
		tension-tension: Aoyagi-Yamada model [8]
	compression after crack	Vecchio-Collins model [9]
	shear thansfer after crack	Li-Maekawa model [10]
RC	tension after crack	tension-stiffening model by Okamura et.al.[5]
concrete	tension after crack	-3 power model [11]
reinforcement	smeared reinforcement	tri-linear model [12]

Results and discussion

Figure 10 shows analytical results of Case 6, 7 and 8, in which individual girders with various level of deterioration were analyzed. Yield load of reinforcement and ultimate load are shown in Table IV. Ultimate state of the member is judged when average compressive strain of concrete reaches 3500 . In any cases, strain of reinforcement has not reached its ultimate strain at this stage. In Figure 10, both yield and ultimate load decreases with increasing of degree of deterioration. The reduction ratio of yield load of the deteriorated girders to the sound one is almost corresponding with the maximum reduction ratio of cross sectional area of reinforcement. The stiffness of the girders, which is the slope of load-deflection curve, is not much different with each other at the initial stage. However, around the yield load, the stiffness decreases with increasing of deterioration. The deflection of the deteriorated girder at the ultimate stage is greater than that of the sound one. These tendencies are, in general, same with those found in the laboratory test of corroded RC beam specimen [1, 2].

Figure 11 shows analytical results of Case 1, 2, 3, 4 and 5, in which superstructures with five main girders were analyzed. Figure 12 shows deflection of five girders in Case 2 at yield load. Yield and ultimate load of Case 2 was the lowest. Case 4, in which only G5 is highly deteriorated, showed the second lowest yield and ultimate load. Yield and ultimate load of Case 3 were not much low though it involves girders with both high and moderate deterioration. It is consequently significant to properly consider the location and the degree of deterioration in the structure as well as enhanced constitutive models for deteriorated materials.

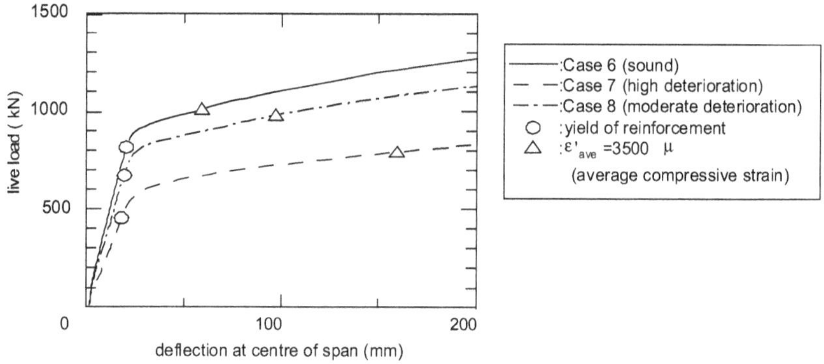

Figure 10. Load-deflection curves of individual girders

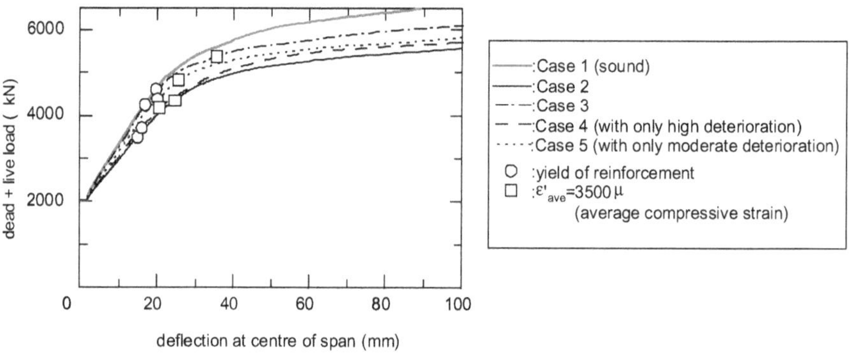

Figure 11. Load-deflection curves of superstructures with five main girders

Table IV. Yield load and ultimate load of each case

	analytical case	yield load (kN)	ultimate load (kN)	relative load	
				yield	ultimate
superstructure with five main girders	Case 1 (sound)	4887	6941	1	1
	Case 2	3768	4612	0.77	0.66
	Case 3	4510	5650	0.92	0.81
	Case 4	3981	4439	0.81	0.64
	Case 5	4652	5100	0.95	0.73
single girder	Case 6 (sound)	822	1010	1	1
	Case 7	461	793	0.56	0.79
	Case 8	678	981	0.82	0.97

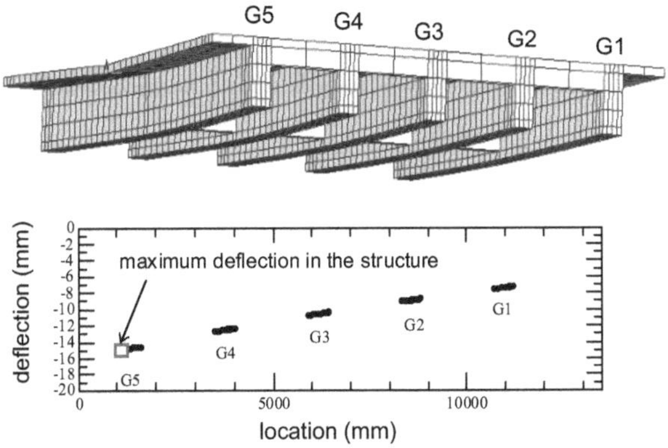

Figure 12. Deflection of girders in Case 2 at yield load

Comparing the results of superstructures and individual girders, it is found that reduction of yield and ultimate load due to corrosion of superstructures is smaller than that of girders. This is because localization of strain does not likely occur in superstructures with five main girders. It is advantageous to analyze whole structure instead of part of the structure when performance of deteriorated structure is assessed by nonlinear FE analysis.

Conclusions

Based on the results of JSCE-331, general procedure of assessment of residual structural performance of deteriorated existing concrete structures by nonlinear FE analysis and integrated case study were introduced in this paper.

(1) In detailed inspection of existing structure in prior to nonlinear FE analysis, it is necessary to collect information about deterioration in the structure enough to represent them in the analysis effectively.

(2) In nonlinear FE analysis of deteriorated concrete structures, it is essential to adequately consider degree and spatial distribution of deterioration such as corrosion of reinforcement and spalling of concrete as well as enhancement of constitutive model for deteriorated materials.

(3) It is advantageous to analyze whole structure instead of part of the structure when performance of deteriorated structure is assessed by nonlinear FE analysis.

Acknowledgements

This work was done as a part of research project by JSCE-331, a committee in Japan Society of Civil Engineers on structural performance of concrete structures with material deterioration. The authors would like to their sincere gratitude to all the members of JSCE-331 who had made contributions to the project.

References

[1] JSCE-331 (2006), *Report of technical committee on structural performance of deteriorated concrete structures*, Vol.1, JSCE, Tokyo (in Japanese).

[2] JSCE-331 (2009), *Report of technical committee on structural performance of deteriorated concrete structures*, Vol.2, JSCE, Tokyo (in Japanese).

[3] JSCE (2003), *Standard specifications for concrete structures-2001 "Maintenance"*, JSCE Guidelines for Concrete, No. 4, JSCE, Tokyo.

[4] R. Takahashi, Y. Sato, K. Konno and T. Ueda (2005), 3D nonlinear punching shear simulation of steel - concrete composite slab, *JCI Journal of Advanced Concrete Technology*, Vol. 3 No. 2, pp. 297-308.

[5] H. Okamura, K. Maekawa and S. Sivasubramaniyam (1985), Verification of modeling for reinforced concrete finite element, in: *Finite Element Analysis of Reinforced Concrete Structures, Proceedings of the Seminar ASCE*, pp. 528-543.

[6] K. Maekawa, J. Takemura and P. Irawan (1993), Triaxial elasto-plastic and fracture model for concrete, *Journal of Materials, Concrete Structures and Pavements, JSCE*, No. 460/V-18, pp. 131-138.

[7] J. Niwa, K. Maekawa and H. Okamura (1981), Nonlinear finite element analysis of deep beams, in: *IABSE Colloquium Delft on Advanced Mechanics of Reinforced Concrete*, pp. 625-638.

[8] Y. Aoyagi and K. Yamada (1984), Strength and deformation characteristics of reinforced concrete shell elements subjected to in-plane forces, *JSCE Concrete Library International* No. 4, pp. 129-160.

[9] M. P. Collins and F. J. Vecchio (1986), The modified compression-field theory for reinforced concrete elements subjected to shear, *ACI Journal*, pp. 219-231, Mar./Apr.

[10] B. Li and K. Maekawa (1987), Contact density model for cracked concrete, in: *IABSE Colloquium Delft on Computational Mechanics of Concrete Structures – Advances and Applications*, pp. 51-62.

[11] R. Takahashi, T. Higai and S. Saito (2008), Study on the influence of crack model in shear behaviour analysis of RC beam, *Proceedings of the JCI*, Vol. 30, No. 3, pp. 55-60 (in Japanese).

[12] K. Maekawa and N. Fukuura (1999), Re-formulation of spatially averaged RC constitutive model with quasi-orthogonal bi-directional cracking, *Journal of Materials, Concrete Structures and Pavements, JSCE*, No. 634/V-45, pp. 157-176 (in Japanese).

Probabilistic Approach to Service Life Prediction of Concrete Structures Subjected to Load and Environmental Actions

Mitsuyoshi Akiyama[1], Dan M. Frangopol[2], Ikumasa Yoshida[3], Hiroaki Tsuruta[4] and Takumi Shimomura[5]

[1]Tohoku University, Japan
[2] Lehigh University, USA
[3]Tokyo City University, Japan
[4]Kansai University, Japan
[5]Nagaoka University of Technology, Japan

Abstract. In the design of reinforced concrete (RC) structures in a marine environment, it is important to consider the effects of this environment on structural long-term performance. In this paper, a time-dependent structural reliability analysis method taking the hazard associated with airborne chlorides into consideration is proposed. Also, a procedure to obtain the failure probabilities of RC structures in a marine environment updated by Sequential Monte Carlo Simulation (SMCS) is indicated. In this procedure, the corrosion crack width is used as observational data. For illustrative purposes, time-dependent reliability analyses are presented for one-way RC slabs in a marine environment. Using SMCS, multiple random variables related to observation information can be updated simultaneously. This is realized by taking into consideration the joint probability density functions of the random variables. The effects of the hazard associated with airborne chlorides and an inspection result of corrosion cracking on the updated estimate of one-way RC slab reliability are discussed in this study.

Introduction

Because of the presence of uncertainties, it is necessary that long-term structural performance be treated based on reliability concepts and methods [1]. Stochastic treatment of structural design problems takes into account the uncertain nature of long-term structural performance making a reliable design of RC structures possible [2].

C. Andrade and G. Mancini (eds.), *Modelling of Corroding Concrete Structures*,
RILEM Bookseries 5, DOI 10.1007/978-94-007-0677-4_19, © RILEM 2011

Unlike the case associated with actions that are usually considered in structural design (e.g., seismic hazard assessment), there has been a lack of research on marine environmental hazard assessment. A methodology for the probabilistic hazard assessment associated with airborne chlorides was established by Akiyama et al. [3]. In this paper, the procedure to integrate the hazard assessment associated with airborne chlorides into long-term performance prediction of RC bridge slabs is provided.

Even though time-dependent reliability for RC bridge slabs subjected to corrosion resulting from chlorides depends on many aleatoric and epistemic uncertainties, for existing RC bridge slabs it is possible to reduce epistemic uncertainties using inspection results. The relationships between observed physical quantities such as inspection results, and the PDFs of related random variables are used in Bayesian updating. When nonlinear relations or non-Gaussian variables are involved, an approximate solution can be found by using several approaches. Monte Carlo Simulation (MCS) approach is in general used because of its versatility. MCS-based methods for non-linear filtering technique have been developed since 1990s. The present paper presents a probabilistic framework for estimating the time-dependent reliability for RC bridge slabs in a marine environment based on a non-linear filtering technique denoted Sequential MCS (SMCS). Visual inspections of corrosion crack width is used as observational information. The emphasis is placed on investigating the effects of the hazard associated with airborne chlorides and inspection results on updated estimates of RC bridge slab reliability.

Structural Performance of RC Bridge Slab in a Marine Environment

Hazard curves of amount of airborne chlorides

Akiyama et al. [3] proposed the probabilistic model of hazard associated with airborne chlorides taking into consideration the spatial-temporal variation effects. The results of hazard assessment associated with airborne chlorides depend on the ratio of sea wind (defined as the percentage of time during one day when the wind is blowing from sea toward land), wind speed in the location analyzed, and distance from coastline. In this paper, the reliabilities of RC bridge slab in Wakkanai City, Japan are estimated. The distance from coastline is assumed to be 0.1 km and 1.0 km. Hazard curves for Wakkanai City are shown in Fig. 1. Based on the hazard assessment of airborne chlorides, the effect of marine environment can be quantified.

Steel weight loss due to corrosion

The procedure to calculate the steel weight loss $W(t)$ (%) due to corrosion resulting

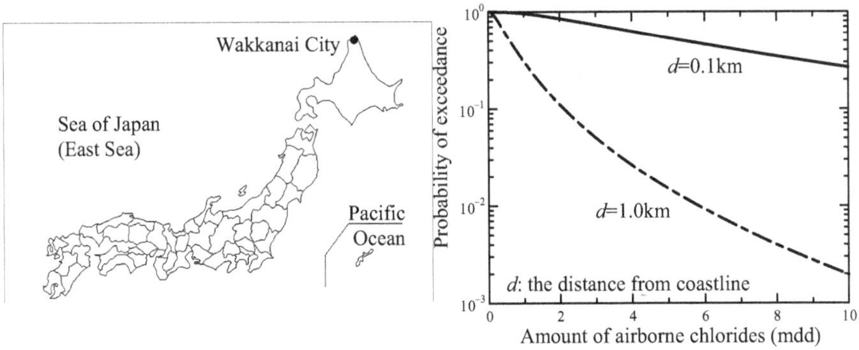

Figure 1. Probability of exceedance of various amounts of airborne chlorides in Wakkanai City at distance of 0.1 km and 1.0 km from coastline

from chlorides at the time t after construction was developed by Akiyama et al. [3]. It can be expressed as

$$W(t)=\begin{cases} 0.0, \quad t_1 \geq t \\ \dfrac{\pi\phi}{(\pi\phi^2\rho_s)/4}(t-t_1)V_1x_9\cdot, \quad t_2 \geq t > t_1 \\ \dfrac{\pi\phi}{(\pi\phi^2\rho_s)/4}(t_2-t_1)V_1x_9+(t-t_2)V_2x_9, \quad t > t_2 \end{cases} \tag{1}$$

where, $$t_1=\frac{1}{4D_c}\left\{\frac{0.1(c+x_6)}{erf^{-1}\left(1-\dfrac{x_4C_T}{x_5C_0}\right)}\right\}^2 \tag{2}$$

$$t_2=t_1+\frac{x_8Q_{cr}(c)}{x_9V_1} \tag{3}$$

$$D_c=x_7\cdot 10^q \tag{4}$$

$$q=-6.77(\mathrm{W/C})^2+10.10(\mathrm{W/C})-3.14 \tag{5}$$

$$C_0=x_3\cdot 0.988{C_{air}}^{0.379} \tag{6}$$

$$C_{air}=x_2\cdot 1.29\cdot r\cdot(x_1\cdot u^{0.386})\cdot d^{-0.952} \tag{7}$$

$$Q_{cr}(c)=\gamma\,(W_{c1}+W_{c2}) \tag{8}$$

$$W_{c1}=\frac{\rho_s}{\pi(\gamma-1)}\left[\alpha_0\beta_0\frac{0.22\{(2(c+x_6)+\phi)^2+\phi^2\}}{F_u(c+x_6+\phi)}{f'_c}^{2/3}\right] \tag{9}$$

$$W_{c2}=\alpha_1\beta_1\frac{\rho_s}{\pi(\gamma-1)}\frac{c-\phi}{5c+3\phi}w_c \tag{10}$$

x_1 = Gaussian random variable related to wind speed; x_2 = lognormal random variable representing model uncertainty of attenuation relationship between the airborne chloride and distance from coastline; u = wind speed, d = distance from coastline, r = the ratio of sea wind, x_3 = lognormal random variable representing model uncertainty associated with the relationship between the surface chloride content C_0 and airborne chloride; C_T = critical threshold of chloride concentration (kg/m^3); c = concrete cover (mm); t = time after construction (year); W/C = ratio of water to cement, erf = error function; D_c = coefficient of diffusion of chloride; x_4 = Gaussian variable associated with the evaluation of C_T; x_5 = lognormal variable representing the model uncertainty associated with the estimation of C; x_6 = normal variable representing the construction error of c; and x_7 = lognormal variable representing the model uncertainty associated with the estimation of D_c; ρ_s = steel density (7.85 (mg/mm^3)); γ = expansion rate of volume of corrosion product =3.0; f'_c = concrete strength (MPa); w_c = crack width due to corrosion of the steel bar = 0.1 mm; E_c = modulus of elasticity of concrete (MPa); ϕ = diameter of the steel bar (mm); V_1 = corrosion rate of the steel bar before the occurrence of corrosion crack (mg/mm^2/year); α_0, β_0, α_1, and β_1 = coefficients taking into account the effects of concrete cover, steel bar diameter, and concrete strength, respectively; η = correction factor; x_8 = lognormal random variable representing the model uncertainty associated with the estimation of Q_{cr}; x_9 = lognormal random variable related to the corrosion rate; C_{air} = the amount of airborne chloride; Q_{cr} = critical threshold of corrosion amount at cover cracking initiation; and V_2 = corrosion rate of steel bars after the occurrence of corrosion cracking.

All parameters of random variables involved in the evaluation of steel weight loss were obtained by survey of existing concrete structure in a marine environment and by comparing experiment results with calculated values, and are summarized in Table I. The study by Akiyama et al. [3] shows the details of how to determine these statistics.

Table I. Parameters of random variables

Variables	Distribution	Mean	COV
x_1	Normal	1.00 (u = 4.65 m/s)	0.0778
x_2	Lognormal	1.06	1.25
x_3	Lognormal	1.43	1.08
x_4	Normal	1.00	0.375
x_5	Lognormal	1.24	0.906
x_6	Normal	specified + 8.5mm	16.6mm / (specified+8.5mm)
x_7	Lognormal	1.89	1.87
x_8	Lognormal	1.00	0.352
x_9	Lognormal	1.00	0.580

Relationship between the steel weight loss and flexural strength loss

It is necessary to evaluate the flexural strength loss based on the steel weight loss provided by Eqn. (1). Recently, experimental studies on RC beams with chloride-induced corrosion of steel reinforcement have been reported [4-9]. The specimens were corroded by electric corrosion or drying/wetting condition using salt water.

In this study, test results of corroded RC beams are used to obtain the relationship between the steel weight loss and flexural strength loss h defined as the ratio of flexural strength of corroded specimen to that of specimen without corrosion. When the steel weight loss of RC bridge slab at time t is given by Eqn. (1) under the condition that the spatial corrosion distributions are ignored and each reinforcing bar has the same steel weight loss, h can be expressed as

$$h = -1.64\times10^{-4}\cdot(W(t))^2 - 9.73\times10^{-3}\cdot(W(t)) + 1.00 \tag{11}$$

Structural Reliability Analysis and Updating Method (SMCS) for Existing RC Structure

Time-dependent reliability analysis

Mori and Ellingwood [10] predicted the life of concrete structures using the probability that the structure survives during the time interval $(0, t_L)$. They assumed that the structure is subjected to two statistically independent load processes with intensities S_1 and S_2, respectively, but only S_1 varies in time and it is modeled as Poisson process. Their approach can be used for structures located in moderate environment, since the variability of degradation function can be ignored. However, if the RC structures using poor concrete quality and having small concrete cover are located in severe marine environment, the variability of degradation function is not negligible. As a result, the failure probability, $F(t_L)$, is

$$F(t_L) = 1 - L(0, t_1)\cdot L(t_1, t_2)\cdots\cdots L(t_{n-1}, t_n) = 1 - L(t_L) \tag{12}$$

where $t_1 < t_2 < \cdots\cdots < t_n : t_L$

$$L(t_{i-1}, t_i) = \int_0^\infty \int_0^\infty \int_0^\infty \exp\left[-\lambda_{S_1}\cdot\left[\Delta t_i - \int_{t_{i-1}}^{t_i} F_{S_1}\{r_0\cdot a_i\cdot h_g(t) - s_2\}dt\right]\right] f_{S_2}(s_2) f_{R_0}(r_0) f_{A_i}(a_i) ds_2 dr_0 da_i \tag{13}$$

λ_{S1} = mean occurrence rate of S_1; $h_g(t)$ = degradation function; F_{S1} = cumulative distribution function (CDF) of S_1; f_{S2} and f_{R0} = PDF of S_2 and initial strength R_0, respectively; $\Delta t_i = t_i - t_{i-1}$, A_i = random variable representing the variability in degradation function; and f_{Ai} = PDF of A_i.

The details of how to calculate the cumulative-time failure probability using Eqns. (1) to (13) are shown in Akiyama et al. [11].

Updating based on inspections results

Using Eqn. (12), time-dependent reliability analysis can be performed for new RC structure in a marine environment. Since inspection results can be obtained for existing structures, the mean and COV of random variables shown in Table I should be updated. This study uses the inspection results of corrosion crack width as observational data. These results depend on several of random variables listed in Table I, and the relationships between the inspection results and these random variables are nonlinear. Therefore, it is impossible to perform the updating of the random variables in Table I by a theoretical closed form solution such as a Kalman Filter algorithm. In this study, the Sequential Monte Carlo Simulation (SMCS) is applied to update these random variables. The detailed procedure of SMCS applied to reliability analysis was given by References [11, 12].

Illustrative Example

Description of RC bridge slab and loads

RC bridge slabs in Wakkanai City (d = 0.1 km and d = 1.0 km) are used in this example. Figure 2 shows the properties of the slab analyzed. The combination of dead and live loads (D+L) is considered and the failure probability of the deteriorated flexural strength is presented. It is assumed that the initial resistance of the bridge slab, R_0, is described by a lognormal distribution with mean value, μ_R = 1.15R_n (where R_n = 2260 kN m), and the coefficient of variation (COV) 0.15. The live load effect is described by a Type I largest value distribution with mean value, μ_L = 0.40 L_n (where L_n = 755 kN m), and the COV of 0.5. The live load is modeled as Poisson process and treated as S_1 in Eqn. (13) and the mean occurrence rate of S_1 is assumed to be 0.5 /year. The dead load is assumed Gaussian variable with mean, μ_D = 1.00 D_n where D_n = 755 kN m, and the COV of 0.07. The dead load is treated as S_2 in Eqn. (13).

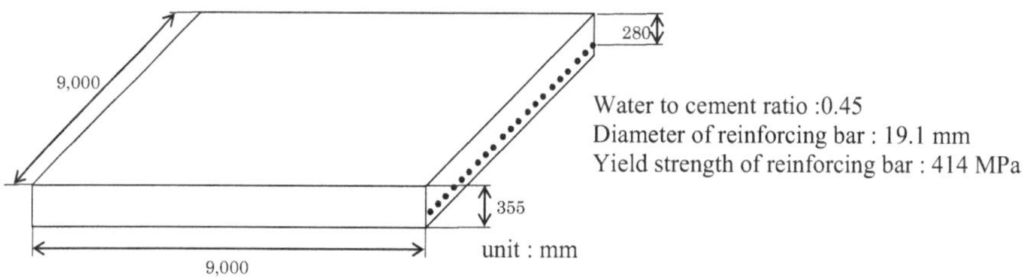

Figure 2. Simply supported RC slab

Modeling of observational data

Visual inspection for corrosion crack width are used as observational information in SMCS. Corrosion products of steel bars cause the corrosion cracks. Based on the survey of marine RC structures the relationships between steel weight loss and corrosion crack width are classified as shown in Table II. Suzuki et al. [13]

presented the experimental results of corrosion crack width and the amount of steel weight loss of corroded RC specimens. The amount of steel weight loss was measured at the location of the crack. Under the condition that the category of steel weight loss is given, the probability of each category of corrosion crack width is calculated using the relationship between the corrosion crack width and steel weight loss based on the test results provided by Suzuki et al. [13]. The observational model of visual inspection of corrosion crack width is shown in Table III. Each probability in Table III is used in the calculation of likelihood in SMCS.

Table II. Relationship between steel weight loss and crack width

Crack width		Steel weight loss	
Category	Width (mm)	Category	Amount (%)
I	0 - 0.1	1	0 - 2.3
II	0.1 - 0.2	2	2.3 - 5.0
III	0.2 - 0.5	3	5.0 - 20
IV	0.5 -	4	20 -

Table III. Observational model of visual inspection of corrosion crack width

Category based on steel weight loss	Category based on crack width			
	I	II	III	IV
1	0.811	0.159	0.031	0.000
2	0.268	0.410	0.313	0.009
3	0.019	0.120	0.600	0.261
4	0.000	0.004	0.219	0.776

Time-dependent reliability analysis of RC slabs

In this case study, two Cases are assumed; RC slab is not inspected for entire lifetime in Case 0; and crack width classified as Category II at 30 years after construction is found by visual inspection in Case 1. Fig. 3 shows the cumulative-time failure probability of RC bridge slab at d = 0.1 km and d = 1.0 km, respectively. Due to the difference of hazard associated with airborne chloride, the failure probability of RC bridge slab at d = 0.1 km is much larger than that at d = 1.0 km. The differences of marine environment need to be considered in long-term structural reliability assessment. The updated failure probability using the visual inspection results of corrosion crack width at 30 years is shown in Fig. 3. Since corrosion crack widths provided by inspection in Case 1 is less than that expected by random variables listed in Table I, the failure probability at 30 years after updating is much smaller than that before updating (Case 0).

Correlation between x_3 and x_5 for RC slab at d = 1.0 km is shown in Figure 4. Figure 4(b) shows x_3 and x_5 after updating. All parameters of random variables related to the observational data can be updated by SMCS simultaneously using the joint PDFs of the random variables (including means, COVs and correlations). x_3 and x_5 are statistically independent before updating. As these two random variables are updated using same inspection results, it is confirmed in Figure 4 that they have to be correlation after updating.

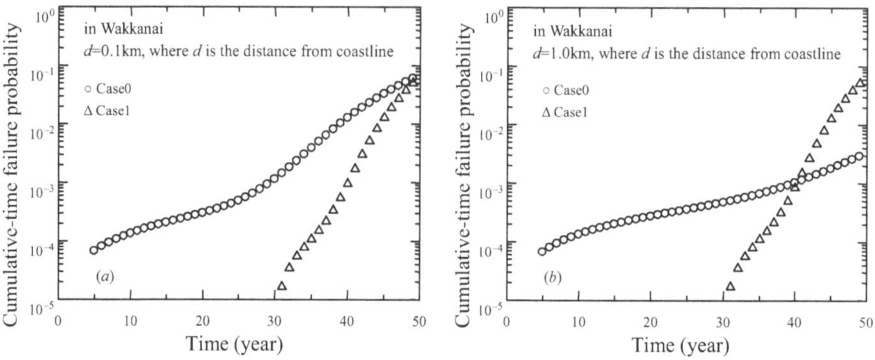

Figure 3. Relationship between year and cumulative-time failure probability

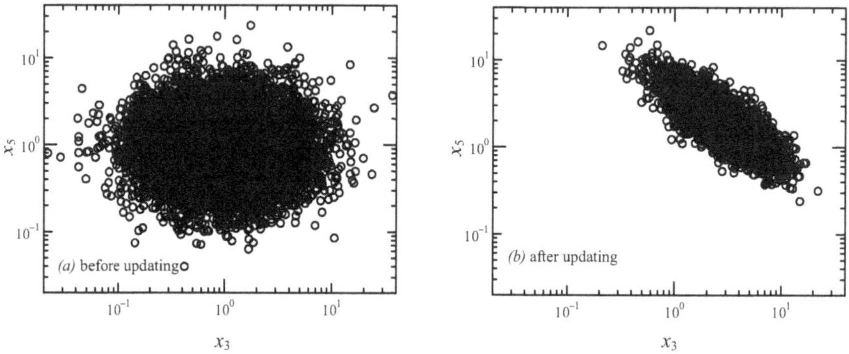

Figure 4. Relationship between x_3 and x_5

Conclusions

1. For the long-term performance prediction and reliability assessment of RC structures in a marine environment it is necessary to take into consideration the hazard associated with airborne chlorides.
2. When multiple random variables are related to observation information, and the relationship between random variables and observation information is non-linear, SMSC can be used successfully.
3. The effects of marine environment and inspection results on updated estimates of RC bridge slab reliability can be quantified using the proposed approach.

References

[1] Ellingwood, B.R. (2005) Risk-informed condition assessment of civil infrastructure: state of practice and research issues. *Structure and Infrastructure Engineering*, vol. 1, n. 1, pp. 7-18.

[2] Frangopol, D.M. and Liu, M. (2007) Maintenance and management of civil infrastructure based on condition, safety, optimization, and life-cycle cost. *Structure and Infrastructure Engineering*, vol. 3, n. 1, pp. 29-41.

[3] Akiyama, M., Frangopol, D.M. and Suzuki, M. (2009) Integration of the effects of airborne chlorides into reliability-based durability design of R/C structures in a marine environment. *Structure and Infrastructure Engineering*, DOI: 10.1080/15732470903363313.

[4] Yamazumi, K., Miyamoto, M., Sato, T. (1990) Deterioration and load capacity of RC beams. *Proc Japan Concr Inst*, vol. 12, n. 1, pp. 557-562 [in Japanese].

[5] Iwanami, M., Yokota, H., Sato, F. (2002) Influence of rebar corrosion on load carrying capacity on RC beams. *Proc Japan Concr Inst*, vol. 24, n. 2, pp. 1501-1506 [in Japanese].

[6] Murakami, Y., Kinoshita, A., Suzuki, S., Fukumoto, Y., Oshita, H. (2005) Study on residual flexural strength of RC beam with corrosion of reinforcement. *Concr Res Technol*, vol. 17, n. 1, pp. 61-74 [in Japanese].

[7] Murakami, Y., Yamauchi, Y., Tsutsumi, T., Oshita, H. (2006) Influence of shear reinforcement on residual flexural strength of RC beams with corrosive rebar. *Proc Japan Concr Inst*, vol. 28, n. 2, pp. 727-732 [in Japanese].

[8] Nakagawa, T., Seshimo, Y., Tsutsumi, T., Yasuda, N. (2002) Maintenance support system of RC structure under the chloride deterioration environment. *Concr J Japan Concr Inst*, vol. 40, n. 3, pp. 53-58 [in Japanese].

[9] Oyado, M., Sato, T. (2005) Evaluation of bending strength of corroded reinforced concrete member. *RTRI Rep*, vol. 19, n. 12, pp. 21-26 [in Japanese].

[10] Mori, Y. and Ellingwood, B.R. (1993) Reliability-based service life assessment of aging concrete structures. *Journal of Structural Engineering, ASCE*, vol. 119, n. 5, pp. 1600-1621.

[11] Akiyama, M., Frangopol, D.M., Yoshida, I. (2010) Time-dependent reliability analysis of existing RC structures in a marine environment using hazard associated with airborne chlorides, Engineering Structures, DOI: 10.1016/j.engstruct.2010.08.021.

[12] Yoshida, I. (2009) *Data assimilation and reliability estimation of existing RC structure*. COMPDYN 2009, CD-281, Rhodes, Greece.

[13] Suzuki, S., Tsutsumi, T., Yoshida, I., Oshita, H. (2009) The model of visual inspection on reinforced concrete structure affected salt damage. *Proceedings of Japan Concrete Institute*, vol. 31, n. 2, pp. 1543-1548 [in Japanese].

Author Index

C. Andrade and G. Mancini (eds.), *Modelling of Corroding Concrete Structures*,
RILEM Bookseries 5, DOI 10.1007/978-94-007-0677-4, © RILEM 2011

REUNION INTERNATIONALE DES LABORATOIRES ET EXPERTS
DES MATERIAUX, SYSTEMES DE CONSTRUCTION ET OUVRAGES

INTERNATIONAL UNION OF LABORATORIES AND EXPERTS IN CONSTRUCTION MATERIALS, SYSTEMS AND STRUCTURES

RILEM Publications – 5 November 2010

The following list is presenting our global offer, sorted by series.

RILEM PROCEEDINGS

PRO 1: Durability of High Performance Concrete (ISBN: 2-912143-03-9); *Ed. H. Sommer*

PRO 2: Chloride Penetration into Concrete (ISBN: 2-912143-00-04); *Eds. L.-O. Nilsson and J.-P. Ollivier*

PRO 3: Evaluation and Strengthening of Existing Masonry Structures (ISBN: 2-912143-02-0); *Eds. L. Binda and C. Modena*

PRO 4: Concrete: From Material to Structure (ISBN: 2-912143-04-7); *Eds. J.-P. Bournazel and Y. Malier*

PRO 5: The Role of Admixtures in High Performance Concrete (ISBN: 2-912143-05-5); *Eds. J. G. Cabrera and R. Rivera-Villarreal*

PRO 6: High Performance Fiber Reinforced Cement Composites - HPFRCC 3 (ISBN: 2-912143-06-3); *Eds. H. W. Reinhardt and A. E. Naaman*

PRO 7: 1st International RILEM Symposium on Self-Compacting Concrete (ISBN: 2-912143-09-8); *Eds. Å. Skarendahl and Ö. Petersson*

PRO 8: International RILEM Symposium on Timber Engineering (ISBN: 2-912143-10-1); *Ed. L. Boström*

PRO 9: 2nd International RILEM Symposium on Adhesion between Polymers and Concrete ISAP '99 (ISBN: 2-912143-11-X); *Eds. Y. Ohama and M. Puterman*

PRO 10: 3rd International RILEM Symposium on Durability of Building and Construction Sealants (ISBN: 2-912143-13-6); *Eds. A. T. Wolf*

PRO 11: 4th International RILEM Conference on Reflective Cracking in Pavements (ISBN: 2-912143-14-4); *Eds. A. O. Abd El Halim, D. A. Taylor and El H. H. Mohamed*

PRO 12: International RILEM Workshop on Historic Mortars: Characteristics and Tests (ISBN: 2-912143-15-2); *Eds. P. Bartos, C. Groot and J. J. Hughes*

PRO 13: 2nd International RILEM Symposium on Hydration and Setting (ISBN: 2-912143-16-0); *Ed. A. Nonat*

PRO 14: Integrated Life-Cycle Design of Materials and Structures - ILCDES 2000 (ISBN: 951-758-408-3); (ISSN: 0356-9403); *Ed. S. Sarja*

PRO 15: Fifth RILEM Symposium on Fibre-Reinforced Concretes (FRC) - BEFIB'2000 (ISBN: 2-912143-18-7); *Eds. P. Rossi and G. Chanvillard*

PRO 16: Life Prediction and Management of Concrete Structures (ISBN: 2-912143-19-5); *Ed. D. Naus*

PRO 17: Shrinkage of Concrete – Shrinkage 2000 (ISBN: 2-912143-20-9); *Eds. V. Baroghel-Bouny and P.-C. Aïtcin*

PRO 18: Measurement and Interpretation of the On-Site Corrosion Rate (ISBN: 2-912143-21-7); *Eds. C. Andrade, C. Alonso, J. Fullea, J. Polimon and J. Rodriguez*

PRO 19: Testing and Modelling the Chloride Ingress into Concrete (ISBN: 2-912143-22-5); *Eds. C. Andrade and J. Kropp*

PRO 20: 1st International RILEM Workshop on Microbial Impacts on Building Materials (CD 02) (e-ISBN 978-2-35158-013-4); *Ed. M. Ribas Silva*

PRO 21: International RILEM Symposium on Connections between Steel and Concrete (ISBN: 2-912143-25-X); *Ed. R. Eligehausen*

PRO 22: International RILEM Symposium on Joints in Timber Structures (ISBN: 2-912143-28-4); *Eds. S. Aicher and H.-W. Reinhardt*

PRO 23: International RILEM Conference on Early Age Cracking in Cementitious Systems (ISBN: 2-912143-29-2); *Eds. K. Kovler and A. Bentur*

PRO 24: 2nd International RILEM Workshop on Frost Resistance of Concrete (ISBN: 2-912143-30-6); *Eds. M. J. Setzer, R. Auberg and H.-J. Keck*

PRO 25: International RILEM Workshop on Frost Damage in Concrete (ISBN: 2-912143-31-4); *Eds. D. J. Janssen, M. J. Setzer and M. B. Snyder*

PRO 26: International RILEM Workshop on On-Site Control and Evaluation of Masonry Structures (ISBN: 2-912143-34-9); *Eds. L. Binda and R. C. de Vekey*

PRO 27: International RILEM Symposium on Building Joint Sealants (CD03); *Ed. A. T. Wolf*

PRO 28: 6th International RILEM Symposium on Performance Testing and Evaluation of Bituminous Materials - PTEBM'03 (ISBN: 2-912143-35-7; e-ISBN: 978-2-912143-77-8); *Ed. M. N. Partl*

PRO 29: 2nd International RILEM Workshop on Life Prediction and Ageing Management of Concrete Structures (ISBN: 2-912143-36-5); *Ed. D. J. Naus*

PRO 30: 4th International RILEM Workshop on High Performance Fiber Reinforced Cement Composites - HPFRCC 4 (ISBN: 2-912143-37-3); *Eds. A. E. Naaman and H. W. Reinhardt*

PRO 31: International RILEM Workshop on Test and Design Methods for Steel Fibre Reinforced Concrete: Background and Experiences (ISBN: 2-912143-38-1); *Eds. B. Schnütgen and L. Vandewalle*

PRO 32: International Conference on Advances in Concrete and Structures 2 vol. (ISBN (set): 2-912143-41-1); *Eds. Ying-shu Yuan, Surendra P. Shah and Heng-lin Lü*

PRO 33: 3rd International Symposium on Self-Compacting Concrete (ISBN: 2-912143-42-X); *Eds. Ó. Wallevik and I. Nielsson*

PRO 34: International RILEM Conference on Microbial Impact on Building Materials (ISBN: 2-912143-43-8); *Ed. M. Ribas Silva*

PRO 35: International RILEM TC 186-ISA on Internal Sulfate Attack and Delayed Ettringite Formation (ISBN: 2-912143-44-6); *Eds. K. Scrivener and J. Skalny*

PRO 36: International RILEM Symposium on Concrete Science and Engineering – A Tribute to Arnon Bentur (ISBN: 2-912143-46-2); *Eds. K. Kovler, J. Marchand, S. Mindess and J. Weiss*

PRO 37: 5th International RILEM Conference on Cracking in Pavements – Mitigation, Risk Assessment and Prevention (ISBN: 2-912143-47-0); *Eds. C. Petit, I. Al-Qadi and A. Millien*

PRO 38: 3rd International RILEM Workshop on Testing and Modelling the Chloride Ingress into Concrete (ISBN: 2-912143-48-9); *Eds. C. Andrade and J. Kropp*

PRO 39: 6th International RILEM Symposium on Fibre-Reinforced Concretes - BEFIB 2004 (ISBN: 2-912143-51-9); *Eds. M. Di Prisco, R. Felicetti and G. A. Plizzari*

PRO 40: International RILEM Conference on the Use of Recycled Materials in Buildings and Structures (ISBN: 2-912143-52-7); *Eds. E. Vázquez, Ch. F. Hendriks and G. M. T. Janssen*

PRO 41: RILEM International Symposium on Environment-Conscious Materials and Systems for Sustainable Development (ISBN: 2-912143-55-1); *Eds. N. Kashino and Y. Ohama*

REUNION INTERNATIONALE DES LABORATOIRES ET EXPERTS
DES MATERIAUX, SYSTEMES DE CONSTRUCTION ET OUVRAGES

INTERNATIONAL UNION OF LABORATORIES AND EXPERTS IN CONSTRUCTION MATERIALS, SYSTEMS AND STRUCTURES

PRO 42: SCC'2005 - China: 1st International Symposium on Design, Performance and Use of Self-Consolidating Concrete (ISBN: 2-912143-61-6); *Eds. Zhiwu Yu, Caijun Shi, Kamal Henri Khayat and Youjun Xie*

PRO 43: International RILEM Workshop on Bonded Concrete Overlays (e-ISBN: 2-912143-83-7); *Eds. J. L. Granju and J. Silfwerbrand*

PRO 44: 2nd International RILEM Workshop on Microbial Impacts on Building Materials (CD11) (e-ISBN: 2-912143-84-5); *Ed. M. Ribas Silva*

PRO 45: 2nd International Symposium on Nanotechnology in Construction, Bilbao (ISBN: 2-912143-87-X); *Eds. Peter J. M. Bartos, Yolanda de Miguel and Antonio Porro*

PRO 46: ConcreteLife'06 - International RILEM-JCI Seminar on Concrete Durability and Service Life Planning: Curing, Crack Control, Performance in Harsh Environments (ISBN: 2-912143-89-6); *Ed. K. Kovler*

PRO 47: International RILEM Workshop on Performance Based Evaluation and Indicators for Concrete Durability (ISBN: 978-2-912143-95-2); *Eds. V. Baroghel-Bouny, C. Andrade, R. Torrent and K. Scrivener*

PRO 48: 1st International RILEM Symposium on Advances in Concrete through Science and Engineering (e-ISBN: 2-912143-92-6); *Eds. J. Weiss, K. Kovler, J. Marchand, and S. Mindess*

PRO 49: International RILEM Workshop on High Performance Fiber Reinforced Cementitious Composites in Structural Applications (ISBN: 2-912143-93-4); *Eds. G. Fischer and V.C. Li*

PRO 50: 1st International RILEM Symposium on Textile Reinforced Concrete (ISBN: 2-912143-97-7); *Eds. Josef Hegger, Wolfgang Brameshuber and Norbert Will*

PRO 51: 2nd International Symposium on Advances in Concrete through Science and Engineering (ISBN: 2-35158-003-6; e-ISBN: 2-35158-002-8); *Eds. J. Marchand, B. Bissonnette, R. Gagné, M. Jolin and F. Paradis*

PRO 52: Volume Changes of Hardening Concrete: Testing and Mitigation (ISBN: 2-35158-004-4; e-ISBN: 2-35158-005-2); *Eds. O. M. Jensen, P. Lura and K. Kovler*

PRO 53: High Performance Fiber Reinforced Cement Composites - HPFRCC5 (ISBN: 978-2-35158-046-2); *Eds. H. W. Reinhardt and A. E. Naaman*

PRO 54: 5th International RILEM Symposium on Self-Compacting Concrete (ISBN: 978-2-35158-047-9); *Eds. G. De Schutter and V. Boel*

PRO 55: International RILEM Symposium Photocatalysis, Environment and Construction Materials (ISBN: 978-2-35158-056-1); *Eds. P. Baglioni and L. Cassar*

PRO56: International RILEM Workshop on Integral Service Life Modelling of Concrete Structures (ISBN 978-2-35158-058-5); *Eds. R. M. Ferreira, J. Gulikers and C. Andrade*

PRO57: RILEM Workshop on Performance of cement-based materials in aggressive aqueous environments (e-ISBN: 978-2-35158-059-2); *Ed. N. De Belie*

PRO58: International RILEM Symposium on Concrete Modelling - CONMOD'08 (ISBN: 978-2-35158-060-8); *Eds. E. Schlangen and G. De Schutter*

PRO 59: International RILEM Conference on On Site Assessment of Concrete, Masonry and Timber Structures - SACoMaTiS 2008 (ISBN set: 978-2-35158-061-5); *Eds. L. Binda, M. di Prisco and R. Felicetti*

PRO 60: Seventh RILEM International Symposium on Fibre Reinforced Concrete: Design and Applications - BEFIB 2008 (ISBN: 978-2-35158-064-6); *Ed. R. Gettu*

PRO 61: 1st International Conference on Microstructure Related Durability of Cementitious Composites 2 vol., (ISBN: 978-2-35158-065-3); *Eds. W. Sun, K. van Breugel, C. Miao, G. Ye and H. Chen*

PRO 62: NSF/ RILEM Workshop: In-situ Evaluation of Historic Wood and Masonry Structures (e-ISBN: 978-2-35158-068-4); *Eds. B. Kasal, R. Anthony and M. Drdácký*

PRO 63: Concrete in Aggressive Aqueous Environments: Performance, Testing and Modelling, 2 vol., (ISBN: 978-2-35158-071-4); *Eds. M. G. Alexander and A. Bertron*

PRO 64: Long Term Performance of Cementitious Barriers and Reinforced Concrete in Nuclear Power Plants and Waste Management - NUCPERF 2009 (ISBN: 978-2-35158-072-1); *Eds. V. L'Hostis, R. Gens, C. Gallé*

PRO 65: Design Performance and Use of Self-consolidating Concrete - SCC'2009 (ISBN: 978-2-35158-073-8); *Eds. C. Shi, Z. Yu, K. H. Khayat and P. Yan*

PRO 66: 2nd International RILEM Workshop on Concrete Durability and Service Life Planning - ConcreteLife'09 (ISBN: 978-2-35158-074-5); *Ed. K. Kovler*

PRO 67: Repairs Mortars for Historic Masonry (e-ISBN: 978-2-35158-083-7); *Ed. C. Groot*

PRO 68: Proceedings of the 3rd International RILEM Symposium on 'Rheology of Cement Suspensions such as Fresh Concrete (ISBN 978-2-35158-091-2); *Eds. O. H. Wallevik, S. Kubens and S. Oesterheld*

PRO 69: 3rd International PhD Student Workshop on 'Modelling the Durability of Reinforced Concrete (ISBN: 978-2-35158-095-0); *Eds. R. M. Ferreira, J. Gulikers and C. Andrade*

PRO 71: Advances in Civil Engineering Materials - The 50-year Teaching Anniversary of Prof. Sun Wei' (ISBN: 978-2-35158-098-1; e-ISBN: 978-2-35158-099-8); *Eds. C. Miao, G. Ye, and H. Chen*

PRO74: International RILEM Conference on 'Use of Superabsorsorbent Polymers and Other New Addditives in Concrete' (ISBN: 978-2-35158-104-9; e-ISBN: 978-2-35158-105-6), *Eds. O.M. Jensen, M.T. Hasholt, and S. Laustsen*

PRO75: International Conference on 'Material Science - 2nd ICTRC - Textile Reinforced Concrete - *Theme 1*' (ISBN: 978-2-35158-106-3; e-ISBN: 978-2-35158-107-0), *Ed. W. Brameshuber*

PRO76: International Conference on 'Material Science - HetMat - Modelling of Heterogeneous Materials - *Theme 2*' (ISBN: 978-2-35158-108-7; e-ISBN: 978-2-35158-109-4), *Ed. W. Brameshuber*

PRO77: International Conference on 'Material Science - AdIPoC - Additions Improving Properties of Concrete' - *Theme 3* (ISBN: 978-2-35158-110-0; e-ISBN: 978-2-35158-111-7), *Ed. W. Brameshuber*

PRO78: 2nd Historic Mortars Conference and RILEM TC 203-RHM Final Workshop – HMC2010 (e-ISBN: 978-2-35158-112-4), *Eds J. Válek, C. Groot, and J. J. Hughes*

RILEM REPORTS

Report 19: Considerations for Use in Managing the Aging of Nuclear Power Plant Concrete Structures (ISBN: 2-912143-07-1); *Ed. D. J. Naus*

Report 20: Engineering and Transport Properties of the Interfacial Transition Zone in Cementitious Composites (ISBN: 2-912143-08-X); *Eds. M. G. Alexander, G. Arliguie, G. Ballivy, A. Bentur and J. Marchand*

Report 21: Durability of Building Sealants (ISBN: 2-912143-12-8); *Ed. A. T. Wolf*

Report 22: Sustainable Raw Materials - Construction and Demolition Waste (ISBN: 2-912143-17-9); *Eds. C. F. Hendriks and H. S. Pietersen*

REUNION INTERNATIONALE DES LABORATOIRES ET EXPERTS DES MATERIAUX, SYSTEMES DE CONSTRUCTION ET OUVRAGES

INTERNATIONAL UNION OF LABORATORIES AND EXPERTS IN CONSTRUCTION MATERIALS, SYSTEMS AND STRUCTURES

Report 23: Self-Compacting Concrete state-of-the-art report (ISBN: 2-912143-23-3); *Eds. Å. Skarendahl and Ö. Petersson*

Report 24: Workability and Rheology of Fresh Concrete: Compendium of Tests (ISBN: 2-912143-32-2); *Eds. P. J. M. Bartos, M. Sonebi and A. K. Tamimi*

Report 25: Early Age Cracking in Cementitious Systems (ISBN: 2-912143-33-0); *Ed. A. Bentur*

Report 26: Towards Sustainable Roofing (Joint Committee CIB/RILEM) (CD 07) (e-ISBN 978-2-912143-65-5); *Eds. Thomas W. Hutchinson and Keith Roberts*

Report 27: Condition Assessment of Roofs (Joint Committee CIB/RILEM) (CD 08) (e-ISBN 978-2-912143-66-2); *Ed. CIB W 83/RILEM TC166-RMS*

Report 28: Final report of RILEM TC 167-COM 'Characterisation of Old Mortars with Respect to Their Repair (ISBN: 978-2-912143-56-3); *Eds. C. Groot, G. Ashall and J. Hughes*

Report 29: Pavement Performance Prediction and Evaluation (PPPE): Interlaboratory Tests (e-ISBN: 2-912143-68-3); *Eds. M. Partl and H. Piber*

Report 30: Final Report of RILEM TC 198-URM 'Use of Recycled Materials' (ISBN: 2-912143-82-9; e-ISBN: 2-912143-69-1); *Eds. Ch. F. Hendriks, G. M. T. Janssen and E. Vázquez*

Report 31: Final Report of RILEM TC 185-ATC 'Advanced testing of cement-based materials during setting and hardening' (ISBN: 2-912143-81-0; e-ISBN: 2-912143-70-5); *Eds. H. W. Reinhardt and C. U. Grosse*

Report 32: Probabilistic Assessment of Existing Structures. A JCSS publication (ISBN 2-912143-24-1); *Ed. D. Diamantidis*

Report 33: State-of-the-Art Report of RILEM Technical Committee TC 184-IFE 'Industrial Floors' (ISBN 2-35158-006-0); *Ed. P. Seidler*

Report 34: Report of RILEM Technical Committee TC 147-FMB 'Fracture mechanics applications to anchorage and bond' Tension of Reinforced Concrete Prisms – Round Robin Analysis and Tests on Bond (e-ISBN 2-912143-91-8); *Eds. L. Elfgren and K. Noghabai*

Report 35: Final Report of RILEM Technical Committee TC 188-CSC 'Casting of Self Compacting Concrete' (ISBN 2-35158-001-X; e-ISBN: 2-912143-98-5); *Eds. Å. Skarendahl and P. Billberg*

Report 36: State-of-the-Art Report of RILEM Technical Committee TC 201-TRC 'Textile Reinforced Concrete' (ISBN 2-912143-99-3); *Ed. W. Brameshuber*

Report 37: State-of-the-Art Report of RILEM Technical Committee TC 192-ECM 'Environment-conscious construction materials and systems' (ISBN: 978-2-35158-053-0); *Eds. N. Kashino, D. Van Gemert and K. Imamoto*

Report 38: State-of-the-Art Report of RILEM Technical Committee TC 205-DSC 'Durability of Self-Compacting Concrete' (ISBN: 978-2-35158-048-6); *Eds. G. De Schutter and K. Audenaert*

Report 39: Final Report of RILEM Technical Committee TC 187-SOC 'Experimental determination of the stress-crack opening curve for concrete in tension' (ISBN 978-2-35158-049-3); *Ed. J. Planas*

Report 40: State-of-the-Art Report of RILEM Technical Committee TC 189-NEC 'Non-Destructive Evaluation of the Penetrability and Thickness of the Concrete Cover' (ISBN 978-2-35158-054-7); *Eds. R. Torrent and L. Fernández Luco*

Report 41: State-of-the-Art Report of RILEM Technical Committee TC 196-ICC 'Internal Curing of Concrete' (ISBN 978-2-35158-009-7); *Eds. K. Kovler and O. M. Jensen*

Report 42: 'Acoustic Emission and Related Non-destructive Evaluation Techniques for Crack Detection and Damage Evaluation in Concrete' - Final Report of RILEM Technical Committee 212-ACD (e-ISBN: 978-2-35158-100-1); *Ed. M. Ohtsu*

REUNION INTERNATIONALE DES LABORATOIRES ET EXPERTS
DES MATERIAUX, SYSTEMES DE CONSTRUCTION ET OUVRAGES

INTERNATIONAL UNION OF LABORATORIES AND EXPERTS IN CONSTRUCTION MATERIALS, SYSTEMS AND STRUCTURES

RILEM COMPENDIUMS

COMP 01: Trilingual Dictionary for Materials and Structures (English-French-German) (CD01) (1970)

COMP 02: 1947-1997: 50 years of evolution of Building Materials and Structures (e-ISBN: 2-912143-86-1); *Ed. F. Wittmann*

COMP 03: General Conference of RILEM TCs' Chairmen and RILEM Seminar 'Advancing the Knowledge in Materials and Structures' (CD10) (e-ISBN: 2-912143-85-3)

COMP 06: Concrete Science and Engineering Journal – Vols. 1, 2, 3, 4 (1999-2002) (CD05)

COMP 13: RILEM Technical Day, Moscow (e-ISBN: 2-35158-045-1)

FSC
www.fsc.org
MIX
Papier | Fördert
gute Waldnutzung
FSC® C083411

Kolibri 360 GmbH
Ferdinand-Jühlke-Straße 7
99095 Erfurt, Deutschland
produktsicherheit@kolibri360.de